The Direct Detection of Microorganisms in Clinical Samples

Contributors

William L. Albritton
Neil R. Blacklow
John B. Brooks
Sarah Cheeseman
J. Donald Coonrod
Billy H. Cooper
George Cukor
Adnan S. Dajani
Mary Jane Ferraro
Sherwood L. Gorbach
James H. Jorgensen
Jane D. Keathley
George E. Kenny
Richard B. Kohler
Lawrence J. Kunz
Kenneth McIntosh
William J. Martin
James W. Mayhew
E. Richard Moxon
Patrick R. Murray
Richard H. Parker
Lauren Pierik
Jon E. Rosenblatt
Lorry G. Rubin
Edward Tabor
Richard Tilton
Marc H. Weiner
L. Joseph Wheat
H. David Wilson
Washington C. Winn, Jr.
Robert H. Yolken

The Direct Detection of Microorganisms in Clinical Samples

Edited by

J. DONALD COONROD

Veterans Administration Medical Center, and Department of Medicine
University of Kentucky
Lexington, Kentucky

LAWRENCE J. KUNZ

Francis Blake Bacteriology Laboratories
Massachusetts General Hospital, and Harvard Medical School
Boston, Massachusetts

MARY JANE FERRARO

Francis Blake Bacteriology Laboratories
Massachusetts General Hospital, and Harvard Medical School
Boston, Massachusetts

1983

ACADEMIC PRESS, INC.

(Harcourt Brace Jovanovich, Publishers)

Orlando San Diego San Francisco New York London
Toronto Montreal Sidney Tokyo São Paulo

ACADEMIC PRESS, INC.
Orlando, Florida 32887

United Kingdom Edition published by
ACADEMIC PRESS, INC. (LONDON) LTD.
24/28 Oval Road, London NW1 7DX

Library of Congress Cataloging in Publication Data

Main entry under title:

The Direct detection of microorganisms in clinical samples.

Includes index.
1. Micro-organisms, Pathogenic--Identification.
2. Diagnosis, Laboratory. 3. Diagnostic specimens.
I. Coonrod, J. Donald. II. Kunz, Lawrence J. III. Ferraro, Mary Jane. [DNLM: 1. Communicable diseases--Diagnosis. 2. Diagnosis, Laboratory--Methods. WC 100 D598]
QR67.D6 1983 616.07'582 83-11862
ISBN 0-12-187780-9

PRINTED IN THE UNITED STATES OF AMERICA

83 84 85 86 9 8 7 6 5 4 3 2 1

Contents

Contributors xv
Foreword—William B. Cherry xix
Preface xxiii

I. Visualization of Organisms in Clinical Samples

1. Microscopy in the Detection of Bacteria
LAWRENCE J. KUNZ

I. Introduction 3
II. General Uses for Gram's Stain 5
III. Usefulness of Diagnostic Microscopy in Particular Types of Specimens or Infectious Processes 5
IV. Wounds, Abcesses, and Exudates 9
V. Conclusions 11
References 11

2. Microscopic Preparations for Detecting Fungi in Clinical Materials
BILLY H. COOPER

I. Role of Microscopic Techniques in Detecting Fungi in Clinical Specimens 13

II. Methods for Visualizing Fungi in Clinical Materials 19
III. Examples of Fungi in Clinical Materials 31
References 33

3. Direct Immunofluorescence Identification of Bacteria in Clinical Specimens

WASHINGTON C. WINN, JR., and JANE D. KEATHLEY

I. Introduction 35
II. Important Factors in the Performance of a Test 36
III. Identification of Bacteria in Clinical Specimens 41
IV. Selection of Rapid Diagnostic Tests 52
References 53

4. Immunofluorescence in Viral Diagnosis

KENNETH McINTOSH AND LAUREN PIERIK

I. Introduction 57
II. Theory of the Method 58
III. Reagents 61
IV. Staining Method 65
V. Fluorescence Microscope 65
VI. Collection and Preparation of Specimens: Reading of Slides 67
VII. Use of Controls 77
VIII. Interpretation of Findings: Practical Aspects 78
References 79

II. Immunologic Methods for Detecting Soluble Antigens

Introduction *85*

J. DONALD COONROD

5. Procedures for the Detection of Microorganisms by Counterimmunoelectrophoresis

RICHARD TILTON

I. Introduction 87
II. Principle 88
III. Variables 88

IV. Procedure for Counterimmunoelectrophoresis of Spinal Fluid, Serum, Urine, and Other Body Fluids 91
V. Clinical Applications of CIE 94
References 95

6. *Application of Counterimmunoelectrophoresis to the Diagnosis of Meningitis*

RICHARD H. PARKER

I. Introduction 97
II. Meningococcal Meningitis 98
III. *Haemophilus influenzae* Meningitis 99
IV. Pneumococcal Meningitis 100
V. Group B Streptococcal Meningitis 101
VI. *Escherichia coli* Meningitis 102
VII. Conclusion 102
References 102

7. *Counterimmunoelectrophoresis for the Diagnosis of Pneumococcal Respiratory and Other Infections*

GEORGE E. KENNY

I. Introduction 105
II. Summary of the Principles of Counterimmunoelectrophoresis 106
III. Counterimmunoelectrophoresis Methods in Pneumococcal Infections 106
IV. Detection of Antigen in Clinical Samples 107
V. Perspective 110
References 110

8. *Counterimmunoelectrophoresis for the Diagnosis of Intrapleural Empyema*

H. DAVID WILSON

I. Introduction 113
II. Methods 114
References 116

9. *Problems with Precipitin Methods for Detecting Antigenemia in Bacterial Infections*

J. DONALD COONROD

I. Introduction 117
II. Variables in the Detection of Antigenemia 118
III. Future Trends 123
References 123

10. *Evaluation of Counterimmunoelectrophoresis in the Diagnosis of Infectious Diseases*

WILLIAM L. ALBRITTON

I. Introduction 125
II. Counterimmunoelectrophoresis as the Test of Choice 126
III. Use of Counterimmunoelectrophoresis in Early Diagnosis 129
IV. Culture-Negative Diagnosis 130
V. Identification of Pathogens in the Presence of Mixed Flora 130
VI. Discussion 131
References 133

11. *Agglutination Techniques for the Detection of Microbial Antigens: Methodology and Overview*

J. DONALD COONROD

I. Principles of Agglutination 135
II. Advantages and Disadvantages of Agglutination for Antigen Detection 139
III. Future Prospects 141
References 142

12. *Agglutination Tests for the Diagnosis of Meningitis*

ADNAN S. DAJANI

I. Introduction 143
II. Antigen Determinants 144
III. Antibody Determinants 150
IV. Specificity of Agglutination Tests 150
References 152

13. *Diagnosis of Pneumonia by Agglutination Techniques*
J. DONALD COONROD

I. Introduction 153
II. Coagglutination in Pneumonia 154
III. Comparison of Coagglutination and Counterimmunoelectrophoresis 155
IV. Specificity 157
References 157

14. *Immunoassays in Meningitis*
LORRY G. RUBIN AND E. RICHARD MOXON

I. Introduction 159
II. Problems in the Diagnosis of Meningitis 160
III. Detection of Bacterial Antigens in Meningitis 161
IV. Fungal Meningitis: *Cryptococcus neoformans* 168
V. Conclusions and Future Prospects 170
References 172

15. *Use of Immunoassays in Bacteremia*
L. JOSEPH WHEAT

I. Introduction 175
II. Problems of Immunoassays for the Detection of Antigenemia 177
III. Applications to the Rapid Diagnosis of Bacteremia 181
References 184

16. *Diagnosis of Legionnaires' Disease by Radioimmunoassay and Enzyme-Linked Immunosorbent Assay*
RICHARD B. KOHLER

I. Introduction 187
II. Conventional Diagnostic Tests for Legionnaires' Disease 188
III. Radioimmunoassay and Enzyme-Linked Immunosorbent Assay for the Detection of Legionnaires' Antigen 191
IV. Summary 203
References 205

17. *Detection of Fungal Antigens in Clinical Samples*

MARC H. WEINER

I. Introduction 207
II. Candidiasis 208
III. *Aspergillus* 215
References 221

18. *Prospects for Solid-Phase Immunoassays in the Diagnosis of Respiratory Infections*

ROBERT H. YOLKEN

I. Introduction 225
II. Antibody Labels 226
III. Assay Formats 229
IV. Support Systems 233
V. Reagents 235
VI. Collection of Specimens 236
VII. Sensitivity of Assay Systems 237
Appendix 238
References 239

19. *Diagnosis of Hepatitis B and Non-A, Non-B Hepatitis*

EDWARD TABOR

I. Hepatitis B 241
II. Non-A, Non-B Hepatitis 254
References 261

20. *Immunoassays for the Diagnosis of Rotavirus and Norwalk Virus Infections*

GEORGE CUKOR, SARAH H. CHEESEMAN, AND NEIL R. BLACKLOW

I. Introduction 267
II. Rotavirus 268
III. Norwalk Virus 274
References 277

III. Nonimmunologic Detection of Microbial Products

Introduction *281*

MARY JANE FERRARO

21. Current Uses of the Limulus *Amoebocyte Lysate Test*

JAMES H. JORGENSEN

I. Background and Mechanism of the *Limulus* Amoebocyte Lysate Test 283
II. *Limulus* Amoebocyte Lysate Testing in the Pharmaceutical Industry 285
III. Clinical Applications of the *Limulus* Amoebocyte Lysate Test 285
References 293

22. Laboratory Diagnosis of Antimicrobial-Associated Diarrhea

JON E. ROSENBLATT

I. Introduction 295
II. *Clostridium difficile* and Its Toxin 296
III. Laboratory Procedures 299
References 301

23. Perspective on the Current and Future Role of Gas–Liquid Chromatographic Analysis

JAMES W. MAYHEW AND SHERWOOD L. GORBACH

I. Introduction 303
II. Methodology 304
III. Gas–Liquid Chromatography 305
IV. Goals of Analytical Techniques 307
V. Clinical Correlations 307
VI. Pitfalls in Analysis 308
VII. Future Prospects 309
References 310

24. Gas–Liquid Chromatography as an Aid in Rapid Diagnosis by Selective Detection of Chemical Changes in Body Fluids

JOHN B. BROOKS

I. Introduction 313
II. Practical Methods for Recovering Volatile Chemical Compounds from Body Fluids for Derivitization Purposes 314
III. Selective and Sensitive High-Resolution Gas–Liquid Chromatography Systems for Body Fluid Analysis 315
IV. Practical Derivatization Methods for Analysis by FPEC–GLC 318
V. Application of FPEC–GLC to Detection of Chemical Changes in Spent Culture Media and Infected Tissue Culture 319
VI. Application of FPEC–GLC to Detection of Disease-Specific Profiles in Body Fluids 319
VII. Identification of Unknown Peaks Detected by FPEC–GLC 330
VIII. Interpretation of Data Obtained by FPEC–GLC 331
IX. Possibilities for Automation and Computerization of Data Obtained from FPEC–GLC 332
X. Summary 333
References 333

25. Challenges in the Development of Automation for the Clinical Microbiology Laboratory

MARY JANE FERRARO

Text 335
References 342

26. Instrumentation for Detection of Bacteremia

PATRICK R. MURRAY

I. Introduction 343
II. Conventional Methods 344
III. Alternative Modifications of Blood Cultures 345
IV. Summary 350
References 350

27. Rapid Methods and Instrumentation in the Diagnosis of Urinary Tract Infections

WILLIAM J. MARTIN

I. Introduction 353
II. Rapid Methods 354
III. Instrumentation 356
References 360

Index 363

Contributors

Numbers in parentheses indicate the pages on which the authors' contributions begin.

William L. Albritton* (125), Department of Medical Microbiology, University of Manitoba, Winnipeg, Manitoba R3E 0W3, Canada

Neil R. Blacklow (267), University of Massachusetts Medical School, Worcester, Massachusetts 01605

John B. Brooks (313), Centers for Disease Control, Atlanta, Georgia 30333

Sarah H. Cheeseman (267), University of Massachusetts Medical School, Worcester, Massachusetts 01605

J. Donald Coonrod (85, 117, 135, 153), Division of Infectious Diseases, Department of Medicine, University of Kentucky, and Veterans Administration Medical Center, Lexington, Kentucky 40536

Billy H. Cooper (13), Department of Pathology, Baylor University Medical Center, Dallas, Texas 75246

George Cukor (267), University of Massachusetts Medical School, Worcester, Massachusetts 01605

Adnan S. Dajani (143), Department of Pediatrics, Wayne State University School of Medicine, and Division of Infectious Diseases, Children's Hospital of Michigan, Detroit, Michigan 48201

Mary Jane Ferraro (281, 335), Francis Blake Bacteriology Laboratories, Massachusetts General Hospital, and Harvard Medical School, Boston, Massachusetts 02114

*Present address: Sexually Transmitted Diseases Laboratory Program, Centers for Disease Control, Atlanta, Georgia 30333.

Sherwood L. Gorbach (303), Infectious Disease Service, New England Medical Center, Tufts University School of Medicine, Boston, Massachusetts 02111

James H. Jorgensen (283), Department of Pathology, The University of Texas Health Sciences Center, San Antonio, Texas 78284

Jane D. Keathley (35), Department of Pathology, College of Medicine, University of Vermont, Burlington, Vermont 05405

George E. Kenny (105), Department of Pathobiology, School of Public Health and Community Medicine, University of Washington, Seattle, Washington 98195

Richard B. Kohler (187), Wishard Memorial Hospital, and Department of Medicine, Indiana University School of Medicine, Indianapolis, Indiana 46202

Lawrence J. Kunz (3), Francis Blake Bacteriology Laboratories, Massachusetts General Hospital, and Harvard Medical School, Boston, Massachusetts 02114

Kenneth McIntosh (57), Division of Infectious Disease, Children's Hospital Medical Center, Boston, Massachusetts 02115

William J. Martin (353), Clinical Microbiology Laboratories, New England Medical Center, Tufts University School of Medicine, Boston, Massachusetts 02111

James W. Mayhew (303), Infectious Disease Service, New England Medical Center, Tufts University School of Medicine, Boston, Massachusetts 02111

E. Richard Moxon (159), Department of Pediatrics, The Johns Hopkins University School of Medicine, Baltimore, Maryland 21205

Patrick R. Murray (343), Division of Laboratory Medicine, Department of Pathology, Washington University School of Medicine, St. Louis, Missouri 63110

Richard H. Parker (97), Veterans Administration Medical Center, and Howard University College of Medicine, Washington, D.C. 20422

Lauren Pierik (57), Diagnostic Virology Laboratory, Children's Hospital Medical Center, Boston, Massachusetts 02115

Jon E. Rosenblatt (295), Mayo Clinic and Mayo Foundation, Rochester, Minnesota 55901

Lorry G. Rubin* (159), Department of Pediatrics, Division of Infectious Diseases, The Johns Hopkins University School of Medicine, Baltimore, Maryland 21205

Edward Tabor (241), Hepatitis Branch, Division of Blood and Blood Products, Bureau of Biologics, Food and Drug Administration, Bethesda, Maryland 20205

*Present address: Department of Pediatrics, Children's Hospital, Long Island Jewish–Hillside Medical Center, New Hyde Park, New York 11042.

Richard Tilton (87), Department of Laboratory Medicine, John Dempsey Hospital, The University of Connecticut Health Center, Farmington, Connecticut 06032

Marc H. Weiner (207), Audie L. Murphy Memorial Veterans Administration Hospital, and Department of Medicine, University of Texas Health Sciences Center, San Antonio, Texas 78284

L. Joseph Wheat (175), Indiana University School of Medicine, and Indianapolis Veterans Administration Hospital, Indianapolis, Indiana 46202

H. David Wilson (113), Department of Pediatrics, College of Medicine, University of Kentucky, Lexington, Kentucky 40536

Washington C. Winn, Jr. (35), Department of Pathology, College of Medicine, University of Vermont, Burlington, Vermont 05405

Robert H. Yolken (225), Department of Pediatrics, The Johns Hopkins University School of Medicine, Baltimore, Maryland 21205

Foreword

When this writer began his association with the field of clinical microbiology more than 40 years ago, test tubes, petri dishes, and pipettes together with a microscope, an autoclave, and an oven for dry sterilization constituted the major laboratory equipment. Although commercial dehydrated media were available, many laboratories still prepared basic infusion media from fresh ground beef or horse meat. Color comparison with indicator solutions or papers was the usual way of adjusting pH. Electronic pH meters and spectrophotometers were available but uncommon. Anaerobic cultures were done only in relation to trauma and suspected gas-gangrene, and the techniques used were frequently inappropriate for the isolation of anaerobic bacteria. Cultures for fungi were rarely done. Viral examinations, except for rabies, existed in only a few special centers because tissue culture techniques were just being developed for virus diagnosis. Antibiotics had not been discovered and sulfonamides had been in use for only a few years. Bacteremic patients had a high mortality rate, and osteomyelitis, meningitis, pleurisy, acute glomerular nephritis, rheumatic heart disease, and other afflictions were commonplace.

Complement fixation tests for syphilis and febrile agglutination tests for typhoid and paratyphoid fever, brucellosis, tularemia, and infectious mononucleosis were performed in many laboratories. Stools were cultured for *Salmonella* and *Shigella* and examined for ova and parasites. Definitive serotyping of *Salmonella* by agglutination tests and grouping and typing of group A streptococci by precipitin tests were sophisticated procedures available in relatively few laboratories.

The advent of antimicrobic agents stimulated a need for more rapid and precise identification of pathogenic microorganisms to serve as a guide to appropriate therapy. The "cold war" years of the 1950s stimulated interest in civil defense against potential agents of biological warfare. As a result, the fluorescent antibody techniques developed by Albert Coons were extended and applied to the detection or identification of a variety of microbiological disease agents, the goal being to obtain rapid and specific information earlier than by cultural procedures. Bacterial genetics was a new frontier just beginning to be explored by Lederberg and Tatum. In the late 1940s when a graduate student concluded her seminar by stating that "it appears that bacteria have a sexual apparatus but they don't use it all the time," many of the distinguished faculty of her department were left in shock. It was difficult for many microbiologists to accept the fact that bacteria are subject to the same mechanisms of heredity that apply to higher forms of life. Genetic analyses of microbial cells including recombinant DNA, DNA base composition, and plasmid determinations are now indispensable tools of the taxonomist. Led by W. E. C. Moore and his group at Virginia Polytechnic Institute, anaerobic bacteriology has, during the past 20 years, added an important new dimension to the services of the clinical laboratory.

The 1970s and 1980s have provided a virtual explosion of techniques for detection, identification, and quantitation of microbial agents, both in pure culture and in clinical specimens by both direct and indirect means. Rapid progress has been possible for two important reasons: (1) commercial companies have applied the technology of computers and microprocessors to the design and manufacture of instruments that mechanize or automate clinical laboratory analyses, and (2) cooperation in the design, development, and evaluation of useful equipment has existed between industrial and clinical laboratory scientists. Notable examples are gas–liquid chromatography (GLC), enzyme-linked immunosorbent assays (ELISA), radioimmunoassays (RIA), and instruments for testing blood, urine, and antimicrobic susceptibility.

Certain problems that tend to limit the application of direct detection techniques for microorganisms are discussed in this book. These problems usually involve the specificity, sensitivity, and predictive value of a given test and the amount of tolerance that is acceptable. In cost effectiveness, the most successful tests are those that can be applied in a mechanized or automated mode to large-volume clinical specimens such as blood, urine, and serum. Specimens containing mixed flora usually require more processing time because of difficult interpretation, need for isolation of individual components, or possible nonspecific reactions. The weakness of some systems of direct detection is economic, in that manufacturers of diagnostic products do not provide reagents for performing important but low-volume tests. A second problem is the poor quality of some commercial products.

The basic requirements for progress are that laboratory scientists communicate their needs to industry, that the expected volume of tests to be performed are sufficient to support the development of new equipment and reagents, and that the applications are relatively rapid and cost effective. Continued cooperation between clinical microbiologists and the manufacturers is essential for improvement in diagnostic methodology and, indirectly, for better patient care.

The contributors to the present volume—all highly experienced laboratory scientists—have critically reviewed the most practical and widely used procedures for direct detection of microorganisms in clinical specimens. This is, however, a dynamic field that will continue to expand and evolve under the influence of the "information revolution."

William B. Cherry

Preface

Recent years have seen a steady growth in efforts in the clinical laboratory to improve the diagnosis of infectious diseases. There has been a particularly striking renewal of interest in the possibility of detecting microorganisms directly in clinical samples by identifying their soluble antigens or metabolites. Improved diagnostic methods are clearly needed, and numerous investigators have now entered the field of so-called rapid diagnosis in infectious diseases. The word rapid is not a precise one in this context, however, because many of the methods being studied are not rapid, and it is doubtful if any method could be more rapid than the time-honored gram stain. What is sought in infectious diseases is perhaps better labeled "timely" diagnosis, a diagnosis made sufficiently early that the clinician can intercede effectively with therapy or prophylaxis. Whether a test can provide timely diagnosis in a particular situation depends on many variables, including methodologic requirements of the test, the type and stage of the infection, and the nature of the therapeutic possibilities.

As older diagnostic tests are expanded or new ones are developed, it is vital that there be continual assessment of the value of these tests and that a critical attitude be maintained. This book was written to consider for the first time in a critical fashion the multiple methods being used to detect microorganisms directly in clinical samples. Applications to virology and mycology are considered as well as the generally more familiar applications to bacteriology. Interesting work is going on in all these areas, and lessons learned in one prove useful in all. Although certain of the methods discussed here may be applicable to the detection of organisms in pure culture, this subject represents a burgeoning field of research in its own right, and no attempt has been made to include it.

The book is organized into three sections. Established techniques for visualization of intact organisms in clinical samples are considered first. Chapters in the second section deal with immunologic techniques for detecting soluble microbial antigens. Precipitin and agglutination techniques are discussed sequentially, followed by enzymatic and radioimmunoassays. In the third section, diverse nonimmunologic methods for detecting soluble constituents of organisms and their metabolites are discussed. The contents of individual chapters range from purely methodologic discussions to general critiques of the diagnostic process itself. Contributors have been selected because of their extensive personal experience with their subjects. A pragmatic and critical attitude has been encouraged, not only in the area of methodologic "nitty-gritty" but also in terms of the overall diagnostic usefulness of the methods. This attempt at a comprehensive and critically oriented review of methods for detection of microorganisms in clinical samples appears overdue. It is particularly directed at clinical microbiologists and infectious disease clinicians and researchers. Individuals working in analogous areas, such as detection of tumor-specific antigens, detection of unique cell markers, and related fields, may also find sections of interest in this book. Controversial material is presented here as such, and the reader is forewarned that different authors in this text have not always come to the same conclusions. It has not been our goal to attempt a consensus where none currently exists but, rather, to stimulate needed additional research in these areas of controversy.

We especially thank our colleagues whose interests in diagnostic methods in infectious diseases have provided the data that are the basis of this book. We thank also the staff of Academic Press for their patience and kind assistance.

J. Donald Coonrod
Lawrence J. Kunz
Mary Jane Ferraro

I

VISUALIZATION OF ORGANISMS IN CLINICAL SAMPLES

1

Microscopy in the Detection of Bacteria

LAWRENCE J. KUNZ

Francis Blake Bacteriology Laboratories
Massachusetts General Hospital
and
Department of Microbiology and Molecular Genetics
Harvard Medical School
Boston, Massachusetts

I. Introduction 3
II. General Uses for Gram's Stain 5
III. Usefulness of Diagnostic Microscopy in Particular Types of Specimens or Infectious Processes 5
A. Meningitis: Examination of Cerebrospinal Fluid 6
B. Bacteremia: Examination of Peripheral Blood 6
C. Urinary Tract Infections 7
D. Respiratory Tract Infections and Specimens for Examination 7
E. Pharyngitis: Throat Specimens 8
F. Eye, Ear, and Nose Infections and Specimens for Examination 8
G. Gonorrhea: Exudates and Body Sites 8
H. Feces 9
IV. Wounds, Abcesses, and Exudates 9
V. Conclusions 11
References 11

I. INTRODUCTION

The development and practical application of new mechanical, electronic, immunologic, physiologic, and biochemical techniques for the early and rapid detection of microorganisms in clinical specimens seem to have been accompanied by a return of moderate interest in and emphasis on the use of microscopic examination of Gram-stained smears as an aid in diagnosing infectious diseases. It is as if those individuals immediately interested in diagnostic procedures in infection, both physicians and laboratorians, have been moved to reassess this

The Direct Detection of Microorganisms in Clinical Samples

ISBN 0-12-187780-9

century-old procedure and to be aware that this rather primitive, technically unsophisticated, but valuable technique may have fallen into disuse, misuse, or abuse for one or more of several reasons, including the following:

1. Unrewarding use of Gram's stain because of lack of experience or guidance
2. Insufficient opportunities for the examination and interpretation of observations
3. Lack of discrimination in the selection of specimens to be examined
4. Misguided deferring or neglecting of the immediately available, simple procedure in preference to more modern, specific diagnostic techniques

Some or all of these may be attributed to the recent omissions of laboratory segments from microbiology courses in many medical schools and to diminished hands-on experience in the hospital's clinical microbiology laboratory (Griner and Glaser, 1982). Recently, I observed an eager young student-physician request aerobic, anaerobic, mycobacterial, fungal, viral, and chlamydial cultures of a pulmonary biopsy specimen only to discover abundant pneumococcal-like organisms in a sea of polys on a Gram-stained smear performed on the suggestion of a microbiologist in the laboratory!

Only a passing allusion will be made here to the important consideration of costs of laboratory tests in the overall expense of patient care (Martin *et al.*, 1980); the examination of a Gram-stained smear not only is considerably less costly than one or more other conventional microbiological procedures, but provides earlier results.

The use (often somewhat indiscriminately) of the terms "rapid," "early," "specific," "diagnostic," "predictive value," etc., has almost forced practitioners of infectious disease to retrieve the old Gram's stain data and compare them with the data generated by the new techniques. Indeed, most of the investigators who have developed and evaluated the new techniques have themselves compared the results of their diagnostic procedures in actual cases with the useful data of microscopic examination of clinical specimens. In many of the chapters that follow, this type of comparative evaluation will be demonstrated and critically reviewed.

One of the net results of this gathering of new data and comparing them with the data obtained by old, established procedures has been a refocusing on the older use of microscopy for the primary, early assessment of clinical material for purposes of diagnosing infectious diseases. This chapter attempts to outline and summarize, not exhaustively or exhaustingly, some of the principal elements of the use of microscopy in examining clinical specimens for the diagnosis of bacterial infections. To this end, it is simple and useful to employ the use of Gram's stain as a prototype. Certain other helpful procedures for visualizing bacteria are also discussed when appropriate. As already noted, diagnostic microscopic examinations will be compared with other, newly developed and emerging diagnostic techniques when pertinent.

II. GENERAL USES FOR GRAM'S STAIN

As with any microbiologic examination of clinical material, microscopic examination requires that the specimen to be examined be representative of the infectious process. This requirement has been expressed so frequently that it need not be repeated here. Actually, it should be emphasized that the microscopic examination of a Gram-stained smear of the specimen in itself affords a rapid and easy qualitative assessment of the usefulness of the specimen. Microscopic evaluation can also measure the acceptability of specimens for many other diverse microbiological diagnostic techniques, some of which are discussed in other chapters of this volume.

Consideration of the cellular and microbiological composition of the specimen observed by microscopic examination may enable the physician to select the most appropriate diagnostic laboratory procedure. Thus, polymorphonuclear, monocytic, lymphocytic, acellular, or mixed populations of cells and the presence or absence of bacteria may suggest acute or chronic pyogenic, indolent, bacterial, fungal, viral, or other types of infection. Judgments and decisions derived from such observations are the responsibility of the physician; thus, the basic information should be observed by or communicated to the physician so that it can be integrated with other patient-related information for the formulation of diagnostic procedures. Decisions to order special microbiological cultures or other procedures can then be rationally made.

Probably the principal benefit of examining Gram-stained smears is that one can make an educated guess as to the causative bacterial agent of an infectious disease. Thus, many polymorphonuclear leukocytes together with small, thin, coccobacillary, gram-negative rods on a smear of cerebrospinal fluid (CSF) from a 1-year-old child will suggest the presence of *Haemophilus influenzae* on a statistically strong basis.

Finally, critical examination of a Gram-stained smear of a specimen may provide useful guidance to the bacteriologist in the diagnostic laboratory in selecting culture media and procedures as well as suggesting particular organisms to seek on crowded culture plates containing a variety of organisms (Heineman *et al.*, 1977).

III. USEFULNESS OF DIAGNOSTIC MICROSCOPY IN PARTICULAR TYPES OF SPECIMENS OR INFECTIOUS PROCESSES

It is somewhat difficult to write about the microscopic examination of clinical specimens without becoming encyclopedic. It is hoped that the following sections are sufficiently succinct but do not neglect the basic considerations.

A. Meningitis: Examination of Cerebrospinal Fluid

Etiologic diagnosis of acute meningitis is probably the most urgent request made of a clinical bacteriology laboratory because of the necessity of initiating appropriate antimicrobial therapy in the shortest possible time (within minutes, preferably). The Gram-stained smear can be a life-saving test under certain conditions because it can be prepared, examined, and evaluated within minutes of the collection of the CSF. Even allowing time for transport, clerical maneuvers, and reporting of results to responsible physicians, information essential to the design and administration of effective chemotherapy can be acquired well within an hour under ideal circumstances.

The identification of specific etiologies from Gram-stained smears of CSF depends, of course, not only on morphology but also on the statistical likelihood of a particular microbial species in given clinical situations. Given that context, the reporting of the presence of gram-positive diplococci, thin gram-negative rods, gram-negative diplococci, etc., suggests to the physician the possible or probable identity of the organisms observed.

The absence of organisms in a field of several or many inflammatory cells suggests the adoption of one or more of several other possibly helpful procedures:

1. The specimen can be centrifuged and the Gram-stained smear of the sediment reexamined. Concentration can improve the sensitivity of the test.
2. An acid-buffered acridine orange-stained smear (Kronvall and Myhre, 1977) of the CSF or of centrifuged sediment will reveal more clearly the presence of rare gram-negative organisms, which may be obscured or masked by inflammatory cells and background staining with safranin (see Section IV, for a discussion of the acridine orange staining technique).
3. Finally, a negative microscopic search of CSF specimens should lead to a consideration of alternative techniques, as discussed in other chapters, for early specific etiologic diagnosis.

B. Bacteremia: Examination of Peripheral Blood

The microbiological ''load'' of agents of bacteremia varies tremendously but is often low, ranging down to less than one bacterium per milliliter of circulating blood (Finegold *et al.*, 1969). It is not surprising, therefore, that the microscopic examination of peripheral blood for evidence of etiologic agents is not widely used. The examination of buffy coat of centrifuged blood for bacteria seems not to have been productive in experimental studies and has not been successful in routine examinations in human infections (Kostiala *et al.*, 1979).

C. Urinary Tract Infections

Urinary tract infections, if not the most frequently occurring class of infections, nonetheless provoke the largest number of bacteriologic examinations in clinical laboratories (e.g., they constitute one-third of specimens submitted daily in our laboratory). Fewer than 25% of the urine specimens submitted to the laboratory are "positive" for bacterial pathogens. Whether these "negative" urines represent "false" negative specimens is difficult to judge, but it can probably be assumed that the large majority of negative urine cultures represent a nonarduous, but not inexpensive manner of "screening" patients for urinary tract infection. Yet it is rather well established that by microscopic examinations of Gram-stained smears of urine one can detect 10^5 or more bacteria per milliliter with 80% accuracy (Kass, 1956). Positive smears of centrifuged specimens will reveal the magical number of 10^5 or more with even greater assurance (Tilton and Tilton, 1980). Confirmation of the conclusions derived from the examination of smears is easily obtained by cultures.

D. Respiratory Tract Infections and Specimens for Examination

Respiratory tract infections, particularly pneumonia, present special problems and challenges to the diagnostic microbiologist. There are diverse means of acquiring lower respiratory tract specimens (expectorated sputum, transtracheal aspiration, lung puncture, etc.), and the quality of each type of specimen and therefore its usefulness can vary greatly. The examination of Gram-stained smears of sputum has a long history but still a somewhat controversial role in the etiologic diagnosis of pneumonia. As experienced observers know, however, microscopic examination of a "mixed" specimen (sputum and saliva) may require the selection of areas of the smear that contain lower respiratory tract secretions, as evidenced by inflammatory cells, while the surrounding upper tract flora and commensal organisms are ignored. Parallel bacteriologic cultures of such material, unfortunately, may not permit similar discriminative examination. Thus, various viewpoints have been expressed concerning the value of bacteriologic cultures in respiratory tract infection (e.g., see Barrett-Connor, 1970; Drew, 1977; and Heineman *et al.*, 1977). Bacteriologic examination of respiratory tract specimens by microscopy, with or without parallel cultures, remains a laboratory procedure requiring discriminative care in its performance and in its interpretation and reporting to the physician. The potential of microscopy in diagnosing pneumonia appears to be greater than realized; it should be more intensely studied and invites definitive specific recommendations for more discriminating use.

The microscopic examination of respiratory tract specimens subserves a wide range of other uses in the microbiology laboratory. The results may suggest the need for studies for mycobacteria, mycoplasma, agents of systemic mycoses, chlamydia, or viruses. Alternatively, they may indicate the need for a modified acid-fast stain for organisms resembling *Nocardia* or perhaps additional cultures or smears for yeastlike organisms.

E. Pharyngitis: Throat Specimens

With the disappearance of diphtheria from the continental United States, the difficult challenge of diagnosing, presumptively, diphtheria from a smear of a "diphtheritic" membrane has fortunately vanished. However, there continues to be an interest in diagnosing streptococcal pharyngitis by the examination of Gram-stained smears of pharyngeal exudates (Crawford *et al.*, 1979). One wonders whether the considerations now involved in the early presumptive reporting of streptococcal pharyngitis are similar to or different from the considerations formerly involved, considering the changing incidence of nonsuppurative (and especially cardiac) complications. However, the presumptive etiologic diagnosis of streptococcal pharyngitis by microscopic examination of Gram-stained smears would seem to warrant further clinical consideration and study.

F. Eye, Ear, and Nose Infections and Specimens for Examination

The use of nose specimens (not nasopharyngeal exudates) for any purpose other than culture for carrier state can rarely be justified; nose cultures for the accurate etiologic diagnosis of otitis media were discredited by Mortimer and associates many years ago (1956).

Ear and eye specimens, which usually reflect their respective autochthonous bacteria under microscopic examination, may present special problems in interpretation when stained by the method of Gram or with acridine orange. Cultures are preferable to microscopy in this situation.

G. Gonorrhea: Exudates and Body Sites

In a large percentage of patients, especially in heterosexual men, the diagnosis of gonorrhea can be made presumptively by the examination of a Gram-stained smear of urethral exudate. The diagnosis depends on the finding of characteristically shaped gram-negative diplococci within and outside of polymorphonuclear leukocytes. When specimens are obtained from women or from nonurethral mucous membrane surfaces from homosexual men or sexually active women of either persuasion, gram-negative diplococci on smears may not be gonococci and may be other species of *Neisseria,* particularly *N. meningitidis.*

Moreover, when gonococci are few in number, their presence may be overlooked entirely because of the gram-negative background of inflammatory cells and protein, which are usually present (Kronvall and Myhre, 1977; Lauer *et al.*, 1981). In these situations acridine orange stain aids in the visualization of morphologically distinct organisms but, of course, still gives only a presumptive diagnosis. Species identification requires other measures. The problem of mistaking "Herellea" (*Acinetobacter calcoaceticus* var. *anitratus*) for *N. gonorrhoeae* in urethritis (DeBord, 1939; Samuels *et al.*, 1969) is probably no longer a frequent occurrence.

H. Feces

Microscopic examination of fecal smears is only occasionally useful. The detection of polymorphonuclear leukocytes may suggest *Shigella* or *Campylobacter* infection rather than toxinogenic etiologies. Similarly, an overgrowth of staphylococci, yeasts, or plaques in pseudomembranous enterocolitis of diverse etiology may suggest additional microbiological procedures for definitive diagnosis.

In general, the most useful diagnostic preparation for the microscopic examination of feces is a preparation stained with methylene blue or Gram's stain for the presence or absence of polymorphonuclear leukocytes or for the determination of massive overgrowth of a single distinctive morphotype among the normal diverse bacterial population.

Some consideration should be given, for epidemiologic reasons at least, to the utility of examining feces under dark-field or phase microscopy for the "characteristic darting motility" of *Campylobacter* in cases of gastroenteritis (Karmali and Fleming, 1979; Rettig, 1979). The practicality of this procedure has to be measured against the usefulness of the presumptive evidence so derived. Its use does not appear to be warranted for every fecal specimen submitted for examination.

IV. WOUNDS, ABCESSES, AND EXUDATES

For brevity, microscopic examinations of wound, abscess, and exudate specimens are considered together. They often contain myriad tissue cells and inflammatory cells and bacteria are frequently obscured by the distracting background of gram-negative cell debris.

This dilemma was resolved by Kronvall and Myhre (1977), who used the microbiological fluorochrome stain acridine orange for sharpening the distinction between bacteria, tissues, and cells. Acridine orange buffered at pH 4, as used by Kronvall and Myhre, discriminated sharply between bacterial cells, which appeared bright orange, and the greenish-yellowish background of tissue or inflammatory cells. The differentiation of bacterial cells from other components of the smear has been favorably reviewed by others (Lauer *et al.*, 1981; Rosendahl and Valdivieso-Garcia, 1981).

The differential aspects of the acid-buffered staining procedure may be based on color differentiation between DNA and RNA in biological specimens. Whatever the biological basis for differentiation, bacteria seem to be more easily detected in mixed smears with this stain. However, the Gram reaction needed for tentative identification of the bacteria, when discerned, requires examination of a superimposed Gram's stain.

The urgency with which bacteriologic diagnosis should be pursued with exudative materials may depend on conditions not always communicated to the laboratorian. Assuming that it is desirable to make a rapid and presumptive diagnosis, examples can be given to illustrate the ways in which a laboratory can provide key information by direct visualization (which does not exclude other informative procedures and tests by any means):

1. A simply solved diagnostic problem presented by a physician is an exudate (drainage or surgically acquired aspirate) that contains polys and a single bacteriologic morphotype (e.g., gram-positive cocci in clusters). This represents an intellectually and technologically easy presumptive diagnostic response: infection by *Staphylococcus,* probably *S. aureus*. The presumption must be confirmed by follow-up culture.
2. If relatively unique forms (e.g., thin, gram-positive, filamentous organisms) are found in a suggestive site (e.g., brain or lung aspirations), one might suggest important adjuctive procedures, such as staining a duplicate smear with modified Ziehl–Neelsen stain for evidence of nocardial infection, followed by appropriate confirmatory cultures.
3. Examination of a specimen (e.g., pulmonary) that yields no specific evidence of pyogenic bacteria might lead one to suggest to the attending physician the need for additional, more specific procedures, in accordance with the epidemiologic, physical, and clinical condition of the patient. Such special techniques might include diagnostic tests for other bacterial infections (such as *Mycobacterium* and *Legionella*) or for fungi, mycoplasma, chlamydia, or other agents.
4. One can reiterate the special (and not necessarily diagnostic) contribution of microscopic examination of a stained preparation to diagnostic microbiology. Armed with the knowledge of the morphologic types of bacteria that have been observed (e.g., in a Gram-stained smear) and the types and identities of the colonies appearing in cultures of the specimen, one can make rational judgments of what organisms may be "missing" (e.g., anaerobes, dysgonic, or slow growers, fungi). They can then be deliberately sought in reincubated or dupli-

cated cultures. Microscopic examination of the specimen should serve as the ultimate verdict of the acceptability of the results of examination of the culture.

V. CONCLUSIONS

Microscopy is one of the oldest arts in diagnostic microbiology. Its use has been refined and extended not only by improvements in optics and instrumentation, but also by additions to basic staining techniques. Because of the simplicity and rapidity with which preparations can be made for microscopic examination, critically useful (although tentatively or presumptively accurate) observations can be made for diagnosing and treating serious infectious diseases.

The usefulness of Gram's stain in the evaluation of diseases suspected of being infectious is attested to by its century-long history and continuing use (albeit somewhat less frequently). The microscopic examination of a Gram-stained smear remains a procedure of unparalleled rapidity and great technical simplicity. It remains a technique that requires—nay, demands—greater use in this age of medicine in which serious infectious diseases are becoming more and more complicated by the involvement of commensal, autochthonous, and "nonpathogenic" bacteria.

REFERENCES

Barrett-Connor, E. (1970). *Am. Rev. Respir. Dis.* **130,** 845–848.
Crawford, G., Brancato, F. and Holmes, K. K. (1979). *Ann. Intern. Med.* **90,** 293–297.
DeBord, G. G. (1939) *J. Bacteriol.* **38,** 119–120.
Drew, W. L. (1977). *J. Clin. Microbiol.* **6,** 62–65.
Finegold, S. M., White, M. L., Ziment, I., and Winn, W. R. (1969). *Appl. Microbiol.* **18,** 458–463.
Griner, P. F., and Glaser, R. J. (1982). *N. Engl. J. Med.* **307,** 1336–1339.
Heineman, H. S., Chawla, J. K., and Lofton, W. M. (1977). *J. Clin. Microbiol.* **6,** 518–527.
Karmali, M. A., and Fleming, P. C. (1979). *J. Pediatr. (St. Louis)* **94,** 527–533.
Kass, E. H. (1956). *Trans. Assoc. Am. Physicians* **69,** 56–63.
Kostiala, A. A. I., Jormalainen, S., and Kosunen, T. U. (1979). *Am. J. Clin. Pathol.* **72,** 437–443.
Kronvall, G., and Myhre, E. (1977). *Acta Pathol. Microbiol. Scand.* **85,** 249–254.
Lauer, B. A., Reller, L. B., and Mirrett, S. (1981). *J. Clin. Mocrobiol.* **14,** 201–205.
Martin, A. R., Wolf, M. A. Thibodeau, L. A., Dzau, V., and Braunwald, E. (1980). *N. Engl. J. Med.* **303,** 1330–1336.
Mortimer, E. A., and Watterson, R. L. (1956). *Pediatrics* **17,** 359–367.
Rettig, P. J. (1979). *J. Pediatr. (St. Louis)* **94,** 855–864.
Rosendahl, S., and Valdivieso-Garcia, A. (1981). *Appl. Environ. Microbiol.* **41,** 1000–1002.
Samuels, S. B., Pittman, B., and Cherry, W. B. (1969). *Appl. Microbiol.* **18,** 1015–1024.
Tilton, R. E., and Tilton, R. C. (1980). *J. Clin. Microbiol.* **11,** 157–161.

2

Microscopic Preparations for Detecting Fungi in Clinical Materials

BILLY H. COOPER

Director of Mycology
Department of Pathology
Baylor University Medical Center
Dallas, Texas

I. Role of Microscopic Techniques in Detecting Fungi in Clinical Specimens 13
II. Methods for Visualizing Fungi in Clinical Materials 19
 A. Processing Specimens for Microscopic Examination 19
 B. Wet Mounts 28
 C. Routine Stains 29
 D. Special Stains for Fungi 29
 E. Fluorescent Antibody Techniques 30
 F. Direct Fluorescence 30
III. Examples of Fungi in Clinical Materials 31
 References 33

I. ROLE OF MICROSCOPIC TECHNIQUES IN DETECTING FUNGI IN CLINICAL SPECIMENS

Microscopic observation of fungi in stained or unstained clinical materials is usually the most rapid means of detecting a fungus disease (Cooper, 1982). The frustratingly slow rate of growth of many fungi makes their isolation in laboratory culture time-consuming. However, direct microscopic examination of wet mounts or stained smears of clinical materials can be accomplished within minutes of collecting the specimen. Even microscopic examination of formalin-fixed and sectioned tissues stained with special fungus stains can be accomplished within 48 to 72 hr of collection. The detection of fungal antigens with sensitive techniques or of unique fungal metabolites in body fluids using special pro-

The Direct Detection of Microorganisms in Clinical Samples

ISBN 0-12-187780-9

cedures can be carried out with equal speed. Those procedures are described elsewhere in this volume.

With the exception of *Candida* and some other yeasts, fungi do not normally reside in or on body surfaces of humans. For that reason the presence of a particular fungus in a clinical specimen is usually indicative of infection by that fungus. However, some saprophytic molds may transiently reside with *Candida* in the respiratory tract, gastrointestinal tract, and lower genitourinary system without causing overt infection. Yeasts and some saprophytic molds may also exist on skin, hair, and nails without causing invasive disease. However, these transients are usually observed in a different form when they are merely contaminants than when they cause invasive disease. For example, yeast cells and pseudohyphae observed in clinical materials suggest tissue invasion by a *Candida* species, whereas budding yeast cells by themselves suggest only colonization. In like manner, hyphae that penetrate epithelial cells suggest invasive infection by a mold, whereas conidia on the surface of epithelial cells suggest transient residence by a saprophytic mold.

When fungi of any kind are observed in tissues or body fluids that are normally sterile, their clinical significance is immediately recognized. By the same token the presence of the tissue form of a known pathogen that is easily recognized, such as the yeast form of *Blastomyces dermatitidis,* is also immediately significant. This is true even though these forms may be observed in sputum contaminated with saliva or in other contaminated specimens.

Many saprophytic fungi that contaminate clinical materials can also cause invasive disease in patients whose immune defenses are not properly functioning. In such instances it is essential that the invading fungus be seen by microscopic examination of the affected tissue in order to establish a complete chain of evidence for its invasiveness. Mere isolation of a common saprophyte, even in high number, does not rule out the possibility that the specimen could have been contaminated during processing. However, even fungi that have not previously been reported to cause human disease may be documented as the cause of a serious infection when they are first observed microscopically and then isolated from the same specimen. As cytotoxic and immunosuppressing medications are used more aggressively for chemotherapy of malignancies, it becomes even more important that fungi be detected in clinical specimens by direct microscopic examination. This is necessary both for rapid diagnosis and for documentation of unusual infections (Cooper, 1982).

The sensitivity of microscopic detection of fungi depends to a large extent on the number of fungal elements in the specimen and on a high level of suspicion on the part of the observer. The specificity with which fungi can be identified in clinical specimens is dependent on the skill and experience of the observer. Special stains for fungi, such as the periodic acid–Schiff (PAS) stain and the Gomorri methenamine silver (GMS) stain can be used with smears (Forster *et*

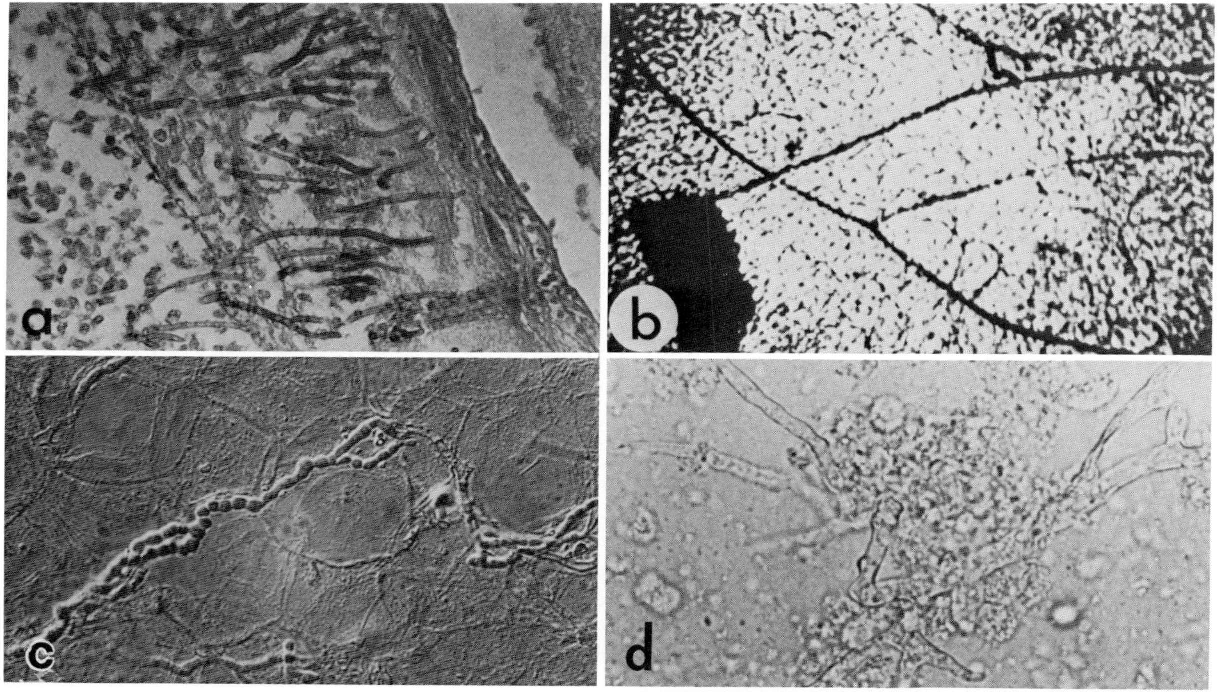

Fig. 1. (a) *Aspergillus* sp. Brain. GPS strain. 400×. (Courtesy Dr. G. Kobayashi.) (b) *Trichophyton rubrum*. Hyphae in skin scraping. KOH and ink (Swartz-Lamkins stain). 400×. (c) *Trichophyton rubrum*. Hyphae and arthroconidia in skin scraping. KOH mount. Nomarski phase-interference illumination. 1000×. (From McCracken and Cawson (eds.), *Clinical and Oral Microbiology* 1983. Hemisphere Publishing Corp.) (d) *Mucor* sp. Sputum. Unstained wet mount. Note hyphae without cross walls (coenocytic). 400×. (Courtesy Dr. G. Kobayashi.)

al., 1976) as well as with tissue sections to good advantage for detecting and identifying fungi. Fluorescent antibody (FA) techniques for detecting fungi enhance both the sensitivity of detection and the specificity of identification (Kaplan and Kaufman, 1961; Gordon *et al.*, 1967; Kaufman and Blumer, 1968; Kaplan, 1973, 1975). However, specific fluorescein-labeled conjugates for detecting fungi are available only at the Centers for Disease Control and at certain other reference laboratories.

In contrast, fungi can frequently be seen in clinical materials that are stained for purposes other than detecting fungi. Alert technologists in bacteriology, cytology, hematology, urinalysis, and parasitology can contribute enormously to the capacity of the laboratory to detect fungi. In our experience the alert detection

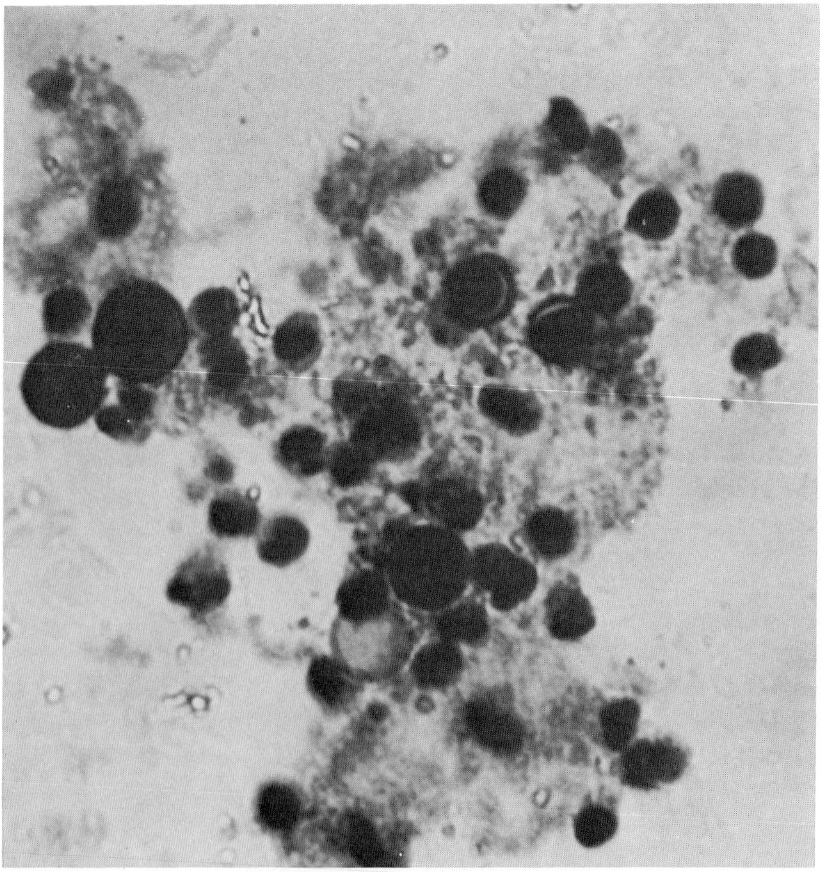

Fig. 2. *Blastomyces dermatitidis*. Sputum. Methylene blue stain. 400×. (From McCracken and Cawson (eds) *Clinical and Oral Microbiology*. 1983. Hemisphere Publishing Corp.).

of fungi in clinical materials where they were not expected has often pointed the way to a correct diagnosis in a confusing clinical situation.

The genus and sometimes even the species of many fungi can be identified by microscopic observation of the tissue phase. One should report the results of microscopic observations as specifically as possible without overdiagnosing. A report such as "encapsulated yeast cells typical of *Cryptococcus* observed" is much more informative than a vague report such as "many fungal elements observed." Fungi are seen in clinical specimens as either (*a*) hyphae (Fig. 1), (*b*) budding yeast cells (Fig. 2–6), (*c*) pseudohyphae (Fig. 7), or (*d*) spherules (Fig. 8). Additional fine characteristics such as the size, shape, and number of buds in yeasts, the presence or absence of septation in hyphae, and the presence of capsules surrounding budding yeasts help to characterize individual fungi. Spherules are seen only in tissues infected with *Coccidioides immitis* or *Rhinosporidum seeberi*. The latter species is rarely seen in the United States. Fungal elements that have a natural brown pigmentation are characteristic of members of the family Dematiaceae (black molds and black yeasts). These fungi cause tinea nigra, mycetoma, chromoblastomycosis, and phaeohyphomycosis. In chromoblastomycosis the forms seen in affected tissues are brown, sclerotic cells (Medlar's cells), which divide by forming transverse septations (Fig. 9). This unique tissue form is only seen in chromoblastomycosis.

Microcolonies of fungi called grains or granules are seen in mycetoma (Fig. 10c) and in actinomycosis ("sulfur granules," Fig. 10b). The size, shape, color, and consistency of these grains provide important clues to their identity. The characteristics of fungi as seen in infected tissues and clinical specimens are summarized in Table I.

II. METHODS FOR VISUALIZING FUNGI IN CLINICAL MATERIALS

A. Processing Specimens for Microscopic Examination

Concentrating fungi that are present in clinical specimens aids in their detection, especially when the number of fungi is small. Centrifugation of fluid specimens at 2500 *g* and examination of the sediment is the recommended approach for concentrating fungi in such specimens. Viscous respiratory secretions can be liquified with *N*-acetyl-L-cysteine (without NaOH) or with dithiothreitol and then centrifuged (Haley *et al.*, 1980). However, the yeasts that are normally present in respiratory secretions will also be concentrated by this

TABLE I

Selected Fungi in Clinical Materials

Species	Description	Illustration
I. Fungi that produce hyphae		
A. *Aspergillus*	Septate hyphae 3–12 μm wide; dichotomous (acute-angle) branching; occasionally, conidiophores and conidia characteristic of the genus are seen	Figure 1a
B. Dermatophytes	Septate branching hyphae 3–15 μm in diameter; keratinized epithelial cells of skin or in nails; arthroconidia on outside of hair (ectothrix) or inside hair (endothrix); not usually seen in deeper tissues	Figure 1b, 1c
C. *Geotrichum*	Septate; arthroconidia	—
D. *Pseudoallescheria* (*Petriellidium*) *boydii*	Septate[a]	—
E. *Trichosporon*	Septate; arthroconidia	—
F. Zygomycetes: *Mucor, Rhizopus, Absidia, Cunninghamella, Saksenaea*	Large hyphae without cross walls (coenocytic) 3–25 μm in diameter; branching at right angles	Figure 1d
II. Fungi that are yeastlike		
A. *Blastomyces dermatitidis*	Large yeast cells, usually 8–15 μm in diameter; thick-walled, single bud with a broad base	Figure 2
B. *Cryptococcus neoformans*	Encapsulated yeast cells usually 4–7 μm in diameter, but quite variable in size and shape; stain positively with mucicarmine and alcian blue	Figure 3
C. *Histoplasma capsulatum*	Small, oval yeast cells 2–5 μm in diameter; usually seen in clusters inside macrophages	Figure 4
D. *Paracoccidioides brasiliensis*	Large, spherical yeast cells 10–30 μm in diameter; multiple buds with a narrow attachment to the parent cell	Figure 5

TABLE I (*Continued*)

Species	Description	Illustration
E. *Sporothrix schenckii*	Small yeast cells 2–5 μm in diameter; variable in size and shape; round, oval, cigar-shaped cells with single or multiple buds	Figure 6
F. *Torulopsis* (*Candida*) *glabrata*	Small, round to oval yeasts, 2–5 μm in diameter; budding is more prevalent in tissue than with *H. capsulatum*, which it closely resembles; may appear in clusters inside macrophages, or individual cells may be scattered in the specimen	Figure 4d
G. Yeasts (*Saccharomyces, Hansenula,* etc.)	—	—
III. Fungi that produce pseudohyphae		
Candida[b]	Pseudohyphae, thin-walled budding yeasts 3–5 μm in diameter; hyphae may be seen in deeper tissues	Figure 7
IV. Fungi that produce spherules		
A. *Coccidioides immitis*	Spherules 10–80 μm in diameter, filled with endospores; septate hyphae may be seen in necrotic tissue or in cavitary lesions of the lungs	Figure 8
B. *Rhinosporidium seeberi*	—	—
V. Fungi that produce sclerotic cells (Medlar's cells)		
A. *Cladosporium carrionii* B. *Fonsecaea compactum* C. *Fonsecaea pedrosoi* D. *Phialophora verrucosa* E. *Wangiella dermatitidis*	Brown spherical cells with transverse septations (sclerotic cells) in chromoblastomycosis; brown septate hyphae, budding yeast cells, and sclerotic cells in phaeohyphomycosis	Figure 9

[a] *Petriellidium boydii* also produces grains in cases of eumycotic mycetoma.

[b] *Candida* species can produce yeast cells, pseudohyphae, and/or true hyphae in tissue. Usually, pseudohyphae and budding yeast cells are observed.

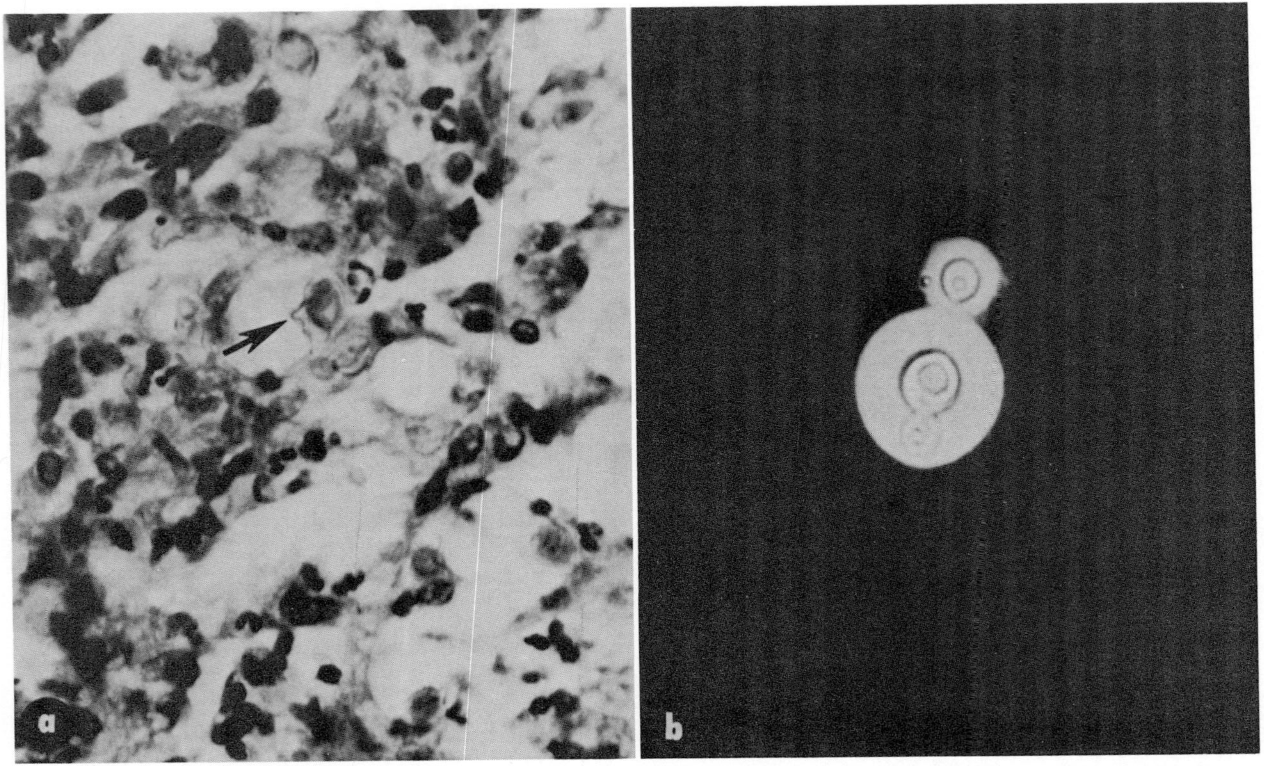

Fig. 3. (a) *Cryptococcus neoformans* (arrow). Lung tissue. Mayer's mucicarmine stain. 400×. (b) *Cryptococcus neoformans*. Cerebrospinal fluid. India ink mount. 400×.

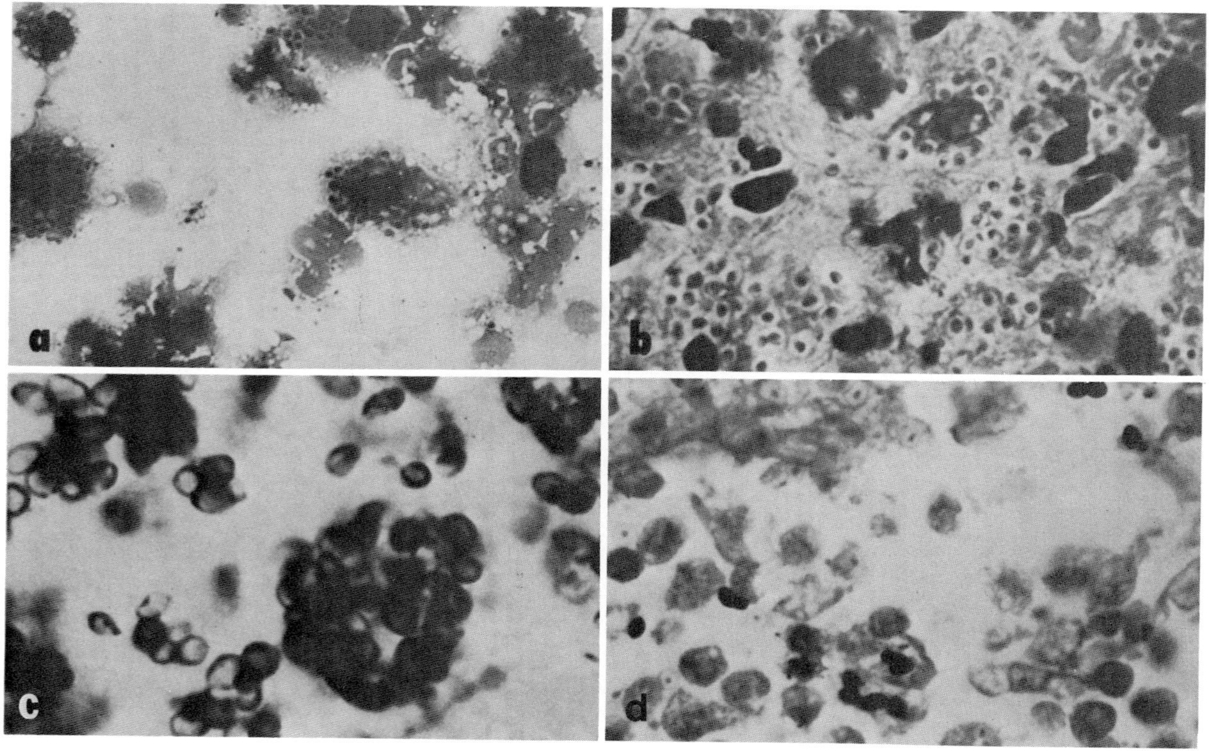

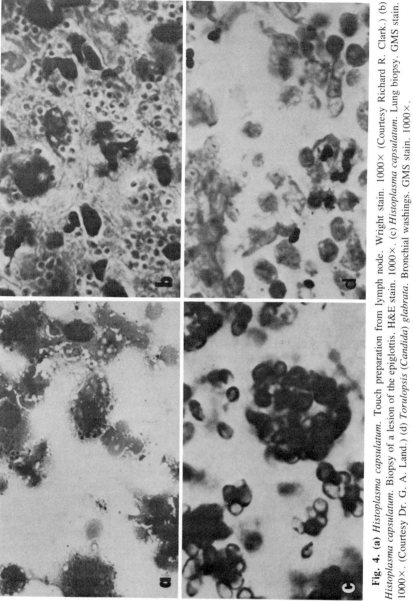

Fig. 4. **(a)** *Histoplasma capsulatum*. Touch preparation from lymph node. Wright stain. 1000× (Courtesy Richard R. Clark.) (b) *Histoplasma capsulatum*. Biopsy of a lesion of the epiglottis. H&E stain. 1000×. (c) *Histoplasma capsulatum*. Lung biopsy. GMS stain. 1000×. (Courtesy Dr. G. A. Land.) (d) *Torulopsis* (*Candida*) *glabrata*. Bronchial washings. GMS stain. 1000×.

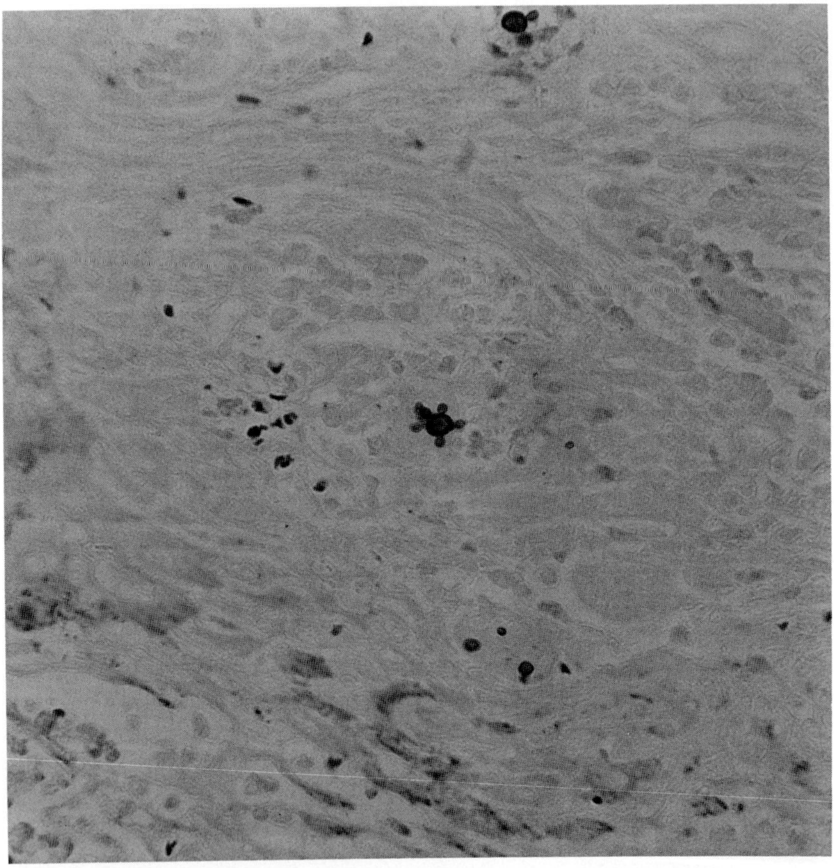

Fig. 5. *Paracoccidioides brasiliensis.* Biopsy of a lesion of the tongue. GMS stain. 400×.

procedure and may obscure other fungi. For that reason it is just as efficient to examine the specimen carefully and pick out flecks of pus, blood, mucous plugs, or other formed elements that can be observed with the unaided eye for further observation with a microscope. This can be accomplished by using a sterile loop or applicator stick. Wet mounts or thin smears should be made with blood, pus, sputum, vaginal discharges, exudates, and other fluid specimens. Thick pus or sputum can be treated with 10 to 20% KOH in order to clear away cellular debris that can obscure any fungi that are present. Tissues should be minced or ground and then treated with KOH. Alternatively, a thin smear can be made by touching the cut surface of a piece of tissue to a microscope slide. When dried in air, this type of smear is very satisfactory for staining by a variety of methods. Tissues fixed with formalin, embedded in paraffin, and sectioned with a microtome are

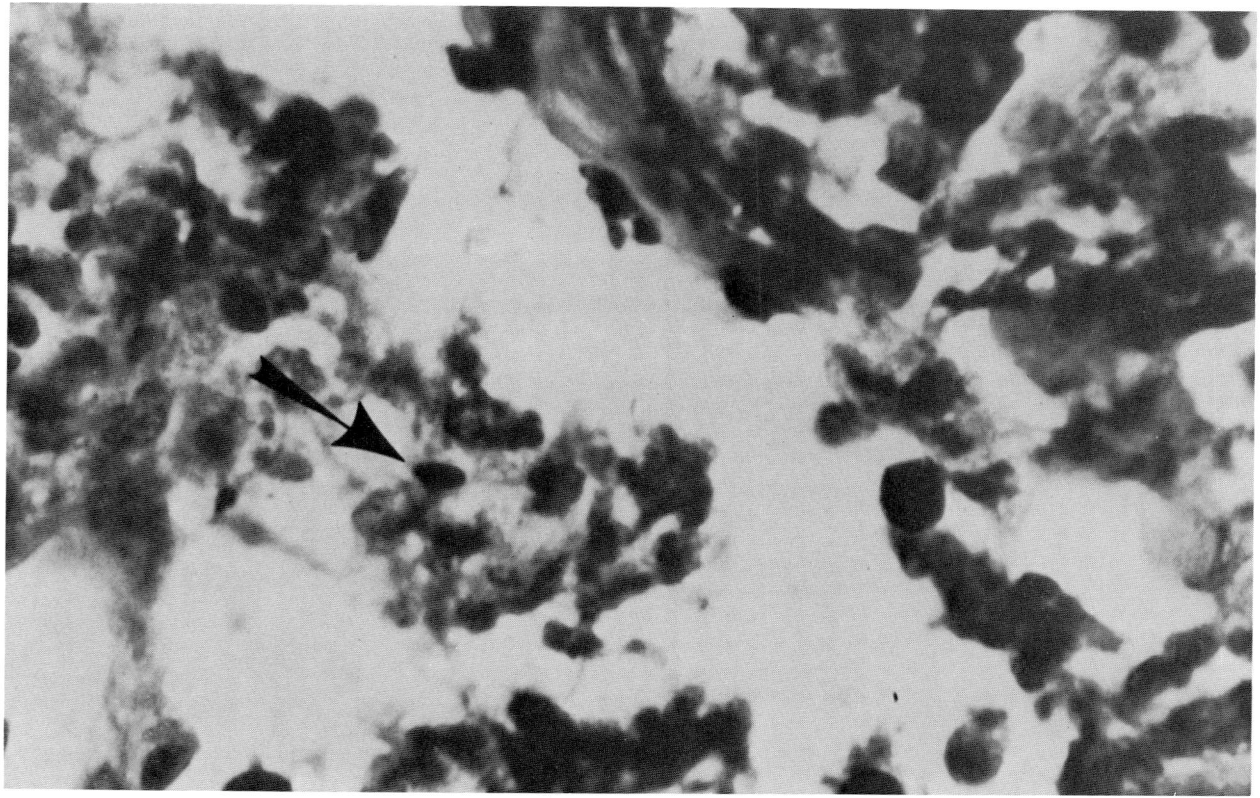

Fig. 6. *Sporothrix schenckii* (arrow). Biopsy of a skin lesion. PAS stain following diastase digestion. 400×. (Courtesy Richard R. Clark.)

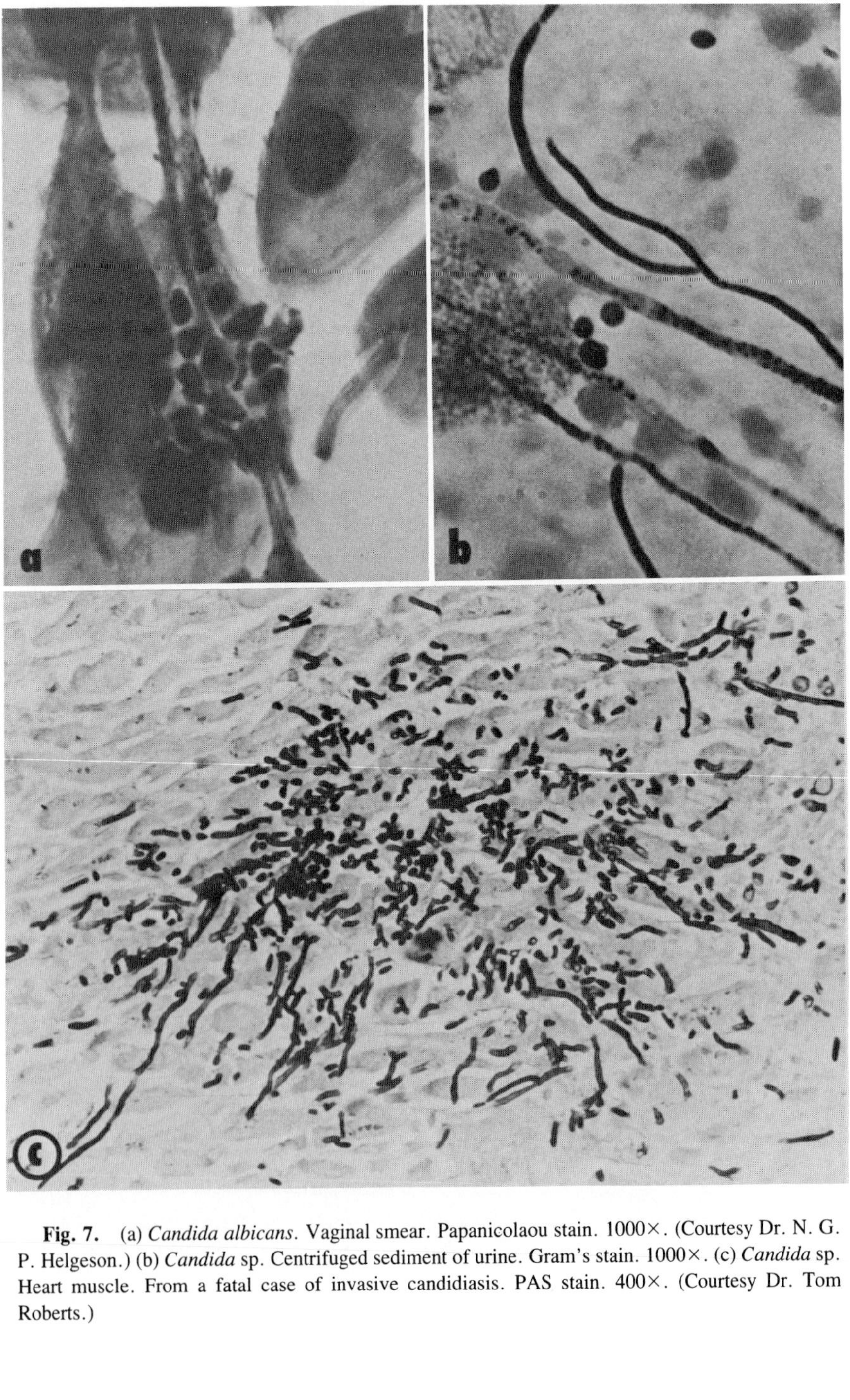

Fig. 7. (a) *Candida albicans*. Vaginal smear. Papanicolaou stain. 1000×. (Courtesy Dr. N. G. P. Helgeson.) (b) *Candida* sp. Centrifuged sediment of urine. Gram's stain. 1000×. (c) *Candida* sp. Heart muscle. From a fatal case of invasive candidiasis. PAS stain. 400×. (Courtesy Dr. Tom Roberts.)

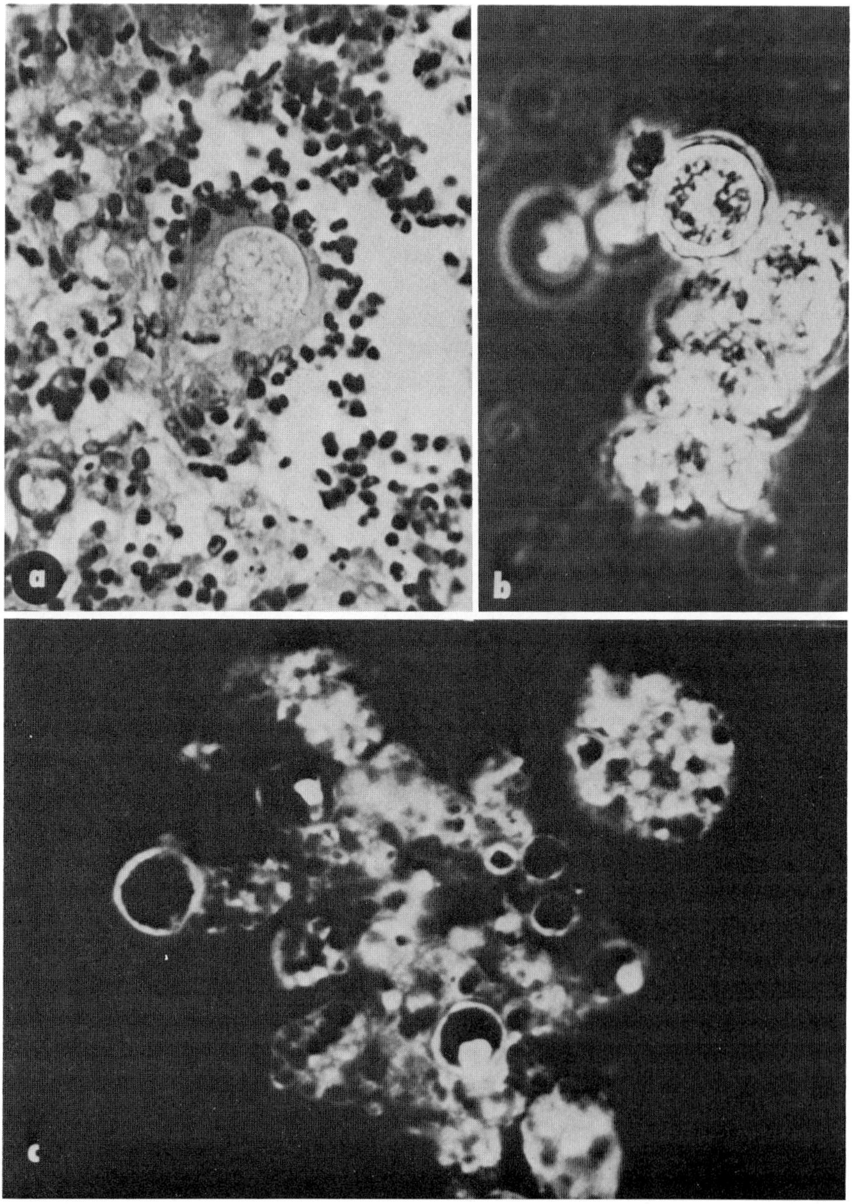

Fig. 8. (a) *Coccidioides immitis*. Lung biopsy. H&E stain. 400× (b) *Coccidioides immitis*. Touch preparation from a lung resection. Unstained. Phase-contrast illumination. 400×. (c) *Coccidioides immitis*. Lung biopsy. Fluorescent antibody. 400×. (Courtesy Dr. W. Kaplan, Centers for Disease Control.)

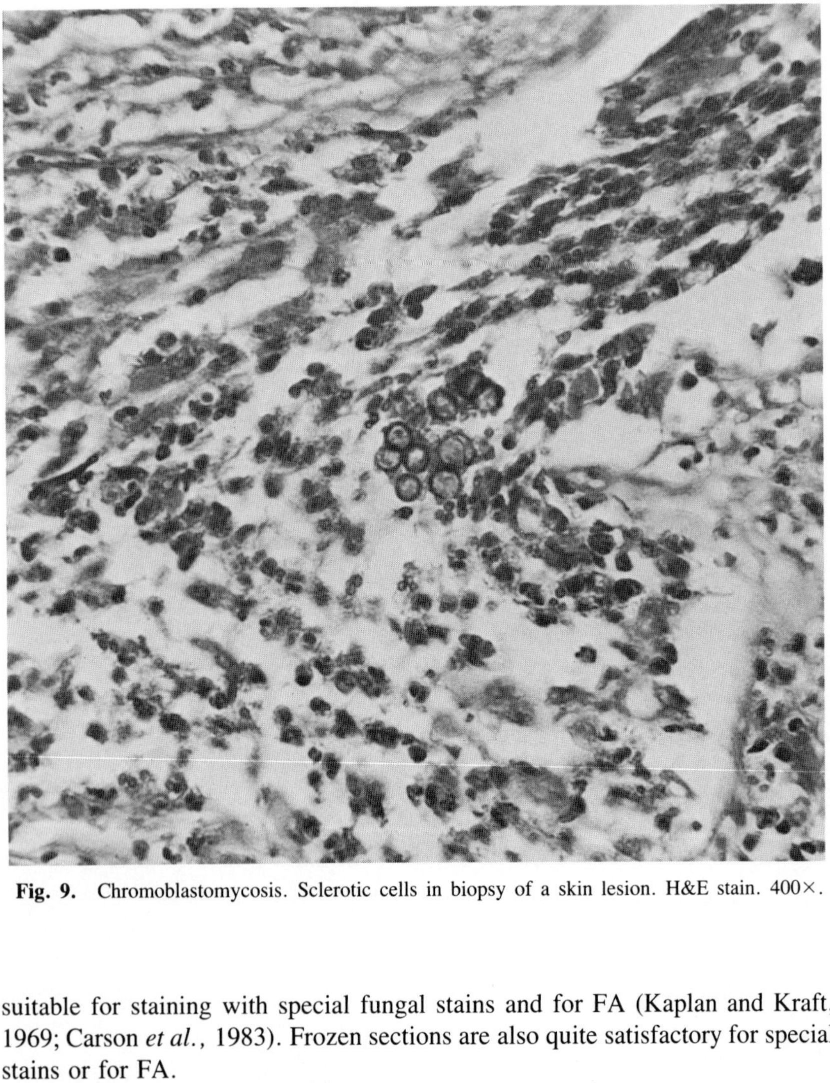

Fig. 9. Chromoblastomycosis. Sclerotic cells in biopsy of a skin lesion. H&E stain. 400×.

suitable for staining with special fungal stains and for FA (Kaplan and Kraft, 1969; Carson *et al.*, 1983). Frozen sections are also quite satisfactory for special stains or for FA.

Scrapings of skin, nail clippings, and hairs are best handled by direct placement onto a microscope slide and addition of KOH. If these materials are collected in a physician's office and sent to a distant laboratory for microscopic examination and culture, they should be placed in a sterile petri dish or similar sterile container or put into a clean envelope for transport to the laboratory.

Pus from a draining subcutaneous lesion or sinus tract should be carefully inspected with a hand lens for the presence of granules or grains. These should be picked out with a sterile loop, washed with sterile saline, and placed on a microscope slide for direct observation or staining.

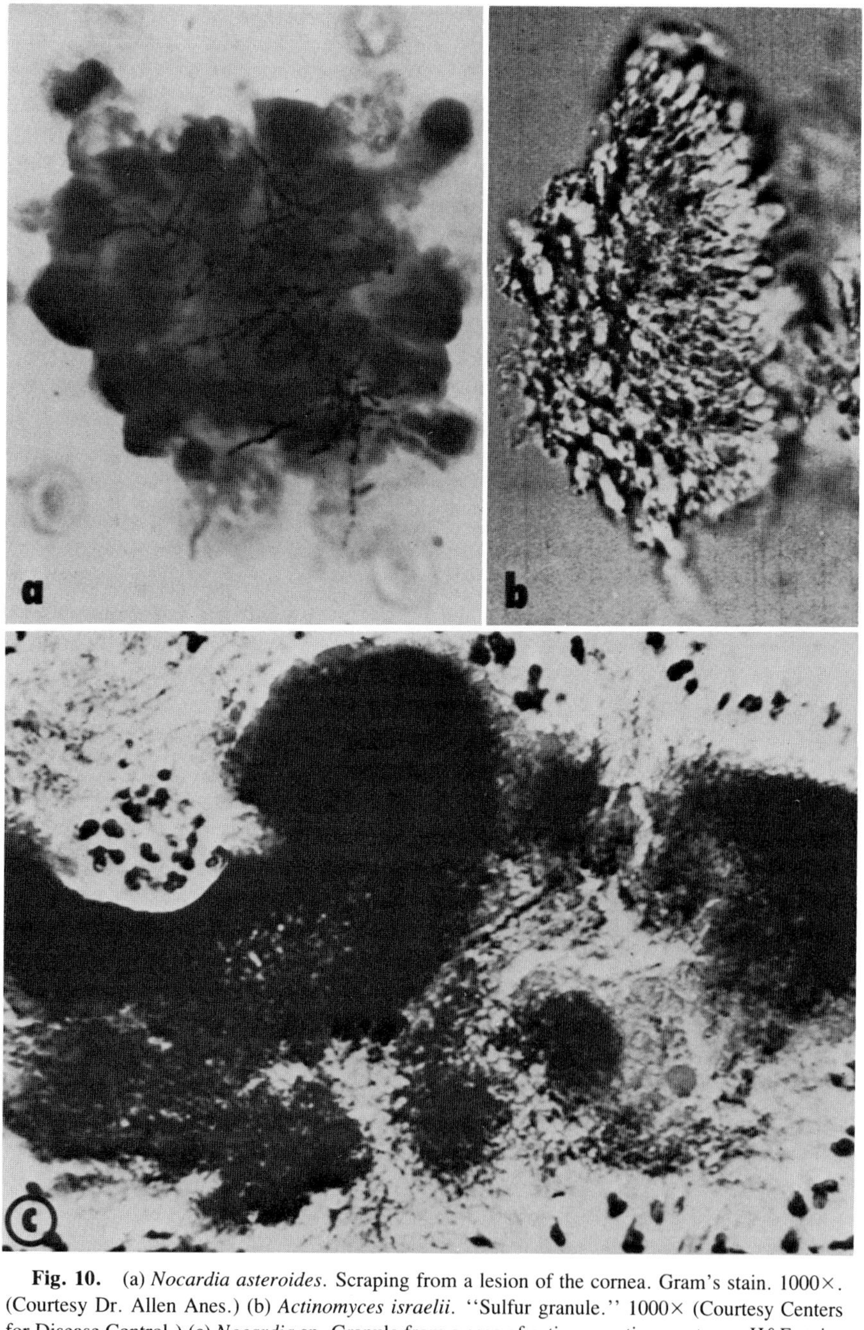

Fig. 10. (a) *Nocardia asteroides*. Scraping from a lesion of the cornea. Gram's stain. 1000×. (Courtesy Dr. Allen Anes.) (b) *Actinomyces israelii*. "Sulfur granule." 1000× (Courtesy Centers for Disease Control.) (c) *Nocardia* sp. Granule from a case of actinomycotic mycetoma. H&E stain. 1000×. (Courtesy Centers for Disease Control.)

B. Wet Mounts

1. Unstained

Fungi can often be detected in unstained wet mounts of fluid specimens such as sputum, pus, urine, peritoneal fluid, cerebrospinal fluid (CSF), and other body fluids. Reduction of the intensity of the light source will heighten the contrast between fungal elements and the specimen and make them more visible. The use of phase-contrast (Fig. 8a) or Nomarski phase-interference illumination (Fig. 1c) increases the resolution of fungal elements in unstained wet mounts (Roberts, 1975).

2. Potassium Hydroxide Preparations

The treatment of skin, hair, and nails as well as thick pus or sputum with 10 to 20% KOH is perhaps the most familiar method for visualizing fungi in clinical materials. Treatment of the specimen with KOH facilitates clearing of the specimen without destroying any fungi that are present. The clearing effect can be accelerated by gently warming the slide or by adding 1% dimethyl sulfoxide to the KOH.

The addition of blue-black ink to the KOH causes any fungi that are present to be stained (Fig. 1b). This modified KOH is known as the Swartz–Lamkins stain (Swartz and Lamkins, 1964) and is commercially available by that name. The blue-black ink used for this stain is no longer available in office supply stores but can be obtained from stores in which drafting and art supplies are sold.

For quality-control purposes the KOH preparation should be treated as a stain. Because KOH-positive clinical materials are not readily available, a suspension of yeast cells or of hyphae may be added to a separate slide and treated with KOH and observed alongside the specimen.

The KOH preparation is not by any means the most satisfactory method for visualizing all fungi in every specimen. It is most useful with skin, hair, and nail specimens.

3. India Ink Mount

The India ink mount is a method for demonstrating yeast cells that have capsules. It is used primarily for detecting *Cryptococcus neoformans* in CSF (Fig. 3b). It can also be used with other fluid specimens. India ink is used as a negative stain; that is, the background is obliterated by the dark carbon particles, and encapsulated yeast cells, which are not penetrated by the carbon particles, stand out in sharp contrast to the dark background. India ink is not a specific stain for *C. neoformans;* however, one is not very likely to observe other encapsulated yeasts in CSF. *Cryptococcus neoformans* is the most frequent cause of fungal meningitis, and to many physicians a positive India ink mount is sufficient for making a diagnosis of cryptococcosis. However, every positive India ink mount should be verified by culture and serology for *Cryptococcus* antigen. The capsule of *C. neoformans* can be stained with alcian blue stain and with Mayer's mucicarmine stain (Fig. 3a) as well as with FA (Chandler *et al.*, 1980; Schwarz, 1982; Carson *et al.*, 1983).

Other fungi can be seen in the India ink mount. Species such as *Candida albicans* that occasionally cause meningitis can be detected but not identified with India ink. White blood cells, red blood cells, and other artifacts seen in CSF can be confused with encapsulated yeasts. For that reason it is convenient to have on hand a formalin-treated suspension of encapsulated yeast cells to use for quality control of the India ink test.

C. Routine Stains

Many fungi can be visualized quite adequately with Gram's stains, Wright or Giemsa stains, Papanicolaou (Pap) stains, and hematoxylin and eosin (H & E) stains (Schwarz, 1982). These routine stains are designed for other purposes and may not be as adequate as special stains and wet mounts for detecting fungi; however, fungi observed in smears stained by these methods should not be ignored. Virtually all fungi can be stained with a simple methylene blue stain (Fig. 2). Although some skill is required for identifying fungi stained in this way, methylene blue may be an acceptable alternative to Gram's stain as a routine stain for smears of fluid specimens. *Candida* species (Fig. 7b) stain very well with Gram's stain. Because this is a widely used stain and is familiar to most technologists, it is important to recognize fungi stained with it. *Cryptococcus neoformans* can be stained with Gram's stain or with the Wright stain, although it may not be easily recognized in smears stained with those stains. *Histoplasma capsulatum* can be observed in smears stained with Wright or Giemsa stain (Fig. 4a) and with H & E (Fig. 4b). With these stains the cell wall does not stain, and the fungus appears to be surrounded by a capsule. The importance of recognizing fungi in cytologic preparations stained with the Pap stain (Fig. 7a) has been stressed (Sutliff and Cruthirds, 1973; Saigo *et al.*, 1977; Sanders *et al.*, 1977; Bernard *et al.*, 1980; Trumbull and Chesney, 1981). *Aspergillus* species and Zygomycetes can be stained very adequately with H & E.

D. Special Stains for Fungi

Stains that have been developed for visualizing fungi in tissues react chemically with carbohydrate polymers that make up the cell walls of fungi. The PAS stain reacts with aldehyde groups created by oxidizing hydroxyl groups with periodic acid. With this stain, fungal cell walls stain red (Carson *et al.*, 1983). The GMS (Grocott, 1955) and Gridley fungus stains (GFS) (Gridley, 1953) react in a similar fashion. With the GMS stain, fungal cell walls stain black and, with the GFS, they stain red-purple. Overall, the GMS stain seems to provide the maximum contrast between fungi and the background. Treatment of a smear or tissue section with diastase before staining digests glycogen and other materials that can obscure fungi. This treatment is especially beneficial for detecting *Sporothrix schenckii* in clinical materials and tissue sections. The H & E stain may be used as a counterstain with the GMS stain. This allows one to study the tissue reaction and at the same time provides a sensitive stain for fungi (Chandler *et al.*, 1980; Schwarz, 1982).

Mayer's mucicarmine stain (Carson *et al.*, 1983) and the alcian blue stain (Carson *et al.*, 1983) react with the acid mucopolysaccharide capsule of *Cryptococcus neoformans*. Mucicarmine stains the capsule carmine red (Fig. 3a), and alcian blue stains the capsule blue. Both permit presumptive identification of *C. neoformans* to be made in smears or tissues, although they are not specific for that species.

E. Fluorescent Antibody Techniques

The preparation of specific antibody tagged with fluorescein isothiocyanate requires patience and skill. Fluorescent antibody conjugates with a high degree of specificity and sensitivity are available at the Centers for Disease Control and at certain other reference laboratories. Positive staining permits not only detection but also identification of a fungus (Fig. 8b). Fluorescent antibody procedures can be used both with smears of clinical materials and with sections of formalin-fixed tissues (Kaplan and Kaufman, 1961; Gordon *et al.*, 1967; Kaufman and Blumer, 1968; Kaplan and Kraft, 1969; Hotchi *et al.*, 1972; Kaplan, 1973, 1975; Chandler *et al.*, 1980).

Although FA procedures have some technical drawbacks (Kaufman and Blumer, 1968; Hotchi *et al.*, 1972; Kaplan, 1975), the most important limitation is the lack of widespread availability. At present, no commercial manufacturer has been willing to invest the resources needed to make FA reagents for fungi readily available.

F. Direct Fluorescence

Direct fluorescence staining of fungi using acridine orange has been described. This technique has not achieved widespread usage, although it is quite a sensitive method for demonstrating the presence of fungi in smears or in tissues. It does not have the same degree of specificity as FA because no antibody reaction is involved in the staining process. The more recent availability of fluorochromes that react chemically with fungal cell walls may increase the popularity and usefulness of this technique.

III. EXAMPLES OF FUNGI IN CLINICAL MATERIALS

The morphology and staining reactions of fungi as seen in the tissues of infected patients have been extensively described (Delacretaz *et al.*, 1976; Chandler *et al.*, 1980; Schwarz, 1982). A brief description of representative fungi that are commonly encountered in clinical materials is given in Table I. The numerous figures in this chapter illustrate the most common morphologic forms of fungi that are observed in clinical materials.

Obviously, there are many variations in staining reaction and morphology that could not be presented in this brief discussion. Figure 11 illustrates an unusual yeast form of a *Mucor* species that could easily be confused with other fungi, especially *Paracoccidioides brasiliensis*. Yeast forms of *Mucor* can be produced

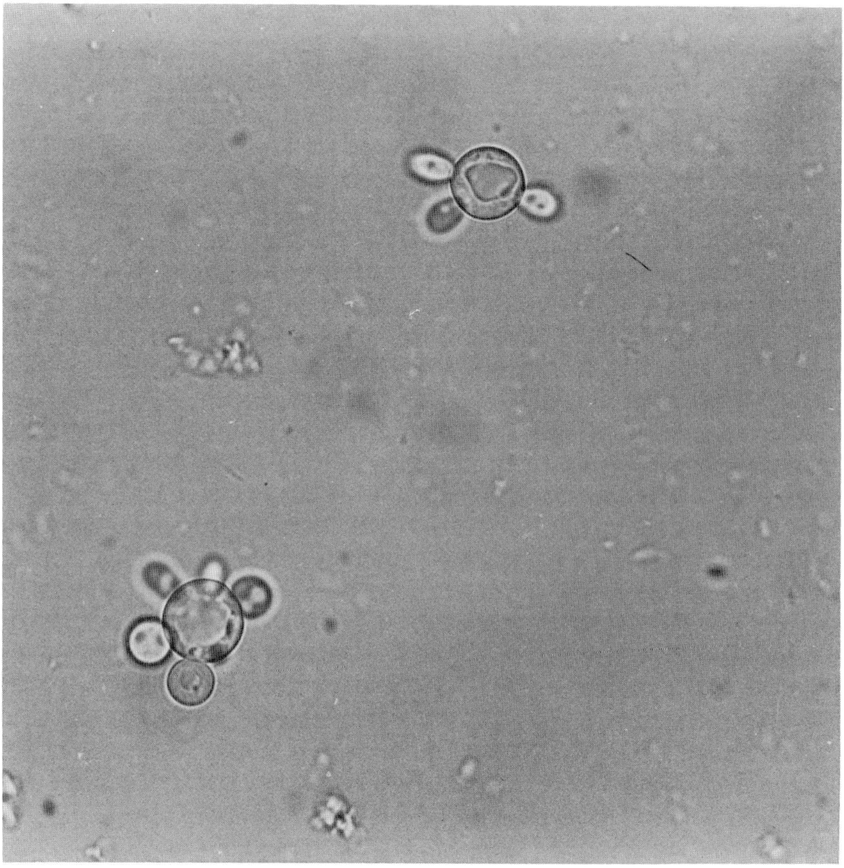

Fig. 11. *Mucor* sp. Yeast form from urine. Note similarity to *P. brasiliensis*. Unstained. 400×.

experimentally by cultivation in a medium containing a hexose carbon source and incubated in an atmosphere of 5 to 10% CO_2. Such forms are rarely seen in clinical specimens. Figure 12 is an illustration of myospherulosis (McClatchie *et al.*, 1969; Rosai, 1978), a condition caused by degenerating red blood cells surrounded by a pseudomembrane. These structures resemble the spherules of *Coccidioides immitis* in H & E-stained tissue but do not stain with any of the special stains for fungi.

A common misconception is that septate hyphae that branch dichotomously, as seen in tissues and clinical specimens, are pathognomonic for *Aspergillus*. Although *Aspergillus* species are perhaps the fungi that most commonly produce septate hyphae in deep tissues, other species can also produce similar structures in clinical materials. *Petriellidium boydii* and *Fusarium* species are but two

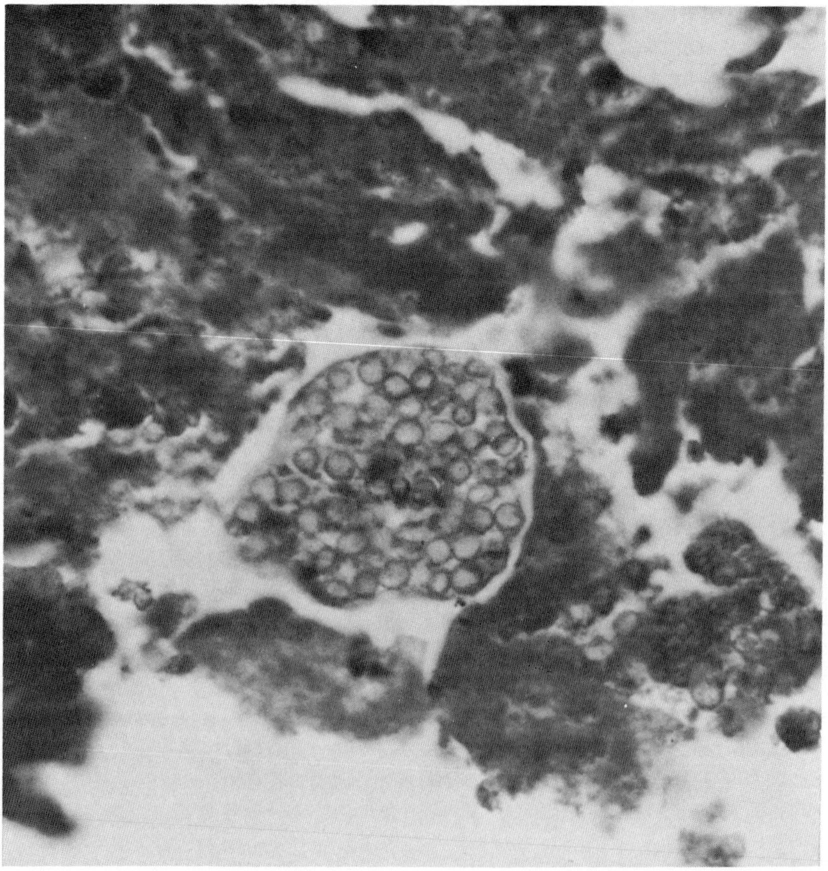

Fig. 12. Myospherulosis. Note resemblance to spherule of *C. immitis*. H&E stain. 400×.

examples of other fungal species that produce septate hyphae in tissues (Schwarz, 1982). *Cocciodioides immitis* can produce septate, branching hyphae in necrotic tissue or in an open cavity of the lung. The cultural isolation and identification of these fungi are essential for accurate diagnosis. When only formalin-fixed tissue is available for study, FA procedures can be very helpful in establishing an accurate diagnosis.

Actinomyces and *Nocardia* are gram-positive, filamentous bacteria. They are not fungi. However, they cause diseases that mimic fungus diseases, and they are encountered in the same kinds of specimens in which fungi are found. For that reason they have been included here. *Nocardia* as well as *Actinomyces* stain with Gram's stain (Fig. 10a). *Nocardia* species are acid-fast, although destaining must be done with aqueous acid instead of acid alcohol because they can be over decolorized more easily than mycobacteria. Members of both genera stain quite well with the GMS stain. *Nocardia* species form granules in mycetomas of the subcutaneous tissues (Fig. 10c), but they do not form granules in deeper tissues. *Actinomyces* species typically form granules in draining sinus tracts ("sulfur granules") that are pale yellow (Fig. 10b). However, they can also be observed as loose aggregates of filamentous bacteria when detected early in the development of an infection.

REFERENCES

Bernard, P. G., Szyfelbein, W. M., Weiss, H. D., and Richardson, E. P., Jr. (1980). *Neurology* **30,** 102–105.

Carson, F. L., Matthews, J. L., and Pickett, J. P. (1983). *In* "Laboratory Medicine" (G. J. Race, ed.), 10th ed., Vol. 3, Chap. 22. Lippincott, Philadelphia, Pennsylvania.

Chandler, F. W., Kaplan, W., and Ajello, L. (1980). "Color Atlas and Text of the Histopathology of Mycotic Diseases." Yearbook Publ., Chicago, Illinois.

Cooper, B. H. (1982). *In* "Rapid Methods and Automation in Microbiology" (R. C. Tilton, ed.), pp. 45–48. Am Soc. Microbiol., Washington, D.C.

Delacretaz, J., Grigoriu, D., and Ducel, G. (1976). "Color Atlas of Medical Mycology." Yearbook Publ., Chicago, Illinois.

Forster, R. K., Wirta, M. G., Solis, M., and Rebell, G. (1976). *Am. J. Ophthalmol.* **82,** 261–265.

Gordon, M. A., Elliott, J. D., and Hawkins, T. W. (1967). *Sabouraudia* **5,** 323–328.

Gridley, M. F. (1953). *Am. J. Clin. Pathol.* **23,** 303–307.

Grocott, R. G. (1955). *Am. J. Clin. Pathol.* **25,** 975–979.

Haley, L. D., Trandel, J., and Coyle, M. B. (1980). "Cumitech 11" (J. C. Sherris, ed.), Am. Soc. *Microbiol.* Washington, D.C.

Hotchi, J., Schwarz, J., and Kaplan, W. (1972). *Sabouraudia* **10,** 157–163.

Kaplan, W., and Kaufman, L. (1961). *Sabouraudia* **1,** 137–144.

Kaplan, W., and Kraft, D. E. (1969). *Am. J. Clin. Pathol.* **52,** 420–432.

Kaplan, W. (1973). *Ann. Clin. Lab. Sci.* **3,** 25–29.

Kaplan, W. (1975). *Sci. Publ. Pan Am. Health Organ.* No. 304, pp. 178–185.

Kaufman, L., and Blumer, S. (1968). *J. Bacteriol.* **95,** 1243–1246.

McClatchie, S., Warambo, M. W., and Bremner, A. D. (1969). *Am. J. Clin. Pathol.* **51,** 699–704.

Roberts, G. D. (1975). *J. Clin. Microbiol.* **2,** 261–265.

Rosai, J. (1978). *Am. J. Clin. Pathol.* **69,** 475–481.

Saigo, P., Rosen, P. P., Kaplan, M. H., Solan, G., and Melamed, M. R. (1977). *Am. J. Clin. Pathol.* **67,** 141–145.

Sanders, J. S., Sarosi, G. A., Nollet, D. J., and Thompson, J. L. (1977). *Chest* **72,** 193–196.

Schwarz, J. (1982). *Human Pathol.* **13,** 519–533.

Sutliff, W. D., and Cruthirds, T. P. (1973). *Am. Rev. Respir. Dis.* **108,** 149–151.

Swartz, J. H., and Lamkins, B. E. (1964). *Arch. Dermatol.* **89,** 89–94.

Trumbull, M. L., and Chesney, T. McC. (1981). *J. Am. Med. Assoc.* **245,** 836–838.

3

Direct Immunofluorescence Identification of Bacteria in Clinical Specimens

WASHINGTON C. WINN, JR, AND JANE D. KEATHLEY

Department of Pathology
College of Medicine
University of Vermont
Burlington, Vermont

I. Introduction 35
II. Important Factors in the Performance of a Test 36
  A. Reagents and Protocol for Preparation of Smears 37
  B. Controls and Interpretation 39
  C. Pitfalls and Remedies 40
III. Identification of Bacteria in Clinical Specimens 41
  A. Bacteria for Which Immunofluorescence Is Most Commonly Employed 41
  B. Miscellaneous Bacteria for Which Immunofluorescence Has Been Employed 51
IV. Selection of Rapid Diagnostic Tests 52
  References 53

I. INTRODUCTION

One of the oldest, if not the most revered, techniques for the expeditious diagnosis of bacterial infection is immunofluorescence microscopy, which has been used both for clinical specimens and for isolated bacteria in the laboratory. Immunofluorescence techniques were applied first to microbiological diagnosis by Coons and associates (1942), who demonstrated antigen of type 3 pneumococcus in experimentally infected mice by reacting frozen tissue sections with rabbit antibody that had been conjugated to fluorescein-4-isocyanate. Improvements in fluorescence microscopes and the introduction of the isothiocyanate

The Direct Detection of Microorganisms in Clinical Samples

ISBN 0-12-187780-9

compound, which facilitated the conjugation of fluorescein to immunoglobulins, resulted in the application of this technique to many microbes and diagnostic problems in the late 1950s and early 1960s. The recognition of problems in interpreting immunofluorescence results on clinical specimens diminished interest and enthusiasm. In the past few years, however, the application of fluorescein-conjugated antibodies to *Legionella* and *Bacteroides* has renewed interest in this approach to rapid diagnosis.

It is important to remember that immunofluorescence is simply another immunologic test, which employs a fluorescent marker as the detection device. The major advantage of immunofluorescence techniques is that the morphology of the object can be evaluated at the same time that the immunologic reactivity is assessed. A strictly immunologic test, such as counterimmunoelectrophoresis, must be interpreted solely on the basis of antigenic reactivity. A careful observer may be able to recognize that the specificity of an immunologic reaction is suspect because the morphology of the organism or the location of the antigenic structure in cells or tissue is atypical. In addition, these morphologic– immunologic procedures can provide important information about disease processes in infected tissues.

The disadvantages of immunofluorescence techniques are related primarily to the subjectivity of the end points and the careful training and attention to detail that are required of the microscopist. Particularly if reagents, microscope, or specimen preparation are suboptimal, an overzealous and uncritical approach to the interpretation of the specimen may result in a false positive diagnosis. Even the morphologic aids may be insufficient to prevent misdiagnoses (Frenkel and Piekarski, 1978). These difficulties are magnified when the number of bacteria that are visualized in positive tests is very small (Tronca *et al.*, 1974; Edelstein *et al.*, 1980a).

This chapter reviews the current status of immunofluorescence tests for the detection of bacteria in clinical specimens. We have concentrated on bacteria for which recent literature suggests a continuing interest in diagnostic immunofluorescence and bacteria for which earlier results suggest that continued evaluation of the technique is desirable. A complete review of early studies was published by Cherry and Moody (1965).

II. IMPORTANT FACTORS IN THE PERFORMANCE OF A TEST

For most laboratories the development of specific antisera and the conjugation of immunoglobulins to fluorescent dyes are not possible. It is important to use commercial antisera that have been evaluated thoroughly in the literature or in the local laboratory. For individuals who wish to prepare their own reagents, a manual has been compiled by investigators at the Centers for Disease Control (Hebert *et al.*, 1972).

Excellent discussions of fluorescence microscopy are provided by Goldman (1968) and Jones *et al.* (1978). Of the several fluorescent dyes available, fluorescein isothiocyanate (FITC) has been evaluated most extensively. Incident-light (episcopic) illumination has many advantages over traditional systems that employ transmitted light through a dark-field condenser. Because immunofluorescence is a technically difficult procedure, it is extremely important to use a high-quality microscope. A check list for evaluating problems with the fluorescence microscope is provided by Neimeister and associates (1980).

A. Reagents and Protocol for Preparation of Smears

1. *Choice of Immunofluorescence Test*

Direct immunofluorescence (DFA) refers to the one-step staining procedure that is employed when the specific antibody is conjugated directly to the fluorescent dye. If the primary antibody is unlabeled and a second antispecies globulin is conjugated to the fluorescent marker, a two-step incubation with rinses between the two stages is described as indirect immunofluorescence (IFA).

The advantages of IFA are that a single conjugated reagent can be employed in multiple systems; this procedure is probably more sensitive than the one-step DFA. Direct immunofluorescence has advantages for a rapid diagnostic test because only one incubation and rinsing step is required. Although less sensitive than IFA, the likelihood of nonspecific reactions is lower and components of the reaction are simpler to control for appropriateness of reaction. When isolated organisms in the laboratory are tested, the choice of tests is not critical. The direct test is probably preferable for clinical specimens, particularly if commercial reagents are used, because the evaluation of performance and problems is less complicated.

2. *Quality of Specimen*

Probably the most important factor in obtaining a successful result is the quality of the specimen itself. If the specimen is of insufficient volume or does not include inflammatory material, the clinician should be notified and the collection of an adequate specimen should be attempted. Portions of specimens that appear to be purulent should be selected and thin smears made on clean glass slides. The importance of specimen collection and processing was emphasized by the finding that sections through the middle of a petechial lesion were required to demonstrate *Rickettsia rickettsii* with maximum sensitivity (Walker *et al.*, 1978).

3. Fixation of Specimen

Fixation of the specimen on the glass slide is usually achieved with acetone, ethanol, or methanol. Gentle heat or formalin may be used with some antigens, but it is important to test each separately. The antigenicity of *Legionella* is maintained in formaldehyde and other fixative solutions (Thomason *et al.*, 1979), but many antigens are masked by aldehyde fixation. Coons noted that the intensity of fluorescence decreased within several days after tissue that had been infected with type 3 pneumococcus was fixed in formalin (Coons *et al.*, 1942). Excellent preservation of antigenic *Treponema pallidum* and *T. pertenue* was achieved by fixation of tissue in Bouin's fixative, but immunofluorescence staining was poorly demonstrated after formalin fixation (Mote *et al.*, 1982). Bouin's fixative was also more satisfactory than formalin alone for the preservation of immunologic reactivity of lymphoid tissue (Dorsett and Ioachim, 1978). Glutaraldehyde distorts or masks antigenic reactivity more completely than formaldehyde. Procedures have been developed for evaluating antigens that have been treated with aldehyde fixatives. The most frequently used procedure employs mild trypsin digestion of tissues (Huang *et al.*, 1976), but chymotrypsin has also been proposed as a useful agent (Brozman, 1980).

4. Performance of Test

The incubation of fluorescence conjugate with substrate is usually accomplished at room temperature or at 37°C for 20 to 60 min. It is important to incubate the specimen in a humidified chamber and ensure that the reagents do not dry on the slide. Rinses are performed in buffered saline, and the specimen is mounted in buffered glycerol. Details of the procedures can be found in several manuals (Jones *et al.*, 1978; Anhalt, 1981). The fluorescence of fluorescein is markedly diminished below pH 7.0, so buffers and mounting fluid should be alkaline. The glycerol solution for mounting coverslips is maintained at pH 9.0 because the solution becomes more acidic with storage. An alkalinized modification of polyvinyl alcohol mounting medium (Lennette, 1978) and the addition of *p*-phenylenediamine to buffered glycerol (Johnson *et al.*, 1981) have been suggested for use with fluorescein immunofluorescence. FITC-conjugated antisera to bacteria have a very high optimum fluorescein/protein ratio (Hebert *et al.*, 1981). The pH of these reagents is also important, because precipitates may form below pH 9.0–9.5. Dialysis against buffer will restore the correct pH.

Scrupulous attention to detail is important to prevent contamination of the smear with bacteria from reagents or other positive specimens. For *Legionella* immunofluorescence we follow a rigorous protocol. Smears are prepared from specimens in an area of the laboratory in which *Legionella* is not cultivated. All reagents (including formaldehyde solutions) are filter-sterilized before use. Reagent bottles are never reentered. Smears are fixed and rinsed individually (disposable plastic slide mailers provide useful containers). Containers that have held positive specimens are not reused. Finally, smears are dried by drainage of buffer onto a clean paper towel rather than a drying rack. During an epidemic of Legionnaires' disease in Burlington, Vermont, during the summer of 1980, we found that the predictive value of a positive smear for *Legionella* could be increased from 33 to 96% if the positive result could be confirmed when the specimen was retested. Edelstein and associates documented false positive results for *Legionella* when distilled water that was used for rinse buffers was contaminated with a cross-reacting strain of *Pseudomonas fluorescens,* when dead bacteria were transferred from serotyping slides to clincial specimens in a drying rack, and by transfer of legionellae from commercially purchased prewashed glass slides and plastic rinse containers (Edelstein *et al.,* 1982). It is clear, therefore, that scrupulous attention to the details of technique is essential for the optimal performance of this test. Similar care in the approach to other bacteria seems prudent.

B. Controls and Interpretation

The initial characterization of the antisera is an important part of the quality control of the test, because the maximum extent of cross-reactivity that might be encountered must be known. Controls for immunologic specificity should be tailored to each antigen on the basis of suggestions from the manufacturer of the antiserum or on recommendations in the literature (Jones *et al.,* 1978).

When tests are performed on clinical specimens, especially from nonsterile sites, false reactions are most likely to be troublesome and controls should be rigorous. If polyvalent conjugates are employed, it is impractical to test every component for reactivity each time the test is performed; the evaluation of all constituents at the time the serum is first used may be followed by testing of a single component with each test. The specificity of the reaction should be tested with each specimen. At the least the smear should be incubated with a normal serum that has been conjugated to the fluorescent dye. Animals differ in their exposure to infectious agents, however, and this control does not rule out the possibility that natural antibodies in the specific serum are producing unwanted fluorescence. For this purpose a preimmune serum from the same animal that produced the specific serum is a superior control, but it is difficult to obtain this reagent in adequate volume to supply a large number of users. Alternatively, the specific serum may be absorbed with antigen. Particularly when contamination of smears with bacteria from the environment is a concern (Edelstein *et al.,* 1982); Winn and Pasculle, 1982), a smear that is known to be negative should be incubated with the specific conjugate in each test. Ideally, this smear should be a clinical specimen of the type that is most frequently tested. Alternatively, a proteinaceous material, such as albumin, may be employed. Saline is probably a less satisfactory reagent, because contaminating bacteria are more likely to be washed free. When a small number of bacteria are present in the clinical specimen, no control is absolute and there is no way to identify a true cross-reaction (unwanted specific fluorescence). An attempt should be made, however, to be rigorous both in controls and in the interpretation of morphology (Frenkel and Piekarski, 1978).

The criteria for determining a positive result differ with each procedure; it is important to state the criteria exactly and analyze the performance of the test on the basis of those parameters. The intensity of fluorescence is usually graded from 1+ to 4+; 3+ to 4+ reactions are considered positive in most studies. The morphologic quality of the staining is an indication that the reaction is of sufficient intensity and an important clue that the fluorescent structure is truly a bacterium. Typically, the perimeter of the bacterium stains intensely, and the center is relatively nonfluorescent. When morphologically typical *Legionella* organisms are present, we have found that the presence of other bacteria that appear to be "moth-eaten" or to have a cloud of antigen at the periphery of the organism is a useful clue that the bacteria have been obtained from an inflammatory process. Although the presence of perfectly intact bacilli does not rule out a true reaction, we have not encountered such "battle-scarred" bacteria in specimens that we later recognized to be false positive tests.

The application of morphologic criteria to fluorescent objects is important, and potentially troublesome false positive interpretations can be minimized. A conservative approach in interpreting positive reactions is justified, however, and one should not overextend morphologic characterization (Frenkel and Piekarski, 1978).

C. Pitfalls and Remedies

Correlating test results with clinical history, response to therapy, and the results of other laboratory tests will help to identify problems early. The reading of smears may be facilitated by the inclusion of a counterstain, such as Evans blue or rhodamine-conjugated bovine serum albumin. An additional advantage of using rhodamine-conjugated normal IgG or whole serum is that receptors on staphylococci and streptococci that bind the Fc fragment of many immunoglobulin classes nonspecifically will be saturated with the rhodamine conjugate. Even after counterstaining, the nonspecific fluorescence of tissues may remain troublesome. In an inflammatory exudate the polymorphonuclear neutrophils may exhibit refractory fluorescence; this difficulty may be diminished by the observation of control smears and use of strict morphologic criteria for a positive interpretation.

III. IDENTIFICATION OF BACTERIA IN CLINICAL SPECIMENS

Increased interest in the application of DFA to the diagnosis of bacterial infections has been sparked mainly by the application of this technique to *Legionella* and *Bacteroides*, but many other agents have been investigated. The following sections summarize the status of those agents for which the literature is sufficient to assess the current performance or future prospects of the test.

A. Bacteria for Which Immunofluorescence Is Most Commonly Employed

1. *Bacteroides*

Both DFA techniques, using commercially available conjugates, and IFA tests, using locally prepared anti-*Bacteroides* antisera, have been evaluated (Kasper *et al.*, 1979: Holland *et al.*, 1979; Slack *et al.*, 1981; De Girolami and Mepani, 1981; Wills *et al.*, 1982). The results of these investigations are summarized in Table I for the *Bacteroides fragilis* group and in Table II for the *Bacteroides melaninogenicus* group. Most investigators selected specimens because anaerobic infection was suspected clinically and appropriately collected material was available for anaerobic culture. The sensitivity of the IFA technique was very high for the detection of the *B. fragilis* group, although only a small number of specimens were evaluated. A commercial direct fluorescence conjugate also performed with moderate to very high sensitivity. *Bacteroides fragilis* was detected in at least 80% of the specimens from which the organism was isolated in all but one study (Wills *et al.*, 1982). The reasons for the poor performance of the commercial conjugate in that study are unclear; a locally prepared antiserum demonstrated fewer *Bacteroides* organisms than it had in a previous trial (Kasper *et al.*, 1979). In the largest series (Slack *et al.*, 1981) only 25% of the specimens were submitted to the laboratory as pus, and most of the remainder were collected and transported on dry swabs. The effect of collection on culture results was not stated; the detection of *Bacteroides* by DFA was adversely affected, however, when the specimen was submitted on a dry swab.

The specificity of the commerical antiserum was also high, exceeding 85% in all studies with *B. fragilis*. The culture was appropriately considered the final arbiter of bacterial etiology. Problems with specimen collection, transport, and selection of colonies from plates that contained mixed bacterial species were recognized as potential contributors to an underestimation of specificity. The relatively high proportion of unsupported DFA examinations for the *B. melaninogenicus* group may be related to the fastidiousness of certain members

TABLE I

Immunofluorescence Identification of *Bacteroides fragilis* Group in Clinical Specimens

Number of specimens	Source of antiserum	Type of test	Method of diagnosis	Sensitivity[a] (%)	Specificity[b] (%)	Comment	Reference
15 positive 28 negative	Local	Indirect	Culture and gas–liquid chromatography	15/15 (100)	18/28 (64)	Antiserum to multiple somatic antigens	Kasper *et al.* (1979)
12 positive 31 negative	Local	Indirect	Culture and gas–liquid chromatography	12/12 (100)	28/31 (90)	Antiserum to *B. fragilis* capsular polysaccharide	
29 positive 20 negative	Commercial	Direct	Culture	28/29 (97)	18/20 (90)	—	Holland *et al.* (1979)
22 positive 97 negative	Commercial	Direct	Culture	20/22 (91)	84/97 (87)	—	Labbe *et al.* (1980)
152 positive 858 negative	Commercial	Direct	Culture	123/152 (81)	841/858 (98)	Pus more frequently positive (30%) than swabs (15%) by DFA	Slack *et al.* (1981)
14 positive 130 negative	Commercial	Direct	Culture	13/14 (93)	127/130 (98)	—	De Girolami and Mepani (1981)
26 positive 41 negative	Commercial	Direct	Culture	13/26 (50)	40/41 (98)	Local antiserum same as polyvalent somatic antiserum in Kasper *et al.* (1979)	Wills *et al.* (1982)
26 positive 41 negative	Local	Indirect	Culture	23/26 (88)	36/41 (88)		

[a] Sensitivity is the number of specimens truly positive by immunofluorescence divided by the number of specimens positive by culture and/or gas–liquid chromatography.

[b] Specificity is the number of specimens truly negative by immunofluorescence divided by the number of specimens negative by culture and/or gas–liquid chromatography.

TABLE II

Direct Immunofluorescence Identification of *Bacteroides melaninogenicus* Group in Clinical Specimens

Number of specimens	Source of antiserum	Method of diagnosis	Sensitivity[a] (%)	Specificity[b] (%)	Reference
15 positive 34 negative	Commercial	Culture	15/15 (100)	29/34 (85)	Holland *et al.* (1979)
9 positive 18 negative	Commercial	Culture	9/9 (100)	13/18 (72)	Labbe *et al.* (1980)
22 positive 988 negative	Commercial	Culture	21/22 (95)	967/988 (98)	Slack *et al.* (1981)

[a] Sensitivity is the number of specimens truly positive by immunofluorescence divided by number of specimens positive by culture.

[b] Specificity is the number of specimens truly negative by immunofluorescence divided by the number of specimens negative by culture.

of this group. In two studies (Labbe *et al.*, 1980; Slack *et al.*, 1981) cross-reactions with aerobic and other anaerobic bacteria were recognized; such cross-reactions were sought, but not found, in other studies (Holland *et al.*, 1979; De Girolami and Mepani, 1981; Wills *et al.*, 1982). Some of these cross-reactions were with bacteria that are morphologically different from *Bacteroides (Staphylococcus, Streptococcus,* and *Escherichia coli*). Such distinctions are more difficult, however, in clinical specimens that contain few bacteria, and the coccobacillary morphology of the *B. melaninogenicus* group might accentuate the problem.

2. *Legionella*

Direct immunofluorescence techniques were employed for the diagnosis of infections by *Legionella pneumophila* soon after the isolation of the bacterium (Cherry *et al.*, 1978; Lattimer *et al.*, 1978). Until recently the only direct fluorescence conjugates available have been produced at the Centers for Disease Control; details of the test are provided in a manual for the laboratory diagnosis of *Legionella* infections (Cherry and McKinney, 1979b). The application of these antisera to the diagnosis of infection by *L. pneumophila* is summarized in Table III. Data are most extensive for epidemic pneumonia caused by serogroup 1 *L. pneumophila* (Edelstein *et al.*, 1980a; Winn *et al.*, 1981). The sensitivity of DFA of respiratory secretions is only approximately 50% in this setting, although the yield is increased to approximately 75% in patients whose disease is confirmed by culture (Winn *et al.*, 1981). Expectorated sputum and tracheal aspirates provide a higher yield than transtracheal aspirates in patients with documented infection, perhaps because of scanty material and the dilution of the specimen

TABLE III

Direct Immunofluorescence Detection of *Legionella pneumophila* in Clinical Specimens[a]

Number of patients	Criterion for infection	Sensitivity[b] (%)	Specificity[c] (%)	Comment	Reference
21 positive 13 negative	Seroconversion or single high titer	5/21 (24)	13/13 (100)	Smears had been stored for up to 6 months at room temperature	Broome *et al.* (1979)
7 positive 8 negative	Seroconversion or single high titer	6/7 (86)	8/8 (100)	—	Winn *et al.* (1980)
19 positive 22 negative	Seroconversion or culture positive	13/19 (68)	22/22 (100)	Includes lung biopsies from 14 patients with *Legionella* infection	Saravolatz *et al.* (1981)
29 positive 24 negative	Seroconversion or culture positive for serogroups 1–4	12/29 (41)	21/24 (88)	Sensitivity 62% if disease defined by positive culture only	Edelstein *et al.* (1980a)
62 positive 22 negative	Seroconversion or culture positive	29/62 (47)	20/22 (91)	Sensitivity increased in patients with positive cultures	Winn *et al.* (1981)

[a] Direct fluorescence performed with antisera produced at the Centers for Disease Control.
[b] Sensitivity is the number of patients truly positive by DFA divided by number of patients positive by culture and/or serology.
[c] Specificity is the number of patients truly negative by DFA divided by the number of patients negative by culture and/or serology.

that may accompany transtracheal aspiration. Direct immunofluorescence and radioimmunoassay for the excretion of serogroup 1 *L. pneumophila* antigen in the urine were complementary in a selected group of patients from the 1980 epidemic of Legionnaires' disease in Burlington (Kohler *et al.*, 1982).

The very low sensitivity (24% of patients) reported earlier may have been affected by prolonged storage of smears at room temperature before testing (Broome *et al.*, 1979). The reasons for the apparently increased sensitivity of DFA in the remaining studies (Winn *et al.*, 1980; Saravolatz *et al.*, 1981) are unclear. Both of these studies include a small number of patients with endemic disease. It is possible that patients who are infected in an endemic setting have more severe disease and/or mobilize alveolar exudate more effectively. Alternatively, physicians may suspect the disease and order the test in patients with mild disease more frequently during epidemic infection.

If appropriate controls and precautions are employed (see Section IIB), the specificity of DFA examination for serogroup 1 *L. pneumophila* is above 90% (Edelstein *et al.*, 1980a; Winn *et al.*, 1981). The importance of cross-contamination of specimens and of contamination of specimens with environmental bacteria is particularly dramatic with this genus (Edelstein *et al.*, 1982; Winn and Pasculle, 1982). Additional false positive results may be obtained because of immunologic cross-reactions with aerobic (Cherry *et al.*, 1978; Cherry and McKinney, 1979a) and anaerobic (Edelstein *et al.*, 1980b) bacteria. Colonization of the airways with legionellae might result in positive DFA tests (and cultures) in the absence of infection. Although *L. pneumophila* has been recovered from patients without radiologically evident pneumonia (Meyer *et al.*, 1980), colonization of the oropharynx has not yet been established. In specimens that have been collected from more than 200 patients in Burlington since the 1980 epidemic, legionellae have been demonstrated by DFA in less than 1% of patients who did not have infection documented by culture or serology.

Information on the demonstration of legionellae other than serogroup 1 *L. pneumophila* is scant, although all of the serogroups and species that have produced human disease have been visualized in lung tissue by DFA. *Legionella micdadei* has been identified by DFA techniques in respiratory specimens (Wing *et al.*, 1981). Data are not available for the other *Legionella* species.

3. *Bordetella pertussis*

Direct immunofluorescence has been used sporadically in the diagnosis of whooping cough for over 20 years (Table IV). The initial studies were based on a clinical diagnosis only (Whittaker *et al.*, 1960; Donaldson and Whittaker, 1960). Subsequently, comparison of DFA testing with culture has produced estimates of sensitivity that range from 18 to 97%. The effect of changes in technique for obtaining specimens was demonstrated dramatically during an outbreak of pertussis in Atlanta in 1977 (Broome *et al.*, 1978). When smears and cultures were

TABLE IV

Immunofluorescence Identification of *Bordetella pertussis* in Clinical Specimens

Number of specimens	Source of antiserum	Method of diagnosis	Sensitivity[a] (%)	Specificity[b] (%)	Comment	Reference
128 positive 36 normal 86 respiratory infection	Local	Clinical pertussis	100/128 (78)	36/36 (100) 84/86 (98)	Mild pertussis?	Whitaker *et al.* (1960)
36 positive	Local	Untreated clinical pertussis	31/36 (86)	36/36 (100)	—	Donaldson and Whitaker (1960)
25 positive 105 negative	Local	Culture	21/25 (84)	98/105 (93)	—	Kendrick *et al.* (1961)
163 positive 364 negative	Local	Culture	138/163 (85)	329/364 (90)	Specimens submitted to state laboratory	Holwerda and Eldering (1963)
29 positive 71 negative	Local	Culture	28/29 (97)	53/71 (75)	Culture technique later improved	Chalvardjian (1966)
10 positive 5 negative	Local	Culture	4/10 (40)	4/5 (80)	Single false positive from treated patient	Field and Parker (1977)
184 positive 254 negative	Commercial	Culture	33/184 (18)	254/254 (100)	Specimens submitted to central health laboratory	Regan and Lowe (1977)
20 positive 35 negative	Local	Culture	20/20 (100)	0/35 (0)	Transport media used	Broome *et al.* (1978)
30 positive 6 negative	Local	Culture	16/30 (53)	0/6 (0)	Direct inoculation of plates and smears	

[a] Sensitivity is the number of specimens truly positive by immunofluorescence divided by number of specimens positive by clinical or cultural criteria.
[b] Specificity is the number of specimens truly negative by immunofluorescence divided by the number of specimens negative by clinical or cultural criteria.

prepared from swabs that had been shipped to the laboratory in Stuart's transport medium, only 20 of 55 (36%) positive fluorescence tests were accompanied by positive cultures of the same specimen. Later in the epidemic the protocol was changed to include the direct preparation of smears and inoculation of agar plates by house officers. Subsequently, 16 of 22 (73%) positive florescence tests were confirmed bacteriologically.

With one exception (Broome *et al.*, 1978) the specificity of DFA for *Bordetella* has ranged from 75 to 100%. Inadequate culture technique and antibiotic therapy may have contributed to the apparent nonspecificity of the test in some studies. Positive DFA tests were obtained in 2 of 30 normal children who had negative cultures for *Bordetella* (specificity, 93%) (Chalvardjian, 1966). It is obvious that with careful technique, good reagents, and experience, excellent results can be obtained (Holwerda and Eldering, 1963; Kendrick *et al.*, 1961). Most recent studies, especially those with improved microbiological technique, are less encouraging about the sensitivity of the test (Field and Parker, 1977; Regan and Lowe, 1977; Broome *et al.*, 1978). The only study that employed a commercially available reagent resulted in a very low estimate of sensitivity (Regan and Lowe, 1977); further studies of this reagent are indicated. A disappointingly low frequency of agreement among three observers of the same slides was noted by Broome and co-workers (1978); the reason for the poor agreement is not apparent, but this issue also deserves further evaluation. As these authors indicated, it is clearly not defensible to rely only on immunofluorescence for diagnosing this infection.

4. Bacterial Agents of Meningitis

Direct immunofluorescence was the first rapid diagnostic technique applied to the diagnosis of bacterial meningitis. The results of many evaluations over the past 20 years are consistent (Page *et al.*, 1961; Biegeleisen *et al.*, 1965; Fox *et al.*, 1969; Koshi and Chacko, 1971; Olcen, 1978; Forre and Gaustad, 1977) and are summarized in Table V. In these studies specimens with evidence of meningeal inflammation were selected, and an etiology was defined if any test was positive. It is impossible to determine the frequency with which a positive immunologic test was not confirmed by culture in several instances, but the relative performance of the tests can be assessed. Results have been consistently best with *Haemophilus influenzae*. In general, the percentage of cases that were identified by each technique was similar. In two studies counterimmunoelectrophoresis was less sensitive than DFA, but it is instructive that the simplest, least expensive, and most rapid test of all, Gram's stain, was as frequently positive as DFA. Although Gram's stain evaluation of cerebrospinal fluid sediment is not immunologically specific, a skilled evaluator should be able to minimize errors in interpreting bacterial morphology.

TABLE V

Diagnosis of Bacterial Meningitis[a]

Antigen	Source of conjugate	Method of diagnosis	Number of patients	Gram-stain positive (%)	Culture-positive (%)	Counterimmuno-electrophoresis-positive (%)	Direct immuno-fluorescence positive (%)	Coagglutination-positive (%)	Immuno-fluorescence-positive only (%)	Reference
Haemophilus influenzae	Commercial	Any test positive	26	88	92	ND	100	ND	8[b]	Page *et al.* (1961)
Streptococcus pneumoniae	Not stated	Culture or fluorescence	19	ND	79	ND	89	ND	21[b]	Biegeleisen *et al.* (1965)
H. influenzae	Not stated	Culture or fluorescence	52	ND	92	ND	94	ND	8[b]	
Neisseria meningitidis	Not stated	Culture or fluorescence	22	ND	91	ND	95	ND	9[b]	
S. pneumoniae	Not stated	Any test positive	83	89	83	ND	92	ND	ND	Fox *et al.* (1969)
H. influenzae	Not stated	Any test positive	151	93	99	ND	96	ND	ND	
N. meningitidis	Not stated	Any test positive	52	90	94	ND	87	ND	ND	

S. pneumoniae	Commercial	culture	13	85	100	ND	100	ND	0	Koshi and Chacko (1971)
H. influenzae, type b	Commercial	culture	11	100	100	ND	100	ND	0	
S. pneumoniae	Commercial and local	Any test positive	15	87	93	53	100	67	ND	Olcen (1978)
H. influenzae	Local	Any test positive	30	83	100	80	93	67	ND	
N. meningitidis	Local	Any test positive	30	60	83	30	67	33	ND	
S. pneumoniae, H. influenzae, or *N. meningitidis*	Commercial and local	Any test positive	59	67	75	49[c]	79	ND	ND	Forre and Gaustad (1977)

[a] ND, not done. [b] Antibiotic therapy. [c] 70% positive after sonication of spinal fluid.

5. *Rickettsia*

Fluorescein-conjugated antisera have been produced against the typhus and spotted fever groups of *Rickettsia* and against *Coxiella burnetii* (Hebert *et al.*, 1980). Clinical studies have been performed with FITC-conjugated antisera to *R. rickettsii* (Woodward *et al.*, 1976; Walker *et al.*, 1978, 1980; Fleisher *et al.*, 1979). By careful selection and cryostat sectioning of petechial lesions, *Rickettsia* organisms were identified in punch skin biopsies from 7 of 10 patients with acute Rocky Mountain spotted fever; false positive tests have not been encountered (Walker *et al.*, 1980). If cryostat sectioning cannot be performed, trypsin digestion procedures have been described for demonstrating rickettsial antigen in formalin-fixed tissue (Walker and Cain, 1978; Hall and Bagley, 1978). Antirickettsial chemotherapy for as little as 24 to 48 hr may produce a false negative test, and this method of diagnosis is not applicable before the onset of the skin rash. At present, the DFA test is the only means of rapidly diagnosing rickettsial infections. Appropriate chemotherapy must be instituted if Rocky Mountain spotted fever is suspected clinically, however, because of the frequent occurrence of false negative examinations.

6. *Treponema pallidum*

Indirect immunofluorescence was first applied to the diagnosis of early syphilis 20 years ago (Edwards, 1962), and DFA was used shortly thereafter (Yobs *et al.*, 1964). The two approaches were found to be comparable by Elsas (1971). Little attention has been devoted to these techniques in more recent years. The identification of *Treponema pallidum* and *T. pertenue* in experimentally infected rabbit tissues has been accomplished with fluorescein and rhodamine-conjugated antisera (Mote *et al.*, 1982), and the technique could presumably be applied to human tissues. A comparison of DFA of exudate from suspected syphilitic lesions with dark-field microscopy of the lesions indicated similar sensitivity and specificity for the two tests (Daniels and Ferneyhough, 1977). Syphilis was diagnosed (criteria not stated completely) in all patients from whom a positive DFA test was obtained. A small number of false positive results were obtained with the dark-field examination of primary lesions. The sensitivity of the DFA examination was greater than that of the dark-field test and was comparable to the FTA–ABS test for the diagnosis of early syphilis. Direct immunofluorescence with appropriately controlled antisera has the advantage of immunologic specificity. Misinterpretations of morphology (Montenegro *et al.*, 1969) should be minimized by the centralization of diagnostic facilities to laboratories that can develop special competence in the test. Further study of this application of early diagnosis is clearly indicated.

7. *Chlamydia trachomatis*

Both DFA and IFA techniques have been applied to the diagnosis of acute and chronic infections with *Chlamydia trachomatis* (reviewed by Schachter and

Dawson, 1978). In several studies DFA was more sensitive than Giemsa staining for the detection of a small number of inclusions in ocular and genital material. When a large number of infected cells were present, as in neonatal inclusion conjunctivitis, the differences between the two techniques were less obvious. An investigation of the specificity of immunofluorescence for diagnosing chlamydial infection was hampered in early studies by the insensitivity of culture techniques. (Gordon *et al.*, 1969). Careful evaluation of smears that were stained consecutively by DFA and by Giemsa stain suggested that the specificity of the reaction was high (Nichols *et al.*, 1963), but further studies comparing DFA with more sensitive cell culture techniques would be of interest.

8. Yersinia pestis

Direct immunofluorescence for demonstrating the presence of *Yersinia pestis* in clinical specimens and in infected animals has proved successful at the Vector Borne Diseases Laboratory of the Centers for Disease Control (T. Quan, personal communication). False negative tests have occurred in some laboratories, however (Counts *et al.*, 1976), and familiarity with the technique is important if a laboratory employs immunologic diagnosis.

B. Miscellaneous Bacteria for Which Immunofluorescence Has Been Employed

Immunofluorescence tests for a variety of bacteria have been applied more successfully to organisms after isolation than directly to clinical specimens. Although one research investigation documented the presence of *Neisseria gonorrhoeae* in normally sterile tissues (Tronca *et al.*, 1974), this procedure has not been tested extensively, and the use of immunofluorescence for demonstrating the presence of gonococci in nonsterile sites has been abandoned. Similarly, the identification of group A beta hemolytic streptococci in clinical specimens is no longer attempted, although the procedure works well after a brief enrichment incubation in the laboratory (Moody *et al.*, 1963). Cross-reactions with other enteric bacteria have limited the usefulness of immunologic methods for identifying *Salmonella, Shigella,* and enteropathogenic serotypes of *Escherichia coli.* False positive reactions are minimized if antisera against well-defined antigens are employed and if small numbers of fluorescent bacilli are ignored (Svenungsson *et al.*, 1979). Even when antiserum to the Vi antigen was used to detect *Salmonella typhi* during an epidemic, however, it was necessary to confirm all positive tests bacteriologically because of apparently nonspecific reactions (Bissett *et al.*, 1969). Many of the early studies were reviewed by Cherry and Moody (1965).

A small number of reports suggest that immunofluorescence may be of use for the demonstration of *Listeria monocytogenes* (Eveland, 1963), *Actinomyces* (Slack *et al.*, 1966; Blank and Georg, 1968; Valicenti *et al.*, 1982), *Mycobac-*

terium tuberculosis (Jones *et al.*, 1966), and *Staphylococcus aureus* (Yum *et al.*, 1978). Further studies are required to establish the utility of DFA techniques for these agents.

IV. SELECTION OF RAPID DIAGNOSTIC TESTS

The choices of infectious agents and methods depend on many factors. Reliable supplies of well-characterized antisera, well-trained personnel, and sufficient clinical material to maintain proficiency in the technique are essential for any test. For fluorescence microscopy of clinical specimens a high-quality fluorescence microscope with appropriate filter combinations is also important.

Most rapid diagnostic tests are designed to document the single etiologic agent of a systemic infection. A test that has very high specificity, even at the expense of sensitivity, is important to prevent one's attention from being diverted from the true etiologic agent. In some situations, however, nonspecificity may be tolerated if all infections are detected and if the treatment of some patients who may be free of infection is not unduly harmful. It is important to evaluate the likely incidence of the disease in the local environment, because variations in the frequency of an infection have a major impact on the performance of any test. Even if the specificity of a test is very high (99%), the predictive value of a positive test will be 50% if only 1% of the population is infected.

A realistic assessment of the frequency and immediacy with which the test can be offered is also important. Immunofluorescence techniques as well as counterimmunoelectrophoresis, radioimmunoassay, and enzyme-linked immunosorbent assay procedures are rapid but are not easy to establish as ''stat'' tests. In such a situation it may be difficult to provide on a regular basis results that will affect initial therapeutic decisions. It may be more feasible to provide expedited information that will form part of the data base for establishing the final diagnosis and evaluating the response to initial therapy.

Careful ongoing evaluation of the results of any immunodiagnostic test is important, particularly if microbiological confirmation of the diagnosis is difficult and if specimens that contain indigenous flora are being tested. Scrupulous attention to technique and controls are particularly important for immunofluorescence tests because of the frequency with which nonspecific reactions have plagued these tests in the past. Evaluation of the morphology of the organisms is important, but undue reliance on morphologic discrimination of cross-reacting organisms should be discouraged.

No immunologic test can replace the microbiological evaluation of an infection. Immunofluorescence continues to play an important role in the expansion of the capabilities of the clinical laboratory for the rapid diagnosis of infections that are clinically important, especially those infections for which other diagnostic techniques are unable to provide expeditious answers.

ACKNOWLEDGMENTS

We thank Cindy Wright and Nancy Green for excellent secretarial assistance.

REFERENCES

Anhalt, J. P. (1981). *In* "Laboratory Procedures in Clinical Microbiology" (J. A. Washington, ed.), pp. 249–265. Springer-Verlag, Berlin and New York.

Biegeleisen, J. Z., Mitchell, M. S., Marcus, B. B., Rhoden, D. L., and Blumberg, R. W. (1965). *J. Lab. Clin. Med.* **65,** 976–989.

Bissett, M. L., Powers, C., and Wood, R. M. (1969). *App. Microbiol.* **17,** 507–511.

Blank, C. J., and Georg, L. K. (1968). *J. Lab. Clin. Med.* **71,** 283–293.

Broome, C. V., Fraser, D. W., and English, W. J., II (1978). *In* "International Symposium on Pertussis" (C. R. Manclark and J. C. Hill, eds.), pp. 19–22, U. S. Department of Health, Education and Welfare, Bethesda, Maryland.

Broome, C. V., Cherry, W. B., Winn, W. C., Jr., and MacPherson, B. R. (1979). *Ann. Intern. Med.* **90,** 1–4.

Brozman, M. (1980). *Acta Histochem.* **67,** 80–85.

Chalvardjian, N. (1966). *Can. Med. Assoc. J.* **95,** 263–266.

Cherry, W. B., and McKinney, R. M. (1979a). *N. Engl. J. Med.* **301,** 1242.

Cherry, W. B., and McKinney, R. M. (1979b). *In* "Legionnaires: The Disease, the Bacterium, and Methodology" (G. L. Jones and G. A. Hebert, eds.), pp. 92–103, U. S. Department of Health, Education and Welfare, Atlanta, Georgia.

Cherry, W. B., and Moody, M. D. (1965). *Bacteriol. Rev.* **29,** 222–250.

Cherry, W. B., Pittman, B., Harris, P. P., Hebert, G. A., Thomason, M., Thacker, L., and Weaver, R. E. (1978). *J. Clin. Microbiol.* **8,** 329–338.

Coons, A. H., Creech, H. J., Jones, R. N., and Berliner, E. (1942). *J. Immunol.* **45,** 159–170.

Counts, J. M., Vernon, T. M., Hodgin, U. G., Horvarth, A. A., Johns, W., Berg, N., Burkhart, M., Mann, J. M., Miller, B., Weber, N., and Gaskin, J. (1976). *Morbid. Mortal. Weekly Rep.* **25,** 189.

Daniels, K. C., and Ferneyhough, H. S. (1977). *Health Lab. Sci.* **14,** 164–171.

De Girolami, P. C., and Mepani, C. P. (1981). *Am. J. Clin. Pathol.* **76,** 78–82.

Donaldson, P., and Whitaker, J. A. (1960). *AMA J. Dis. Child.* **99,** 423–427.

Dorsett, B. H., and Ioachim, H. L. (1978). *Am. J. Clin. Pathol.* **69,** 66–72.

Edelstein, P. H., Meyer, R. D., and Finegold, S. M. (1980a). *Am. Rev. Respir. Dis.* **121,** 317–327.

Edelstein, P. J., McKinney, R. M., Meyer, R. D., Edelstein, M. A. C., Krause, C. J., and Finegold, S. M. (1980b). *J. Infect. Dis.* **141,** 652–655.

Edelstein, P. H., Deboynton, E., Bridge, J., and Cox, N. (1982). *Ann. Meet., Am. Soc. Microbiol.,* Abstract No. C43.

Edwards, E. A. (1962). *Public Health Rep.* **77,** 427–430.

Elsas, F. J. (1971). *Br. J. Vener. Dis.* **47,** 255–258.

Eveland, W. C. (1963). *J. Bacteriol.* **85,** 1448–1450.

Field, L. H., and Parker, C. D. (1977). *J. Clin. Microbiol.* **6,** 154–160.

Fleisher, G., Lennette, E. T., and Honig, P. (1979). *J. Pediat.(St. Louis)* **95,** 63–65.

Forre, O., and Gaustad, P. (1977). *Scand. J. Infect. Dis.* **9,** 285–288.

Fox, H. A., Hagen, P. A., Turner, D. J., Glasgow, L. A., and Connor, J. D. (1969). *Pediatrics* **43,** 44–49.

Frenkel, J. K., and Piekarski, G. (1978). *J. Infect. Dis.* **138,** 265–266.

Goldman, M. (1968). "Fluorescent Antibody Methods." Academic Press, New York.

Gordon, F. B., Harper, I. A., Quan, A. L., Treharne, J. D., Dwyer, R. St. C., and Garland, J. A. (1969). *J. Infect. Dis.* **120,** 451–462.
Hall, W. C., and Bagley, L. R. (1978). *J. Clin. Microbiol.* **8,** 242–245.
Hebert, G. A., Pittman, B., McKinney, R. M., and Cherry, W. B. (1972). "*The Preparation and Physicochemical Characterization of Fluorescent Antibody Reagents.*" U.S. Dept. of Health, Education and Welfare, Atlanta, Georgia.
Hebert, G. A., Tzianabos, T., Gamble, W. C., and Chappell, W. A. (1980). *J. Clin. Microbiol.* **11,** 503–307.
Hebert, G. A., Pittman, B., and McKinney, R. M. (1981). *J. Clin. Microbiol.* **13,** 498–502.
Holland, J. W., Stauffer, L. R., and Altemeier, W. A. (1979). *J. Clin. Microbiol.* **10,** 121–127.
Holwerda, J., and Eldering, G. (1963). *J. Bacteriol.* **86,** 449–451.
Huang, S. N., Minassian, H., and More, J. D. (1976). *Lab. Invest.* **35,** 383–390.
Johnson, G. D., and de Nogueire, C., and Araujo, G. M. (1981). *J. Immunol. Methods* **43,** 349–350.
Jones, W. D., Jr., Beam, R. E., and Kubica, G. P. (1966). *Am. Rev. Respir. Dis.* **95,** 516–517.
Jones, G. L., Hebert, G. A., and Cherry, W. B. (1978). "Fluorescent Antibody Techniques and Bacterial Applications." U.S. Department of Health, Education and Welfare, Atlanta, Georgia.
Kasper, D. L., Fiddian, A. P., and Tabaqchali, S. (1979). *Lancet* Feb. 3, pp. 239–242.
Kendrick, P. L., Eldering, G., and Eveland, W. C. (1961). *Am. J. Dis. Child.* **101,** 149–154.
Kohler, R. B., Wheat, L. J., Winn, W. C., Jr., and Girod, J. C. (1982). *J. Infect. Dis.* **146,** 444.
Koshi, G., and Chacko, J. (1971). *Indian J. Med. Res.* **59,** 996–1001.
Labbe, M., Delamare, N., Pepersack, F., Crockaert, F., and Yourassowsky, E. (1980). *J. Clin. Pathol.* **33,** 1189–1192.
Lattimer, G. L., McCrone, C., and Galgon, J. (1978). *N. Engl. J. Med.* **299,** 1172–1173.
Lennette, D. A. (1978). *Am. J. Clin. Pathol.* **69,** 647–648.
Meyer, R. D., Edelstein, P. H., Kirby, B. D., Louis, M. H., Mulligan, M. E., Morgenstein, A. A., and Finegold, S. M. (1980). *Ann. Intern. Med.* **93,** 240–243.
Montenegro, E. N. R., Nicol, W. G., and Smith, J. L. (1969). *Am. J. Ophthalmol.* **68,** 197–205.
Moody, M. D., Siegel, A. C., Pittman, B., and Winter, C. C. (1963). *Am. J. Public Health* **53,** 1083–1092.
Mote, P. T., Hunter, E. F., VanOrden, A. E., Crawford, J. A., and Freely, J. C. (1982). *Arch. Pathol. Lab. Med.* **106,** 295–297.
Neimeister, R. P., Iampietro, L., Bartola, J., and Pidcoe, V. (1980). *Lab. Med.* **11,** 744–745.
Nichols, R. L., McComb, D. E., Haddad, N., and Murray, E. S. (1963). *Am. J. Trop. Med.* **12,** 223–229.
Olcen, P. (1978). *Scand. J. Infect. Dis.* **10,** 283–289.
Page, R. H., Caldroney, G. L., and Stulberg, C. S. (1961). *Am. J. Dis. Child.* **101,** 155–159.
Regan, J., and Lowe, F. (1977). *J. Clin. Microbiol.* **6,** 303–309.
Saravolatz, L. D., Russell, G., and Gvitkovich, D. (1981). *Chest* **79,** 566–570.
Schachter, J., and Dawson, C. R. (1978). "Human Chlamydial Infections." PSG Pub., Littleton, Massachusetts.
Slack, J. M., Moore, D. W., Jr., and Gerencser, M. A. (1966). *W. Va. Med. J.* **62,** 228–231.
Slack, M. P., Griffiths, D. T., and Johnston, H. H. (1981). *J. Clin. Pathol.* **34,** 1381–1384.
Svenungsson, B., Jorbeck, H., and Lindberg, A. A. (1979). *J. Infect. Dis.* **140,** 927–936.
Thomason, B. M., VanOrden, A., Chandler, F. W., and Hicklin, M. D. (1979). *J. Clin. Microbiol.* **10,** 106–108.
Tronca, E., Handsfield, H. H., Wiesner, P. J., and Holmes, K. K. (1974). *J. Infect. Dis.* **129,** 583–586.
Valicenti, J. F., Jr., Pappas, A. A., Graber, C. D., Williamson, H. O., and Willis, N. F. (1982). •*J. Am. Med. Assoc.* **247,** 1149–1152.

Walker, D. H., and Cain, B. G. (1978). *J. Infect. Dis.* **137,** 206–209.

Walker, D. H., Cain, B. G., and Olmstead, P. M. (1978). *Am. J. Clin. Pathol.* **69,** 619–623.

Walker, D. H., Burday, M. S., and Folds, J. D. (1980). *South. Med. J.* **73,** 1443–1447.

Whitaker, J. A., Donaldson, P., and Nelson, J. D. (1960). *N. Engl. J. Med.* **263,** 850–851.

Wills, A., Taylor, E., Pantosti, A., Phillips, I., and Tabaqchali, S. (1982). *J. Clin. Pathol.* **35,** 304–308.

Wing, E. J., Schafer, F. J., and Pasculle, A. W. (1981). *Am. J. Med.* **71,** 836–840.

Winn, W. C., Jr., and Pasculle, A. W. (1982). *Clin. Lab. Med.* **2,** 343–369.

Winn, W. C., Jr., Cherry, W. B., Frank, R. O., Casey, C. A., and Broome, C. V. (1980). *J. Clin. Microbiol.* **11,** 59–64.

Winn, W. C., Girod, J. C., Pelletier, R., Reichman, R. C., and Dolin, R. (1981). *Interscience Conf. Antimicrob. Agents Chemother.*, Abstract No. 296.

Woodward, T. E., Pedersen, C. E., Jr., Oster, C. N., Bagley, L. R., Romberger, J., and Snyder, M. J. (1976). *J. Infect. Dis* **134,** 297–301.

Yobs, A. R., Brown, L., and Hunter, E. F. (1964). *Arch. Pathol.* **77,** 220–225.

Yum, M. N., Wheat, L. J., Maxwell, D., and Edwards, J. L. (1978). *Am. J. Clin. Pathol.* **70,** 832–835.

4

Immunofluorescence in Viral Diagnosis

KENNETH McINTOSH AND LAUREN PIERIK

Department of Pediatrics
Children's Hospital Medical Center
Boston, Massachusetts

I.	Introduction	57
II.	Theory of the Method	58
III.	Reagents	61
IV.	Staining Method	65
V.	Fluorescence Microscope	65
VI.	Collection and Preparation of Specimens: Reading of Slides	67
	A. Respiratory Secretions	68
	B. Skin and Conjunctival Scrapings	73
	C. Biopsy and Autopsy Specimens	75
VII.	Use of Controls	77
VIII.	Interpretation of Findings: Practical Aspects	78
	References	79

I. INTRODUCTION

Immunofluorescence is one of the most widely used methods for the rapid detection of antigen in viral diagnosis. A number of features of the method have contributed to its popularity and acceptance. The first is its high sensitivity, which derives largely from the requirement that antigens be intracellular to be detected by immunofluorescence. The method thus takes advantage of the natural concentration of antigens in very specific compartments of biological tissues. The second is its specificity, which derives largely from the particular intracellular morphology displayed by the antigens of each virus. Nonspecific fluorescence, from whatever source, can be rapidly assessed and ignored by the skilled microscopist, and false positive readings are thus minimized. Another major advantage of immunofluorescence is that the antigens detected are often intra-

The Direct Detection of Microorganisms in Clinical Samples

ISBN 0-12-187780-9

cellular and can therefore be detected even after the host has started to produce antibody. This is because intracellular antigens are in a privileged position until prepared for staining by fixation. In contrast, viruses in clinical samples are often rendered undetectable by culture or solid-phase immunoassays when the infection has reached the stage at which an immune response has begun (Gardner *et al.*, 1970; Gardner and McQuillin, 1978; McIntosh *et al.*, 1982). Finally, the preparation and staining of specimens is in general faster and in some instances easier than the detection of antigens by solid-phase systems. Incubation periods tend to be shorter. The equipment, including fluorescence microscopes, which are generally present in most microbiology or pathology laboratories, is simple and widely available. The use of immunofluorescence for viral diagnosis has been described in detail by Gardner and McQuillin (1980).

The quality of immunofluorescence diagnosis depends on the quality of the reagents—antisera and conjugates—the capacity and condition of the fluorescence microscope, the quality of the specimen, and the skill of the laboratory worker, both in preparing the specimen and in interpreting the findings. The fact that the technique is not more widely practiced stems largely from the sparsity of high-quality reagents and the unavailability of suitable training facilities for those who are making and interpreting the slides. These deficiencies may diminish as the need for viral diagnosis is more widely recognized.

II. THEORY OF THE METHOD

The theory of fluorescence microscopy is as follows. A fluorescent dye absorbs light at one wavelength and emits it at a different (usually higher) wavelength. It is theoretically possible, therefore, to construct a microscope in which the incoming light is exactly of the dye's absorbance wavelength and the light reaching the observer's eye is only of the emitted wavelength. In this system only molecules with the specific required absorption and emission spectra would be seen, and everything else on the slide would be black. In practice, of course, these very narrow spectrum openings are not possible, and indeed some visible background is desirable so that the morphology of the objects seen can be properly reconstructed.

Both fluorescein and rhodamine, the two fluorescent dyes most widely used for immunofluorescence, have the absorption and emission properties just described. Both are often used in their isothiocyanate forms: fluorescein isothiocyanate (FITC) and tetramethylrhodamine isothiocyanate (TRITC). In addition and with increasing frequency, fluorescein is conjugated to proteins in the form of dichlorotriazinylaminofluorescein (DTAF) (Blakeslee and Baines, 1976). Fluorescein in either form absorbs light most efficiently at a wavelength of 490 nm and emits it at 517 nm, producing apple-green fluorescence. Rhodamine absorbs light maximally at a wavelength of 575 and emits it at 595 nm, producing a reddish orange color (Hiramoto *et al.*, 1958; Riggs *et al.*, 1958).

The other desirable quality of the dyes used for fluorescence is that they should be easily attached (conjugated) to macromolecules of all sorts (particularly immunoglobulins), with no alterations or minimal alterations in the structure and function of those macromolecules. With both FITC and TRITC great care must be used to ensure optimal conjugation and subsequent purification of tagged molecules. Conjugation with DTAF and subsequent purification steps are simpler and are therefore preferred by many (Blakeslee and Baines, 1976).

For practical purposes the molecules used for conjugation to such dyes are species antiglobulins (antibodies made against the Fc portions of immunoglobulins from a particular species of animal), antiviral antibodies, anticomplement antibodies, and staphylococcal protein A (SPA). These are used, respectively, for indirect immunofluorescence, direct immunofluorescence, anticomplement immunofluorescence, and SPA immunofluorescence (a form of the indirect method).

In direct immunofluorescence (Fig. 1) the conjugated antibody is directed against the virus antigens that are sought. This reagent is reacted directly with the tissue on the slide. Excess unreacted material is then washed off. The advantage of direct immunofluorescence is that staining requires only a short time (usually about 1 hr) and, because there is only one antiserum involved, nonspecific reactions are minimized. The disadvantages are that more antiviral antibody is required to achieve a given intensity of staining than with the indirect method and that a separate conjugate is necessary for each virus sought. In practice, very high titered antibodies are now being produced for certain viruses, and conjugation in large batches by commerical firms has been shown to be cost-effective. For this reason high-quality direct conjugates are becoming available for many viruses, and it is likely that direct fluorescence will become widely practiced.

In indirect fluorescence the antiviral antibody is unconjugated, and the fluorescent conjugate is either an antiglobulin or SPA. There are thus two successive incubation steps, with appropriate washings in between. The antiviral antibody, unlabeled, is reacted with the tissue on the slide, excess antibody is washed off, and then the antiglobulin (or SPA), conjugated to a fluorescent dye, is incubated on the slide, attaching itself to the antiviral antibody, which is in turn attached to the viral antigen in the specimen. In this two-stage process there is augmentation of the fluorescence by virtue of the fact that several conjugated antiglobulin molecules can attach to a single antiviral antibody molecule. Thus, in comparison with the direct method, less antiviral antibody is necessary to produce the same degree of fluorescence. Moreover, the technique is more flexible in that a single antiglobulin conjugate can be used with several antiviral sera made in the same species of animal. Because the manufacture and testing of conjugates are tedious, this is an advantage when a laboratory is making its own reagents.

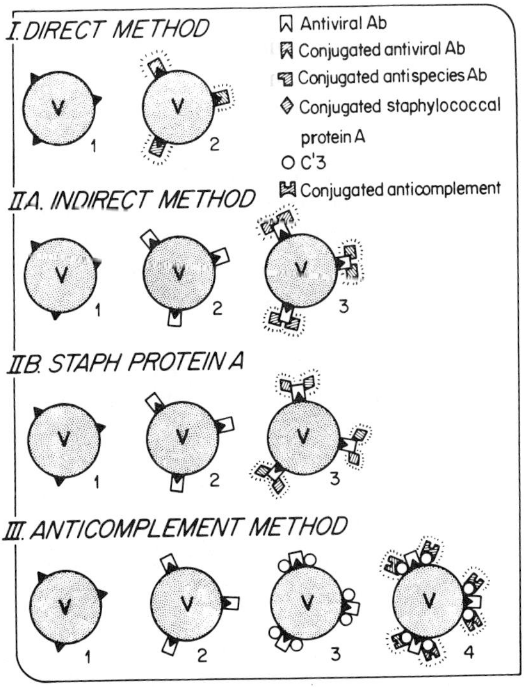

Fig. 1 Diagrammatic depiction of the four general types of immunofluorescence. Washes are carried out between all steps after the second.

However, the indirect method requires more staining time, and the possibilities for nonspecific reactions are greater because more sera are involved. More control systems are also required to rule out such nonspecific reactions. Finally, if human materials are being examined, immunoglobulins are sometimes present in the tissues (in plasma cells or deposited as immune complexes) against which the antispecies globulin conjugate may cross-react. This is particularly a problem when primate, and particularly human, antiviral sera are used in the indirect method.

It is important to note also that, although less antiviral antibody is needed in the indirect method to produce a given level of fluorescence, it is not therefore intrinsically more sensitive as a diagnostic test than the direct method. Sensitivity depends on many factors, only one of which is the number of antiviral antibodies attaching to the antigen in question. Comparisons of the two methods may show either similar sensitivities or even higher sensitivities for the direct method (Minnich and Ray, 1980).

Conjugated SPA may be used in place of antispecies conjugates in the indirect method (Wood and Corbitt, 1982). The theoretical advantages of this conjugate are that SPA is inexpensive and can be produced in a highly purified form with a consequent reduction in nonspecific fluorescence. Staphylococcal protein A also reacts with immunoglobulins from a number of animal species and thus eliminates the need for multiple antispecies conjugates.

However, even under the most favorable conditions the affinity of SPA for globulins is lower than that of a good antispecies antibody. It binds very poorly to the immunoglobulins of some very useful species, such as sheep, cattle, and birds (Goudswaard *et al.*, 1978; Richman *et al.*, 1982). It also binds to only certain subclasses of IgG. For all these reasons higher concentrations of antiviral antisera must be used. Thus, in certain situations it can be useful but probably should not replace normal conjugates for routine indirect immunofluorescence.

Anticomplement immunofluorescence is sometimes used to augment sensitivity still more or to bypass the problems presented by certain viruses (particularly members of the herpes family) that produce Fc receptors on cells during growth. The technique, illustrated in Fig. 1, involves the use of antiviral antibody, a source of complement, and conjugated antibody to the third component of complement (C3). Heat-inactivated serum (often of human origin because of the usefulness of this method in detecting cytomegalovirus) is incubated with the fixed tissues or tissue culture. After the appropriate washing, a source of complement (fresh human serum lacking antibody to the viruses in question may be used at a dilution of 1 : 40) is overlayed. After another washing procedure, conjugated antiserum to human C3 is added (Goldwasser and Shepard, 1958). The procedure would theoretically also pick up immune complexes of any type, but this has apparently not been a problem of great importance with cytomegalovirus in human tissues (Stagno *et al.*, 1980).

III. REAGENTS

The best antiviral antisera are made in experimental animals, are of high titer and high avidity, are free of antibody to other viruses and other microorganisms such as bacteria, yeasts, and fungi, and do not react with normal tissues or the antigens in such common reagents as calf serum (Johansson *et al.*, 1976). There are many ways of producing such sera, but none of them is easy. Whenever possible, purified antigens or viruses should be used in the immunization process. Clean, even germfree, animals are best. Schedules should be constructed with the use of adjuvants so that the antibody is plentiful and avid. Preimmunization sera should always be obtained (in adequate, which usually means large, quantities) so that possible nonspecific reactions can later be controlled for.

It is often not practical or possible to use purified antigens for immunization. In such instances several methods can be used to avoid nonspecific reactions. Antisera may be absorbed with uninfected tissue cultures. Sometimes the problem can be avoided by growing viruses in cell systems unrelated to the tissues ultimately to be examined. For example, embryonated eggs are often an excellent source of high-titered viruses, and antibody against egg antigens usually poses no problem when human tissues are being examined. A second method of avoiding antitissue antibody is to grow the virus with the use of tissue culture and sera obtained from the animal to be immunized. Even under the best circumstances, however, sera usually have to be absorbed before use to remove even low levels of nonspecifically reacting materials.

A practical system for making antisera against a number of virus species is outlined by Gardner and McQuillin (1980). Rabbits are used in all instances, and the antigens are only crudely purified, reliance being placed on several later absorptions with human heteroploid tissue culture lines to remove antitissue antibodies and other nonspecifically reacting substances. The reader is referred to this book for the details of immunization schedules. The great advantages of this system are that a single species of animal is used for all antiviral sera, a large volume of serum can be collected from a single animal, and quite high titered sera can be made. The disadvantages are that multiple time-consuming and expensive absorptions are necessary, and all reagents are indirect, requiring the extra staining time mentioned earlier.

It is often tempting to use high-titered human sera as reagents for certain viruses that are difficult to grow in tissue culture or that fail to infect laboratory animals. The best examples are varicella-zoster virus, cytomegalovirus, and Epstein–Barr virus. Human sera have been used with success, but there are many potential pitfalls. It is sometimes difficult to prove specificity and to avoid completely antibodies against other important viruses. As already mentioned the indirect method with human sera also necessitates staining with antihuman globulin, which may pick up immunoglobulin in the tissues (Fulton and Middleton, 1975; Olding-Stenkvist and Grandien, 1976).

Many of these problems may be avoided when monoclonal antibody becomes available. It would appear to be only a matter of time before such reagents are tested in immunofluorescent antigen detection systems. If they are of moderate to high avidity (Phillips *et al.*, 1982), they can be excellent reagents that demonstrate desirable characteristics of antigenic specificity and freedom from unwanted reactions (Russell *et al.*, 1981; Cepko *et al.*, 1981). Because monoclonal antibodies react with a single antigenic determinant, they can be successfully used to distinguish among viruses that show cross-reactivity with polyclonal antisera such as herpes simplex types 1 and 2 (Balachandran *et al.*, 1982). However, this specificity may limit their usefulness in recognizing a wide range of field strains (Russell *et al.*, 1981). These problems must clearly be examined with care before the use of these reagents becomes widespread.

Few laboratories find it practical to conjugate their own antiviral sera. However, a useful method of doing so with FITC is given in Gardner and McQuillin's book (1980). Methods involving DTAF are somewhat simpler and more reliable (Blakeslee and Baines, 1976).

In the past, the supply of both direct and indirect reagents for rapid diagnosis by immunofluorescence has been variable and undependable, and many reagents have not always been of high titer and specificity. It is hoped that these problems will be solved as the consuming laboratories become more sophisticated and critical. Fortunately, many companies now produce antiglobulin conjugates of high quality and consistency. Whereas only a decade ago all conjugates required absorption before use, this is no longer necessary with currently ''acceptable'' products.

Whether reagents are homemade or store-bought, nonspecific reactions are a potential problem. The types of problems are listed in Table I, along with potential solutions. If the reagent in question was purchased, such reactions are sometimes best dealt with by simply returning the product to the manufacturer. Some, however, are unavoidable or removed only at considerable cost. Under these circumstances the inclusion of proper controls in each staining procedure will sometimes substitute adequately for tedious absorptions. The proper use of controls is discussed in detail in Section VII. Another solution to problems of nonspecific staining, particularly when body tissues (either exfoliated cells or tissue sections) are being examined, is the use of a counterstain such as amido black or Evans' blue. Most counterstains reduce background nonspecific fluorescence but only with some concomitant loss of specific staining. Under circumstances in which unavoidable backgrounds are high, as in work with exfoliated respiratory tract cells, the final result is better when a counterstain is used. This is because the losses of both specific and nonspecific fluorescence tend to be roughly equal, in absolute terms, rather than proportional to the amount of each. Thus, if specific staining without the counterstain is much brighter than nonspecific staining, as it usually is, use of the dye can completely remove the latter but only dampen somewhat the former.

When a new serum or conjugate is obtained, it should be titrated on tissue-culture-grown virus before use with clinical specimens. Twofold dilutions should be made in phosphate-buffered saline (PBS) at pH 7.2 to 7.4 and, if amido black is to be used as a counterstain, it should be added to the diluent (0.5–1%). The serum must give strong specific staining with minimal or no staining of the background or control cells and must retain this strong specific staining for at least a fourfold dilution beyond where it is first seen. Thus, if background staining first disappears entirely at a dilution of, say, 1 : 50, strong, 4+ viral antigen fluorescence should persist at least through a dilution of 1 : 200. The proper dilution of antiviral serum to use on clinical specimens is one that is two to four times as concentrated as the highest one that still gives strong staining with tissue-culture-grown virus. In practice, clinical specimens tend to stain less strongly than tissue cultures, and the two- to fourfold additional strength of the

TABLE I

Nonspecific Reactions in Immunofluorescence of Clinical Samples

Type of reaction	Color	Morphology	Possible source	Potential solution
Hazy glow throughout	Greenish yellow	None	Immersion oil, dirty lenses, or unclean slide	Use glycerol; clean lenses; clean slide
Large chunks	Any, very bright	Irregular	Precipitation in serum conjugate	Centrifuge or filter reagents
Cytoplasmic staining in uninfected cells	Apple green	Looks specific—punctate	Anticellular antibody	Absorb antiviral serum with uninfected cells
Blue irregular background	Blue	Nonspecific, jagged	Use of high-dry lense without coverslip	Use coverslip or, better, get oil-immersion 40× lens
Diffuse, irregular wash	Apple green	Usually thick in places, thin or absent in others	Mucus	Wash the cells more thoroughly next time
Clusters of tiny circles, ovals	Apple green	Looks specific at first; uniform size of circles; extracellular	Bacteria	Ignore
Many cells piled on each other	Yellowish	Dull, not punctate	Specimen too thick	Dilute specimens further before spotting
Cell-sized objects	Dirty green	Dull, not punctate	Dead cells	Ignore
"Positive" appearance of cells, not uniformly distributed on slide, often at edges	Apple green	Can look specific	Drying out of sera or conjugates during staining	Use larger volume of sera or conjugates, maintain proper humidity

antiviral serum is needed. This means that the quality of sera used for rapid diagnosis must be higher than those used for tissue culture work.

IV. STAINING METHOD

Irrespective of the system (indirect, direct, anticomplement, etc.) or the reagents used, the same staining methodology applies. In all instances the slides are placed flat in a humidified box (or simply on a tray if a humidified incubator is available). Moistened towels are a good source of humidity. A practical box is a large petri dish with a cover or a square plastic dish with a tightly fitting lid.

Each properly diluted reagent is placed as a drop, or fraction of a drop, on the square or circle where the cells have been fixed and is spread carefully to cover the area. Platinum loops are good spreaders and can be flamed between squares to avoid possible carry-over of cells or sera.

The slides are then incubated at 37°C for 30 min and then washed. We remove most of the serum drops by tapping the slides sideways on a towel and then immerse them completely in three successive 10-min baths of PBS at pH 7.2 to 7.4.

If several successive stains are needed (as with all but the direct method) either the slides should be allowed to dry completely after the third wash or, short of this, the portions of the slide outside the cell-covered "spots" should dry or be dried thoroughly. This is necessary to prevent the next staining reagent from running between spots.

After the final 30-min stain and three 10-min washes the slide should be dipped briefly (about 1 min) in distilled water, drained, and allowed to dry completely. It is now ready for reading. If a "high-dry" objective lens is to be used, a coverslip with mounting medium is added at this point. This is not necessary for the oil-immersion 40× objective we recommend (see next section).

V. FLUORESCENCE MICROSCOPE

A fluorescence microscope is made up of a light source, filters, and lenses. Each of these will be briefly treated in turn.

Two practical light sources were available at the time of this writing: mercury arc burners and quartz–iodine–tungsten (halogen) lamps. The spectrum emitted by the mercury burner is broader than that of the halogen lamp, and the amount of light produced is greater so that the images are brighter. The broader spectrum also allows the use of fluorochromes other than fluorescein, particularly rhodamine. Rhodamine is a valuable fluorescent dye. Moreover, it has a particular function in the practice of "double staining": The wavelength of light it emits is sufficiently different from that of fluorescein that individual slides can be stained with different antibodies tagged with each dye, allowing the simultaneous visualization of two different antibody species on the same tissue specimen.

However, mercury bulbs are expensive and have a very limited life span (100–200 hr, depending on their use; turning the lamp on and off frequently will wear down the bulb more than leaving it on constantly). Although the quartz–halogen bulbs emit less light and cannot be used for rhodamine staining because of their limited spectrum, they are inexpensive. Moreover, frequent switching off and on does not wear them down, and they are thus preferable in laboratories where the use of the fluorescence microscope tends to be intermittent.

Most light sources are now mounted in such a way that the light path strikes the sample from above, and light emitted from the specimen comes back along the same path, passing through a series of filters and a half-silvered (dichroic) mirror to the eyes of the microscopist. This system, called epiillumination, or incident light, has the advantages of producing brighter images than the older transmitted light systems and obviating the need for immersion oil between the slide and the condenser, a messy inconvenience. There is nothing intrinsically inferior about the older "through-the-specimen" systems, however, and these should not be denigrated if they are being used to advantage.

Filters have also been much improved in the past decade, and almost all the currently available systems use interference technology. These filters are made by depositing thin layers of metal, which form a molecular sieve and allow narrow bands of light to pass through with total exclusion of wavelengths outside these bands. The results are filters with extremely fine performance characteristics and very few of the problems of nonspecificity formerly encountered. Most workers find that a single filter combination for use with fluorescein and another for rhodamine give all the flexibility needed for their work.

Choosing objectives is more a matter of opinion. We have found that, for the examination of clinical samples, particularly exfoliated or scraped cells from mucosal or skin surfaces, a single oil-immersion 40× objective is preferable. The medium-high power allows the necessary strength to distinguish intracellular detail without the loss of light suffered with 100× objectives. The use of oil (or, more conveniently, glycerol) obviates the need for mounting liquid or medium and a coverslip. We rarely have to use a lower power, although this may be preferable for quick scanning of slides, and we almost never go to a higher power. In fact, low-power scanning is less valuable than one would anticipate because of the importance of cellular detail in judging specific as opposed to nonspecific fluorescence.

We have found that glycerol (glycerine) has for all practical purposes the same optical properties as immersion oil and is less expensive, less irritating to the eyes (should it get first on one's fingers and then on one's face), odor-free in the tight confines of a darkened fluorescence room, easier to wipe off the microscope, and much easier to remove from slides that we want to keep overnight or freeze (dip a few times in distilled water). It may be mixed at a ratio of 7 parts glycerol to 1 of phosphate buffer at pH 9.0 or used straight out of the bottle.

Every good fluorescence microscope should be serviced periodically. If a mercury burner is used, the number of hours it is on should be known and the burner changed at 150 hr or whenever it seems to be weak. In spite of the warnings of the manufacturers, we have found that bulbs are often good for more than 200 hr as long as they have not been frequently switched on or off after only a few minutes' use.

VI. COLLECTION AND PREPARATION OF SPECIMENS: READING OF SLIDES

In Table II are listed the common human viruses in which immunofluorescence appears to be a useful technique in diagnosis. A number of viruses are notable for their absence. Diarrhea viruses, such as rotaviruses, "Norwalk-like"

TABLE II

Human Viruses That Have Been Detected in Clinical Specimens by Useful and Practical Immunofluorescence Systems

Virus	Source[a]	Reference
Influenza A and B	NPS, lung	Liu (1956, 1961); McQuillin *et al.* (1970); Daisy *et al.* (1979)
Parainfluenza 1–4	NPS, lung	Fedova and Zelenkova (1969); Gardner *et al.* (1971); Wong *et al.* (1982)
Respiratory syncytial	NPS, lung	Gardner and McQuillin (1968); McQuillin and Gardner (1968)
Adenovirus	NPS, lung	Kalter *et al.* (1969); Gardner *et al.* (1972)
Measles	NPS, lung, urine, skin, brain	Llanes-Rodas and Liu (1965); Fulton and Middleton (1975); McQuillin *et al.* (1976); Olding-Stenkvist and Bjorvatn (1976)
Herpes simplex	Vesicle, brain	Biegeleisen *et al.* (1959); Olding-Stenkvist and Grandien (1976); Schmidt *et al.* (1980)
Varicella-zoster	Vesicle	Schmidt *et al.* (1965); Olding-Stenkvist and Grandien (1976); Drew and Mintz, (1980); Schmidt *et al.* (1980)
Rabies	Brain, cornea	Lennette *et al.* (1965); Dean and Abelseth (1973)

[a] NPS, Nasopharyngeal secretions.

viruses, or caliciviruses, do not appear here because intact cells are often difficult to observe in stools, and solid-phase immunoassays and electron microscopy have been so much more useful. Picornaviruses are also absent. There have been claims that enteroviruses have been observed in the leukocytes of cerebrospinal fluid (Sommerville, 1966), but this is not a practical test for most laboratories. The use of immunofluorescence for the diagnosis of rhinovirus respiratory disease is not recommended because of the very large number of serotypes involved. Epstein–Barr virus has not been found in diagnostic materials, although immunofluorescence has been used in the study of unusual individual cases (Henle and Henle, 1967; Klein *et al.*, 1967). The same is true of the hepatitis viruses (Mathiesen *et al.*, 1979) and papovaviruses (Hogen *et al.*, 1980).

Viruses must be sought within intact cells. These are best obtained by collecting the secretions of mucosal surfaces such as the respiratory tract, the urine, or serosal surfaces (pleura or synovia), by scraping them from intact surfaces such as the conjunctiva or the skin, or by removing tissue at biopsy or at autopsy. Leukocytes have occasionally been used but are rarely of value. They often have Fc receptors or contain immunoglobulins, both of which interfere with the recognition of specific fluorescence.

A. Respiratory Secretions

Because so many viruses infect the epithelial cells lining the respiratory tract, respiratory secretions are often a useful source of material for immunofluorescence. Many viruses are found in the nasopharyngeal mucosa at the time they are producing disease either there or elsewhere, and for this reason it is rarely necessary to sample other areas as well. This is true not only of the respiratory viruses when they are causing pneumonia, bronchitis, bronchiolitis, or croup, but also of several rash diseases (measles or rubella) and also of mumps (Haire, 1969; Fulton and Middleton, 1974).

Many different types of sample may be used. However, we have found that all are in some ways inferior to nasopharyngeal secretions obtained in children by gentle suction into a trap through a plastic catheter or, in adults, by forceful expulsion from the nose after the instillation of about 0.5 ml of PBS. Swabs may not retrieve a large number of cells from the back of the nose. Sputum is both unnecessary and filled with nonspecifically fluorescing material. We have found that secretions obtained by the method of Hall and Douglas (1975) also tend to contain nonspecific fluorescence. Although tracheal secretions obtained through a suction catheter are a good source of cells, they also are usually unnecessary. In addition, we have found that in children the esophagus and stomach are often sampled instead of the trachea, yielding a sample that is strongly acidic and contains squamous epithelial cells (but no virus).

In practice, children are sampled with the use of a no. 8 French feeding tube or suction catheter with a thumb valve and mucus trap attached. The apparatus is connected to a source of mechanical suction, and the tube is inserted along the floor of the nose to the nasopharynx, from which it is slowly withdrawn while gentle suction is applied. The tube may then be inserted into the other nares and the process repeated. One-tenth to 0.6 ml of secretion is usually obtained in this way. The secretion can be washed from the tube into the trap by suction of several milliliters of saline at the bedside, and the sample can then be placed on wet ice and transported to the laboratory. Clearly, care must be taken in obtaining such specimens from severely ill children, but in most instances the removal of secretions from the nose will relieve symptoms of obstruction and produce an overall beneficial effect if it is done gently.

In some conditions secretions will be scant. This is particularly true of croup or of other conditions during convalescence. If the strength of the vacuum is too great or sometimes if the nasal airway is narrow or the mucosa particularly fragile, blood will be obtained. Although this is clearly not desirable, it does not in any way mar the quality of the specimen for fluorescence diagnosis.

Once the specimen is received in the laboratory, it should be gently mixed with a Pasteur pipette. If tissue cultures are to be inoculated, this should be done at this time. We have found that secretion specimens obtained in this way are superior to swabs for the recovery of infectious viruses as well as for fluorescence and may also be used for the detection of soluble antigens by solid-phase assays such as enzyme-linked immunosorbent assay (McIntosh *et al.*, 1982) or radioimmunoassay (Sarkkinen *et al.*, 1981). The contents of the trap are then transferred to a tube, the tube filled with PBS (pH 7.3), and the whole mixed gently and then centrifuged at room temperature at about 300 *g* for 10 min to deposit the epithelial cells. If there is a great deal of mucus in the specimen, this will be found as a hazy layer above the cells. The supernatant wash fluid is now withdrawn with a Pasteur pipette, care being used at the end to remove as much mucus as possible without disturbing the loose pellet of cells. A few drops of PBS are then added, the pellet gently resuspended, and the tube filled again with PBS. The centrifugation and washing steps are repeated as many times as necessary to clean the cells of mucus (usually two or three times total). Finally, about two to eight drops of PBS are added to the final pellet, the quantity depending on the cellularity of the sample. Experience will dictate how many are sufficient. The resultant suspension should be faintly cloudy but by no means dense. When a drop is spread on a slide to dry, there should be 2 to 10 epithelial cells per 400× field. More than this tends to produce nonspecific fluorescence and trapping of dyes. Less gives a slide difficult to read if positive and inadequate if negative.

Many types of slides may also be used. Thin slides (1.1–1.2 mm) frosted on one end are perfectly satisfactory. They should be removed freshly from the box and marked with two or three 8-mm squares with the use of a glass-marking, carborundum- or diamond-tipped pencil. Then they should be dipped (or even stored) in clean acetone and then immediately before use wiped dry with a lint-free cloth or a Kimwipe. The frosted end can then be marked with a regular lead pencil. Other prepared slides are also available, most containing two or three round or square areas for deposition of the specimen and a Teflon or other coating in between. Extra cleaning is often unnecessary with these slides. The signs of an unclean slide are that the staining liquids slide or spread unnecessarily over their surface and that various degrees of background fluorescent haze or grit are seen.

Onto the squares or spots of the prepared slide are then put drops, or more commonly fractions of drops, of the cell suspension. These are spread evenly over the designated surface and allowed to dry, either in a current of warm air or out in the room. The slides are then fixed in reagent-grade acetone at 4°C for 10 min and allowed to dry. It is important always to make more than one slide, because every test should allow for the possibility of repetition and because often several antigens are being tested for.

At this stage the slide can be immediately stained, or it can be left for a period of several days at 4°C, or longer at freezing temperatures. Our experience has been that slides gradually develop increasing nonspecific background fluorescence during storage and, more slowly, decreasing specific fluorescence. This process is slowed by lowering the temperature of storage. At −20°C, in a frost-free freezer, the slides are still quite serviceable 1 year later, but after 2 or more years they become difficult to interpret. At −40°C slides are still usable 4 or 5 years later. We have not had extensive experience with −70°C, but it would appear that slides might be useful indefinitely if they were kept in a moisture-proof container.

When respiratory epithelial cells are stained, a difficult choice must often be made as to which viruses to stain for in a given specimen. Such a choice must be guided by (*a*) the availability of antisera and conjugates; (*b*) the experience of the reader, because there is little sense in staining for a virus that will not be recognized; (*c*) the clinical picture; (*d*) the time of year; (*e*) the prevalent viruses (this applies particularly to influenza viruses); and (*f*) the number of available cells and slides for staining. The last point is important when there are few cells in a sample. A rational basis for these decisions should emerge from a consideration of the points mentioned concerning the specific viruses discussed in the following subsections.

The possible sources of nonspecific staining in clinical specimens can be broadly divided in two categories: those due to problems with the reagents, as already discussed, and those due to problems in the preparation of the specimen. The most common problem in the latter category is insufficient washing of cells. This may be a more acute problem with some specimens than with others, usually because of the admixture of a large amount of mucus, which tends to be difficult to separate from cells by centrifugation. Nevertheless, with thorough homogenization between centrifuge runs, with proper slow-speed centrifugation, and with careful removal of as much of the mucus from above the cells as possible after each run, mucus can be removed from every specimen. The second most common problem is placing too many cells on too small an area of the slide. This merely requires practice in estimating the number of cells in a specimen by the turbidity of the solution after it has been washed free of mucus.

Finally, some specimens contain a large amount of material, particularly bacteria or yeast, which is stained by the reagents at hand. This is an unavoidable problem. The finding of seas of bacteria in a nasopharyngeal specimen has no clinical significance whatsoever. Gram's stain would reveal that most specimens contain a large number of bacteria, and their occasional interference with immunofluorescence diagnosis is primarily an accident of their particular species and the presence of the matching antibodies in the antiviral serum used.

1. Respiratory Syncytial Virus

Respiratory syncytial virus (RSV) is, in temperate climates, a distinctly seasonal virus, appearing usually in late November or December and producing an outbreak that lasts for about 3 months. However, although this behavior is typical, it is not absolutely dependable. Some communities may have smaller or larger outbreaks, and they may begin late in the winter, less commonly earlier, or, very rarely, not at all in a particular winter. Infections during the summer and fall are uncommon. Serious respiratory illness is most common in infants 1–9 months old, but infections are quite common later in childhood and occur occasionally even in adults. The syndromes produced are bronchiolitis, bronchitis, pneumonia, and asthma, with occasional cases of croup. As with all other respiratory viruses, upper respiratory tract infections are common but are usually not the subject of diagnostic investigation.

The fluorescence found in shed epithelial cells infected with RSV is cytoplasmic, finely punctate with occasional large inclusion bodies, and more intense toward the periphery of the cell. It is uncommon to find ciliated cells infected, for some reason, and, in spite of its name, RSV rarely produces syncytia in clinical specimens (McQuillin and Gardner, 1968; Gardner and McQuillin, 1968).

2. Influenza A or B

The seasonality of influenza viruses is well known. They rarely occur except in cold weather and tend to be associated with large outbreaks of one circulating type. Between outbreaks small groups of sporadic wintertime cases occur. Influenza activity tends to be prominent in the news media, and this is a help to the diagnostic virologist. The virus infects at any age, although the very young and the very old are more frequently hospitalized than other age groups. The predominant respiratory syndromes produced are bronchitis and pneumonia. In children croup is common. Also in children influenza can produce extrarespiratory syndromes such as seizures, abdominal pain, myositis, Reye's syndrome, and others. In all age groups fever and constitutional symptoms such as muscle aches and malaise are very common.

Because the fluorescence of influenza viruses is either cytoplasmic or nuclear or both, this is one of the difficult viruses to be certain about in clinical specimens. In the nucleus the staining is usually uniformly bright, with little definition. In the cytoplasm it is often punctate with larger inclusion bodies. Epithelial cells are usually involved; ciliated cells are less often so (Liu, 1956, 1961; McQuillin *et al.*, 1970; Daisy *et al.*, 1979).

3. *Parainfluenza Viruses Types 1 and 2*

These two viruses produce similar clinical syndromes and have similar seasonal epidemiology. They are often seasonal in the autumn, frequently occurring simultaneously. In recent years there has been a pattern of simultaneous epidemics occurring from September or October until December on alternate years. In such instances type 1 is the predominant pathogen. The major syndrome they produce is croup. However, they can be found in all other respiratory syndromes, particularly pneumonia.

The morphology of these two viruses as determined by fluorescence is slightly different from that of RSV. Although the staining is exclusively in the cytoplasm and may be punctate, it is common to see cytoplasmic strands or stripes of fluorescence that coalesce to form irregular blobs. In addition, ciliated cells are often found to contain viral antigens (Fedova and Zelenkova, 1969; Gardner *et al.*, 1971; Wong *et al.*, 1982).

4. *Parainfluenza Virus Type 3*

Parainfluenza virus type 3 infections are endemic in most populations, with occasional 2- or 3-month rises in incidence that appear to be small epidemics. Infections are most severe, as with RSV, in infants 1–9 months of age, but infections are common in later childhood as well and throughout life. The predominant clinical syndromes are the same as those seen with RSV: bronchiolitis, bronchitis, and pneumonia. Croup is somewhat more common than with RSV.

The fluorescence morphology of cells obtained from the respiratory tract in type 3 infection is very similar to that of RSV. However, perhaps as a reflection of the nature of the parainfluenza group as a whole, cells are also occasionally seen with the strands and coalesced strands that are so common with type 1. Moreover, ciliated cells exhibiting specific fluorescence are more frequently seen (Fedova and Zelenkova, 1969; Gardner *et al.*, 1971; Wong *et al.*, 1982).

5. *Adenovirus*

Adenovirus respiratory infections are endemic in children, with little tendency to occur either seasonally or in outbreaks. The only significant exception to this statement relates to adenovirus types 3, 7, and 14, each of which has on occasion produced outbreaks of severe pneumonia in children, usually in selected racial groups. Most adenovirus infections are due to types 1, 2, and 5, and pneumonia, pharyngitis, or mild upper respiratory illness are common clinical presentations. The virus is shed over periods of weeks to months, and therefore a positive culture during illness may not have an etiologic relationship to that illness.

Adenovirus-infected respiratory epithelial cells show either nuclear or nuclear plus cytoplasmic fluorescence. It has been said that extracellular fluorescing particles are more commonly seen in adenovirus than in other respiratory viral infections (Gardner and McQuillin, 1980). Immunofluorescence tends to be negative in specimens from patients carrying the virus incidentally. Thus, adenovirus identified by immunofluorescence during illness is more likely to be of etiologic importance than adenovirus recovered in culture (Kalter *et al.*, 1969; Gardner *et al.*, 1972).

6. Measles Virus

The epidemiology of measles has changed dramatically in countries where vaccine usage is widespread. However, the laboratory diagnosis may be important for several reasons. Atypical clinical syndromes may be seen in children or adults who have not received vaccine because of immunosuppression or immunodeficiency. Moreover, the very rarity of the syndrome has meant a gradual deterioration of the physician's ability to recognize it by clinical criteria, and laboratory confirmation is often of clinical and epidemiologic value.

There is little question that fluorescence of nasopharyngeal epithelial cells represents the easiest and most sensitive method of making a rapid and accurate laboratory diagnosis. The morphology of the fluorescence is very typical. Largely cytoplasmic (but occasionally also nuclear) fluorescence is found in large multinucleated giant cells, often with accentuation along the line of the limiting cell membrane. These cells can also be seen in the urine sediment (Llanes-Rodas and Liu, 1965), but in our experience this is not as consistent as the fluorescence in a respiratory tract aspirate (Fulton and Middleton, 1975). Respiratory epithelial cells containing measles virus antigen can be found from 4 days before the onset of rash to 4 to 5 days after and occasionally as late as 10 days after the rash first appears (McQuillin *et al.*, 1976).

Virus and antigen can also be found in skin biopsies (Olding-Stenkvist and Bjorvatn, 1976), but this procedure is not necessary if respiratory tract cells are used.

B. Skin and Conjunctival Scrapings

Rapid immunofluorescence diagnosis of viral infections on the surface of the body requires great care in obtaining the maximum number of cells with the minimum amount of nonspecifically fluorescing material. For infections by herpes simplex virus (HSV) or varicella-zoster virus (VZV), fresh vesicles must be sampled, because the crusts of aging lesions produce a large amount of interfering fluorescence. The base of each lesion must be scraped carefully but thoroughly in order to dislodge as many cells as possible. The most practical instruments to use for this are a scalpel blade (for skin vesicles), a platinum spatula (for conjunctival scrapings), or a wire swab tipped with calcium alginate or Dacron (for mucosal lesions).

A clean slide with two or three squares etched on it should be ready for use as the specimen is being taken. When either a scalpel blade or spatula is used, a small drop of saline is placed on each square just before the procedure, and as soon as the cells are scraped from the lesion they are transferred to the drop of saline with the help of a dissecting needle. A second dissecting needle is then used to tease apart any large chunks and disperse the material on the full areas of the etched square. At this point the slide is allowed to dry and immediately fixed in cold acetone at 4°C for 10 min. For cells obtained with a swab, the swab is rolled, touched, or gently rubbed on the surface of the slide and the material then allowed to dry (Biegeleisen *et al.*, 1959; Schmidt *et al.*, 1965; 1980; Drew and Mintz, 1980).

The specimen obtained from such maneuvers, if properly fixed and stained, is equally suitable for cytologic examination (the Tzanck smear), and many dermatologists and ophthalmologists prefer this technique (Blank *et al.*, 1951). However, immunofluorescence has the advantage of identifying the species of virus (HSV or VZV) and could even be used to differentiate HSV type 1 from HSV type 2 further if suitably specific antisera were available. Such fine distinctions may have value when immunosuppressed patients are involved, from the viewpoint of both prognosis and, with increasing frequency, treatment. In addition, fluorescence may be more sensitive than more traditional cytologic methods.

1. Herpes Simplex Virus

For fresh vesicular lesions on the skin, immunofluorescence offers sensitivity and specificity very similar to that achieved by culture. In somewhat older lesions, particularly when they are recurrent and therefore contain both less virus and also extracellular virus–antibody complexes, culture is probably better. This pattern is exactly the opposite of that seen with the respiratory viruses.

In genital herpes the virus may be present in lesions on ordinary skin, on dermal–mucosal junctional areas, and on the labia minora, as well as in cervical secretions. Presumably, the sensitivity of immunofluorescence is similar to that of culture for fresh skin lesions. For lesions at or near the vaginal mucosal border, a swab is usually used to obtain cells, and the sensitivity of the method is about 70–80% that of culture. For cervical specimens the sensitivity drops to 50%, and the technique is not recommended (Moseley *et al.*, 1981).

For corneal lesions, fluorescence appears to be highly sensitive and specific. This is because the level of nonspecific fluorescence is lower than with skin or mucosal lesions.

The fluorescence morphology of HSV is similar in all types of specimens. Staining usually fills the cells and is of high intensity throughout. In these instances, the cells often appear ragged, with torn edges presumably reflecting impending cell disintegration. In some cells the fluorescence is predominantly or exclusively cytoplasmic and of lesser intensity. In all cases the cells are the large, spread-out squamous epithelial cells of the skin or the oral or genital mucosa, very unlike the smaller compact cells found in respiratory secretions.

2. *Varicella-zoster Virus*

The techniques for preparing slides from patients with chicken pox or herpes zoster are the same as those for herpes simplex. Either scrapings with a scalpel blade or cells obtained with a swab may be used. Because VZV is a difficult virus to grow in tissue culture, the diagnosis by immunofluorescence is considerably more sensitive than by virus isolation. For example, of 24 samples examined by Drew and Mintz (1980) and found positive by either immunofluorescence or culture, 14 were found positive only by fluorescence. None were positive as determined only by culture.

As with HSV, the cells stained are large, and the fluorescence is found both in the cytoplasm and in the nucleus. All specimens examined for VZV should also be stained for HSV because the two viruses may rarely produce clinically indistinguishable lesions. For both VZV and HSV, direct conjugates give fewer nonspecific reactions than indirect conjugates, probably owing both to the reduction in Fc receptor staining and to the frequent presence of antibody in clinical lesions that reacts with antispecies conjugates (Olding-Stenkvist and Grandien, 1976).

C. Biopsy and Autopsy Specimens

The third major source of material for diagnosis by immunofluorescence is tissue specimens obtained from either biopsy or autopsy. The purposes of such investigations vary widely. On the one hand, intensive study may be needed to establish structure–pathogen relationships. On the other, a rapid and definitive diagnosis may be required to guide therapeutic choices. The methodologic approach will depend on the setting.

In general, it is possible, and sometimes even preferable, to use "touch preparations" made from fresh tissue fragments for the rapid identification of a virus in a tissue specimen. Frozen or paraffinized sections retain anatomic relationships but are often time-consuming to prepare (and usually involve the services of a pathologist), require specialized equipment, and, most important, tend to have a higher background of nonspecific fluorescence than touch preparations. In addition, most technicians will have obtained their prior experience with fluorescence in cells distributed singly over the surface of a slide, rather than with relatively thick tissue sections. The appearance of intracellular immunofluorescence is different when cells are bounded by neighboring cells and tends to be more difficult to recognize as specific in spite of all the advantages of intact anatomical structure.

If autopsy specimens are being examined, tissue sections may be a necessity because tissues are often sent for study some time after the procedure has been done. In addition, tissues may have been preserved for anatomic study in formalin, necessitating special procedures for immunofluorescence. These techniques are described in detail by Swoveland and Johnson (1979). In brief, formalin-preserved specimens imbedded in paraffin are cut with a microtome, and the sections are deparaffinized with xylene and graded alcohol baths. They are then washed in Ca–Mg-free PBS and incubated for 1 hr at room temperature in 0.25% trypsin. Purified trypsin preparations for some reason work less well than crude ones. Staining is then carried out in the usual fashion.

Touch preparations are simple and rapid to make. Slides are prepared as for respiratory secretions. Squares about 8mm on a side are etched on their surfaces and are washed in acetone. On removal from the acetone they are wiped dry with a lint-free cloth or a Kimwipe and labeled appropriately. The tissue is then, if necessary, cut into pieces 2–4 mm on a side. These are picked up carefully with either fine or, preferably, round-tipped smooth forceps and pressed lightly once or twice within the boundaries of each square in turn. Enough spots are made so that all the necessary sera can be tested, all controls can be performed, and slots remain for repeat staining if this be necessary. This requires making more than one slide. The tissue should touch the slide only once for each area (i.e., repeated daubs in the same place should not be made; enough cells are almost always deposited with a single touch), although if the fragment is small, several touches can be made within a single etched square. As soon as the cells are dried on the slide, they should be fixed for 10 min in acetone at 4°C. They then can be stained as for any other preparation.

Frozen sections should, if possible, be deposited two per slide. They should also be fixed for 10 min in cold acetone, and the area around them scored with a glass-marking pencil to assist in the staining procedures.

1. Viruses in Brain: Rabies, Herpes Simplex, and Others

The viruses most often sought in neural tissue are rabies and herpes simplex virus (Schmidt *et al.*, 1980). In unusual circumstances, such viruses as measles virus, rubella virus, papovaviruses, and others might be sought. The rules for such immunofluorescence are generally the same as those already outlined. There is a particular need for controls under certain circumstances. In general, microscopists have had less experience with such procedures than with respiratory secretions or skin scrapings, and for this reason both positive and negative controls become valuable and sometimes a necessity. Conjugate controls are essential whenever a new or unfamiliar conjugate is being used or whenever a familiar one is being used under unfamiliar circumstances.

Rabies is an important special case that deserves separate consideration. The search for rabies virus is not a task to be undertaken by any but those working in laboratories familiar with both the hazards involved and also the techniques of specimen preparation and virus recognition.

The most rewarding tissue in which to seek this virus is the brain, but salivary glands should be examined in animals suspected of spreading the virus through bites. The hippocampus is the most likely area of involvement, and this tissue should be specifically prepared. Touch preparations, as described in the preceding subsection, are made from cut tissue surfaces and stained (usually by the direct method) with specific antisera. Positive and negative controls should always be run, along with immune and control antisera. In many instances blocking by a preparation of soluble antigen adds to the reliability of the reading.

When experienced technicians are using well-tested specific reagents, immunofluorescence is the most reliable rapid method of diagnosis for rabies. Mouse inoculation should always be used to corroborate findings but, when comparisons have been made, agreement between immunofluorescence and mouse inoculation has been 95–100%, whereas search for Negri bodies by traditional staining techniques has been only 35–75% sensitive (Lennette *et al.*, 1965; Dean and Abelseth, 1973).

2. *Viruses in Pulmonary Tissue: Cytomegalovirus and Others*

The usual viruses sought in pulmonary tissue are the respiratory viruses, measles virus in the immunoincompetent host, and CMV. The last is particularly tricky to work with because of its capacity to induce Fc receptors on the surface of infected cells. These trap immunoglobulins of any idiotype and of many animal species and can produce misleading staining that looks specific. In order to overcome this problem, either Fab or $(Fab)_2$ fragments or anticomplement immunofluorescence must be used. Most laboratories are not set up for such specialized procedures, and they had probably best be avoided if not performed routinely. In our experience culture is considerably more sensitive than immunofluorescence for the detection of CMV in biopsy tissues.

VII. USE OF CONTROLS

The following is a list of possible controls that can be used to prove the specificity of a positive or negative reading in a particular clinical specimen (pcs):

1. Repeat staining of the pcs for the same virus.
2. The pcs stained with control, preimmune antiviral serum.
3. The pcs stained with a different antiviral serum from the same animal species.
4. The pcs stained with PBS and conjugate (omission of the antiviral serum, for indirect fluorescence only).
5. A known positive clinical specimen stained for the virus in question.
6. A known negative clinical specimen stained for the virus in question.
7. A positive infected tissue culture preparation stained for the virus in question.
8. A negative uninfected tissue culture preparation stained for the virus in question.
9. Blocking tests: incubation of the pcs with an antiviral, unlabeled serum (and, in a duplicate slide or square, the appropriate control, preimmune serum) from an unrelated animal species followed by staining in the usual fashion. The unlabeled antiviral serum should block or severely dampen the fluorescence, whereas the control serum should not.

Clearly, all of these controls are not necessary for every positive or negative specimen. In practice, most clinical specimens are controlled by a combination of daily use of the same reagents, the experience of the microscopist, and control system 3. However, a number of circumstances would dictate that one or another additional control procedure be applied: (*a*) unusual intracellular morphology, (*b*) staining in an unfamiliar tissue (e.g., thymus), (*c*) an unfamiliar situation (e.g., measles virus in a brain biopsy), (*d*) unfamiliar sera or conjugates, (*e*) disagreement between immunofluorescence and culture, (*f*) a surprising match of virus and clinical syndrome (e.g., RSV and encephalitis), (*g*) a surprising epidemiologic or seasonal finding (e.g., influenza A in the summertime).

Choosing a control is sometimes difficult and very much a matter of judgment. In many instances a combination of controls 1, 2, 3, and 4 is satisfactory. If the tissue is unfamiliar, the most helpful control is often 6, and, of course, 5 is very useful if such positive tissue is available. In the absence of these controls and in an unfamiliar situation, controls 7 and 8 may be valuable. Finally, blocking is definitive if properly performed. It is used primarily in research and in rabies diagnosis. Nevertheless, any laboratory seriously involved with immunofluorescence should be familiar with it and have used it in known situations so that it can be effectively employed if the need arises.

VIII. INTERPRETATION OF FINDINGS: PRACTICAL ASPECTS

This section contains a number of general comments that apply to all viruses and all types of specimen. Even when all the rules are followed regarding careful specimen collection and handling, technique in making slides, selection of the highest-quality antisera and conjugates, and care and maintenance of the microscope, technologists still face general problems in the interpretation of slides.

First, what are the features required for a positive reading? The issue of controls is discussed in the preceding section, and the comments here are over and above the requirements described there.

1. The fluorescence must be of the right color.
2. The fluorescence must be intracellular.
3. The viral antigens must be in the right kind of cell.
4. The antigen must be in the right portion of the cell, depending on the nature of the virus.
5. The morphology of the fluorescence must be characteristic for that virus in that type of cell.
6. More than one convincingly positive cell must be seen.

The last point deserves some elaboration. If a specimen is truly positive, a number of almost acceptable cells will usually be seen in addition to the acceptable ones. Thus, even if only two cells are ultimately found, there will be a background of borderline positivity which in practice is very important in supporting the positive reading. If one or two strongly positive cells are seen against a uniformly and convincingly negative background, the microscopist should beware. It may be that, through error or carelessness, a few infected cells have been transferred onto the slide from another truly positive slide during preparation or staining.

Confidence comes only with practice. This means constantly checking one's readings against simultaneous tissue culture results, or, less commonly, gathering extensive experience from a large group of known positive and negative slides. In our laboratory, after several years of experience, we feel confident to give a diagnosis without culturing a specimen only when dealing with the few virus systems with which we are most familiar: RSV, parainfluenza 1, and parainfluenza 3. In all other systems, and in most instances with those three viruses as well, we culture the specimens at the same time and keep close track of the sensitivity and specificity of our immunofluorescence. As discussed above, we always try to have available at least one extra slide so that we can repeat our staining if we wish either immediately in the case of an equivocal result, or later if there is disagreement between culture and fluorescence. This is a crucial part of the constant educational process and allows us to gain confidence as we gain experience.

REFERENCES

Balachandran, N., Frame, B., Chernesky, M., Kraiselburd, E., Kouri, Y., Garcia, D., Lavery, C., and Rawls, W. E. (1982). *J. Clin. Microbiol.* **16,** 205–208.

Blakeslee, D., and Baines, M. G. (1976). *J. Immunol. Methods* **13,** 305–319.

Blank, H., Burgoon, C. F., Baldridge, G. D., McCarthy, P. L., and Urback, F. (1951). *J. Am. Med. Assoc.* **146,** 1410–1412.

Biegeleisen, J. Z., Jr., Scott, L. V., and Lewis, V., Jr. (1959). *Science* **129,** 640–641.

Cepko, C. L., Changelian, P. S., and Sharp, P. A. (1981). *Virology* **110,** 385–401.

Daisy, J. A., Lief, F. S., and Friedman, H. M. (1979). *J. Clin. Microbiol.* **9,** 688–692.

Dean, D. J., and Abelseth, M. K. (1973). *In* "Laboratory Techniques in Rabies" (M. M. Kaplan and H. Koprowski, eds.), 3rd ed., pp. 73–84. World Health Organization, Geneva.

Drew, W. L., and Mintz, L. (1980). *Am. J. Clin. Pathol.* **73,** 699–701.

Fedova, D., and Zelenkova, L. (1969). *J. Hyg., Epidermiol., Microbiol., Immunol.* **13,** 13–23.

Fulton, R. E. and Middleton, P. J. (1974). *Infect. Immun.* **10,** 92–101.

Fulton, R. E. and Middleton, P. J. (1975). *J. Pediat. (St. Louis)* **86,** 17–22.

Gardner, P. S., and McQuillin, Jr. (1968). *Br. Med. J.* **3,** 340–343.

Gardner, P. S., and McQuillin, Jr. (1978). *J. Med. Virol.* **2,** 165–173.

Gardner, P. S., and McQuillin, J. (1980). "Rapid Virus Diagnosis. Application of Immunofluorencence," 2nd ed. Butterworth, London.

Gardner, P. S., McQuillin, J., and McGuckin, R. (1970). *J. Hyg.* **68,** 575–580.

Gardner, P. S., McQuillin, J., McGuckin, R., and Ditchburn, R. K. (1971). *Br. Med. J.* **2,** 7–12.

Gardner, P. S., McGuckin, R., and McQuillin, J. (1972). *Br. Med. J.* **3,** 175.

Goldwasser, R. A., and Shepard, C. C. (1958). *J. Immunol.* **80,** 122–131.

Goudswaard, J., van der Donk, J. A., Noordzij, A., van Dam, R. H., and Vaerman, J. P. (1978). *Scand. J. Immunol.* **8,** 21–28.

Haire, M. (1969). *Lancet* **1,** 920–921.

Hall, C. B., and Douglas, R. G., Jr. (1975). *J. Infect. Dis.* **131** 1–5.

Henle, G., and Henle, W. (1967). *Cancer Res.* **27,** 2442–2446.

Hiramoto, R., Engel, K., and Prissman, D. (1958). *Proc. Soc. Exp. Biol. Med.* **97,** 611–614.

Hogan, T. F., Padgett, B. L., Walker, D. L., Borden, E. C., and McBain, J. A. (1980). *J. Clin. Microbiol.* **11,** 178–183.

Johansson, M. E., Bergquist, N. R., and Grandien, M. (1976). *J. Immunol. Methods* **11,** 265–272.

Kalter, S.S., Armour, V., and Reinarz, J. A. (1969). *Arch. Gesamte Virusforsch.* **28** 34–40.

Klein, G., Clifford, P., Klein, E., Smith, R. T., Minowada, J., Kourilsky, F. M., and Burchenal, J. H. (1967). *J. Natl. Cancer Inst.* **39,** 1027–1044.

Lennette, E. H., Woodie, J. D., Nakamura, K., and Magoffin, R. L. (1965). *Health Lab. Sci.* **2,** 24–34.

Liu, C. (1956). *Proc. Soc. Exp. Biol. Med.* **92,** 883–887.

Liu, C. (1961). *Am. Rev. Respir. Dis., Suppl.* **83,** 130–132.

Llanes-Rodas, R., and Liu, C. (1965). *N. Engl. J. Med.* **275,** 516–523.

McIntosh, K., Hendry, R. M., Fahnestock, M. L., and Pierik, L. P. (1982). *J. Clin. Microbiol.* **16,** 329–333.

McQuillin, J., and Gardner, P. S. (1968). *Br. Med. J.* **1,** 602–605.

McQuillin, J., Gardner, P. S., and McGuckin, R. (1970). *Lancet* **2,** 690–695.

McQuillin, J., Bell, T. M., Gardner, P. S., and Downham, M. A. P. S. (1976). *Arch. Dis. Child.* **51,** 411–419.

Mathiesen, L. R., Fauerholdt, L., Møller, A. M., Aldershvile, J., Dietrichson, O., Hardt, F., Nielsen, J. O., and Skinhøj, P. (1979). *Gastroenterology* **77,** 623–628.

Minnich, L., and Ray, C. G. (1980). *J. Clin. Microbiol.* **12,** 391–394.

Moseley, R. C., Corey, L., Benjamin, D., Winter, C., and Remington, M. L. (1981). *J. Clin. Microbiol.* **13,** 913–918.

Phillips, D. J., Galland, G. G., and Reimer, C. B., and Kendal, A. P. (1982). *J. Clin. Microbiol.* **15,** 931–937.

Olding-Stenkvist, E., and Bjorvatn, B. (1976). *J. Infect. Dis.* **134,** 463–469.

Olding-Stenkvist, E., and Grandien, M. (1976). *Scand. J. Infect. Dis.* **8,** 27–35.

Richman, D. D., Cleveland, P. H., Oxman, M. N., and Johnson, K. M. (1982). *J. Immunol.* **128,** 2300–2305.

Riggs, J. L., Seiwald, R. J., Burckhalter, J. H., Downs, C. M., and Metcalf, T. G. (1958). *Am. J. Pathol.* **34,** 1081–1097.

Russell, W. C., Patel, G., Precious, B., Sharp, I., and Gardner, P. S. (1981). *J. Gen. Virol.* **56,** 393–408.

Sarkkinen, H. K., Halonen, P. E., Arstila, P. P., and Salmi, A. A. (1981). *J. Clin. Microbiol.* **13,** 258–265.

Schmidt, N. J., Lennette, E. H., Woodie, J. D., and Ho, H. H. (1965). *J. Lab. Clin. Med.* **66,** 403–412.

Schmidt, N. J., Gallo, D., Devlin, V., Woodie, J. D., and Emmons, R. W. (1980). *J. Clin. Microbiol.* **12,** 651–655.

Sommerville, R. G. (1966). *Arch. Gesamte Virusforsch* **19,** 63–69.

Stagno, S., Pass, R. F., Reynolds, D. W., Moore, M. A., Nahmias, A. J., and Alford, C. A. (1980). *Pediatrics* **65,** 251–257.

Swoveland, P. T., and Johnson, K. P. (1979). *J. Infect. Dis.* **140,** 758–764.

Wong, D. T., Welliver, R. C., Riddlesberger, K. R., Sun, M. S., and Ogra, P. L. (1982). *J. Clin. Microbiol.* **16,** 164–167.

Wood, D. J., and Corbitt, G. (1982). *J. Clin. Pathol.* **35,** 472–475.

II

IMMUNOLOGIC METHODS FOR DETECTING SOLUBLE ANTIGENS

Introduction

J. DONALD COONROD

The chapters in this section deal with immunologic techniques that have been used to detect soluble microbial antigens in clinical samples. Chapters 5–10, on precipitin techniques, deal almost exclusively with counterimmunoelectrophoresis (CIE). An argument can be made (Chapter 8) for the use of double immunodiffusion techniques in situations in which neutral rather than acidic polysaccharides may be present in high concentration (e.g., type 7 or type 14 pneumococcal empyema). However, immunodiffusion and capillary techniques are too insensitive, generally, for routine application to the detection of microbial antigens. Chapter 5 deals with the methodology of CIE and Chapters 6–9 with the specific results obtained with the test. Chapter 10 is an in-depth critique of CIE. Chapters 11–13 cover methods and results obtained with agglutination techniques. The future of these rapid and technically simple diagnostic techniques in clinical microbiology laboratories appears to be more secure than that of CIE, but many aspects of the critique of CIE in Chapter 10 could be applied to agglutination tests as well. Chapter 14 deals with enzymatic assays and radioimmunoassays in meningitis and contains many comparative data obtained by both agglutination methods and CIE. The successful application of agglutination techniques to cryptococcal meningitis is discussed in Chapters 12 and 14. Applications of enzymatic and radioimmunoassays to the diagnosis of a variety of bacterial, fungal, and viral diseases are discussed in Chapters 15–20. Insights into the methodologic challenges of these methods are offered in most of these chapters but are dealt with in greatest detail in Chapters 15, 16, 18, and 20. It does not appear that there will be any immediate end to technologic innovation in this area!

5

Procedures for the Detection of Microorganisms by Counterimmunoelectrophoresis

RICHARD TILTON

Department of Laboratory Medicine
John Dempsey Hospital
The University of Connecticut Health Center
Farmington, Connecticut

I. Introduction 87
II. Principle 88
III. Variables 88
A. Antisera 89
B. Buffers 89
C. Support Systems 90
D. Electrophoresis Chamber and Power Supply 90
IV Procedure for Counterimmunoelectrophoresis of Spinal Fluid, Serum, Urine, and Other Body Fluids 91
A. Preparation of Agarose 91
B. Electrophoresis 92
C. Sample CIE Setup for CSF and Urine 92
D. Concentration of Samples 93
E. Liquefaction of Sputum for CIE 93
F. Staining of CIE Plates 94
V. Clinical Applications of CIE 94
References 95

I. INTRODUCTION

In 1901 Vincent and Bellot described the use of the tube precipitin reaction for the detection of meningococcal antigen in cerebrospinal fluid (CSF). Dochez and Avery identified the capsular polysaccharide of the pneumococcus in patients' urine as early as 1917. Counterimmunoelectrophoresis (CIE) was originally de-

The Direct Detection of Microorganisms in Clinical Samples

ISBN 0-12-187780-9

scribed in 1959 by Bussard and was first used clinically for the detection of "Australia antigen" (Hb_sAg). Radioimmunoassay soon replaced CIE for HB_sAg, but CIE has become a valuable immunologic tool for the rapid detection of other microbial antibodies and antigens.

II. PRINCIPLE

Counterimmunoelectrophoresis is based on the principle of immunodiffusion modified by electrophoretically driving the antigen and antibody toward each other. The antigen is placed in a well on the cathode side and because of its negative charge is drawn toward the anode. The antibody is placed in a well on the anode side and moves toward the cathode by endoosmotic flow. The mobility of the antibody is the sum result of the migration to the anode due to weak negative charges and the rather swift migration toward the cathode by endoosmotic flow. If conditions of voltage, current, pH, buffer, relative concentration of antigen and antibody, and reactivity of antisera are optimal, a precipitin line will form in 30 to 45 min approximately equidistant from the anode and the cathode well. Figure 1 is a diagrammatic sketch of the CIE reaction.

III. VARIABLES

As in any laboratory procedure, a number of variables must be standardized if CIE is to be a rapid and sensitive method of antigen detection.

A. Antisera

The quality of antisera is the most important variable in CIE. Without reactive antisera, quality control of other physicochemical biological parameters is use-

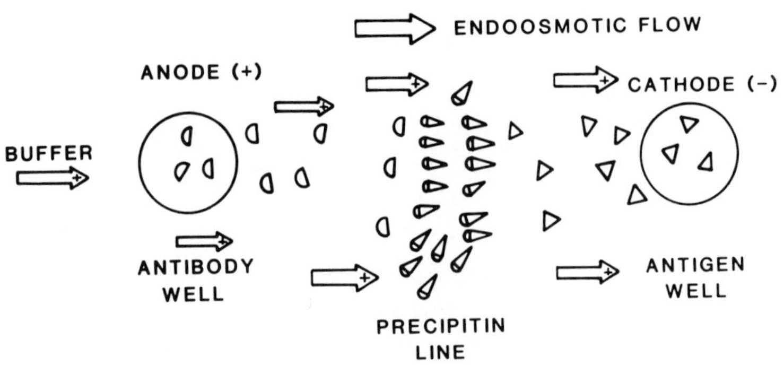

Fig. 1. Principles of counterimmunoelectrophoresis.

TABLE I

Sensitivity of Counterimmunoelectrophoresis for Antisera

Antisera	Limits of detection (μg/ml)
Haemophilis influenzae type B	0.05–0.1[17]
Streptococcus pneumoniae (Omniserum)	0.05[10]
Neisseria meningitidis	
Groups A, B, C, X, Y, Z	0.05[10]
Groups A, B, C, D, G streptococci	0.05–0.1

less. It is difficult to procure antisera that reproducibly have a high precipitin titer, migrate in CIE, and are of moderate cost. Even if the highest-quality antisera are used, the nature of the antigen may preclude a precipitin reaction. The commercial availability of monoclonal antibodies may solve many of the problems of sensitivity and specificity. Anhalt and Yu (1975) reviewed 128 published cases of pneumococcal pneumonia and found that not a single case of *Streptococcus pneumoniae* types 7 and 14 was detected by CIE. These antigens have a neutral charge and do not migrate to the anode. Thus, no precipitin line is formed when a barbital buffer is used. When a borate–barbital buffer is substituted (Anhalt and Yu, 1975), precipitin lines develop. The pneumococcal antibody may be sulfonated so that it migrates anodally. The antibody well is then placed to the right of the antigen well instead of to the left as in most applications. Table I indicates the sensitivity of CIE for the commonly used antisera.

B. Buffers

The buffer in CIE maintains the pH and ionic strength and affects the endo-osmotic flow. Usually, the buffer is alkaline (i.e., barbital buffer, pH 8.2–8.6) and confers a negative charge on the molecules. The buffer should not be at the isoelectric point of the reactants (Rytel, 1979). The ionic strength (u) of a solution is defined as the sum of the products obtained by multiplying the molarity of each ion by the square of its valence and dividing by 2.

$$u = EmX^2/2$$

Where u is ionic strength, m is molarity, and X is valence. This formula enables one to calculate the ionic strength if the molarity of the buffer system is known. The optimal ionic strength for CIE in most systems is 0.05 μ (Tilton, 1978).

Buffer systems may be continuous or discontinuous. A continuous buffer system is one in which the buffer in the electrophoresis chamber is of the same pH and ionic strength as the buffer used to make the agar gel. In a discontinuous system the buffers in the gel and in the wells are different. In the author's experience discontinuous buffer systems offer no advantage.

C. Support Systems

Immunologic reactions occur in gels. Several supports have been used for CIE, including agarose (Moody, 1976), cellulose acetate (Cohn and Kahan, 1976), Noble agar, and bacteriologic agar (Kelkar and Niphadkar, 1974). Certain antibodies such as *Neisseria meningitidis* group B migrate best in the highly charged Noble agar due to the higher endoosmotic flow. However, as the endoosmotic flow increases, the sensitivity of antigen detection decreases. For most applications agarose is the gel of choice. It is a neutral linear polysaccharide, is water soluble, and will form a gel at 0.01% (Rytel, 1979). Agarose gels should be no thicker than 3 mm. Thick gels may result in false negative results. However, if the gels are too thin (<1.0–1.5 mm), they may be unstable due to excessive heat.

Gels may be placed on glass microscopic slides, plastic plates, or Mylar film. The author prefers Mylar film because a permanent stained record of the CIE analysis can be retained.

D. Electrophoresis Chamber and Power Supply

The chamber provides a physical support for the gel, it contains buffer, and it transmits the voltage and current to the gel matrix. Many chambers are commercially available, although the author prefers one that can be cooled. Similarly, many power supplies are available. The user should choose one that offers variable voltage and amperage.

Constant current is the preferred mode of operation for CIE. The relationship between voltage (*V*), amperage (*I*), and resistance (*R*) in the CIE has been described as follows

> At a given voltage, the resistance (*R*) of the gel support will influence the current. In order to increase the electrophoretic migration rate, the voltage (*V*) must be increased in order to increase the current (*I*), assuming constant *R* in the gel bed. However, as *V* is increased, heat is generated, *R* falls, and distillation of water in the system occurs which causes *R* to fall even more. In fact, the heat generated may cause denaturation of the antibodies. Consequently, either *V* or *I* must be stabilized; *I* will increase because *R* is progressively falling. If *I* increases, additional heat will then be generated. If *I* is stabilized, *V* varies as does *R*. However, less heat is generated at a constant current [Anhalt and Yu, 1975].

The CIE procedure presented in this chapter is the one currently in use at the University of Connecticut Health Center. Although other laboratories may prefer their own variations, this procedure is simple, adaptable, and economical. It is not a reference procedure, nor is it presented as being a better procedure than others.

For the routine detection of bacterial antigens, CIE is usually performed at 100 V and 30–40 mA per 12 × 18 cm (5 × 7 in.) Mylar strip. However, in order to generate 30–40 mA at 100 V, the 1% agarose gel strip must be freshly prepared. Edwards (1971) has reported successful results using 3–6 mA per microscope slide and 12–18 mA per lantern slide for 30 to 45 mins.

The end result of ensuring quality antisera and controlling the many variables in CIE is that, when an antigen and antibody are electrophoresed, the precipitin line will be visible and approximately equidistant between the two wells. Barring antigen or antibody excess, the formation of such a line constitutes a positive reaction; the absence of a line indicates a negative test but not necessarily the absence of disease. It has been the author's experience that prozone effects usually occur due to antigen excess. The CIE is then repeated at a 1:10 dilution.

Although some (Rytel, 1979) do not recommend staining, in the author's laboratory all negative CIE plates are stained with Coomassie blue.

IV. PROCEDURE FOR COUNTERIMMUNOELECTROPHORESIS OF SPINAL FLUID, SERUM, URINE, AND OTHER BODY FLUIDS

A. Preparation of Agarose

1. Prepare 30 ml of 1% agarose in barbital buffer (0.3 gm agarose and 30 ml buffer). Use an analytical balance to weigh the agarose to the nearest 0.01 g. Carefully bring the agarose solution to a boil in a foil-covered flask until the agarose is completely dissolved. Do *not* allow the agarose solution to boil over because the concentration will be altered.
2. Adjust the leveling table and label a 5 × 7 in. Mylar sheet in the lower right-hand corner with a black marker. If few tests are to be done, the sheets may be cut and less agarose used.
3. With a 10-ml pipette, apply 25 ml of molten agarose evenly to the sheet. When the agarose has gelled, place the sheet in a humid chamber (plastic freezer containers) at 4°C until ready for use.
4. Using the sheet (Fig. 2) as a guide, cut wells in pairs with a 4-mm well cutter, according to the number of wells needed. Remove the agar plugs by aspiration.
5. Fill the wells with the appropriate antigen, antibodies, and control reactants using either micropipettes or disposable capillary pipettes.

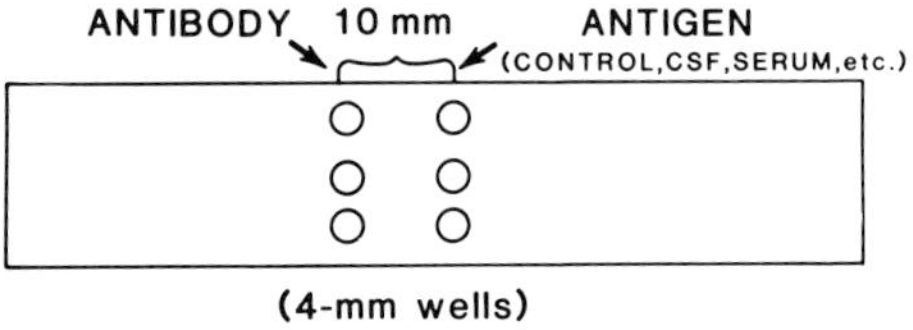

Fig. 2. A CIE template.

B. Electrophoresis

1. Fill the chamber troughs with barbital buffer (pH 8.2).
2. Place the agarose sheets in the chamber. The antigen is *always* placed near the cathode. The polarity can be changed each time a run is done with the same buffer to prevent the accumulation of salts on one side of the chamber. Replace the cover.
3. Connect the electrodes to the power supply and turn power on.
4. After 30 to 45 mins, turn power off, remove the electrodes, and record the results.
5. If precipitin lines are not visible, place in refrigerator (4°C) for 30 mins and reinspect.
6. Issue a preliminary report. Place the agarose sheet between two paper towels and apply pressure with several old textbooks to remove water from the gel. Place in the deproteinizing solution before staining.

C. Sample CIE Setup for CSF and Urine

All patients:

1. *Haemophilus influenzae* type B
2. *Neisseria meningitidis* groups A–C, X, Y, Z, WR-135 or *N. meningitidis* polyvalent plus XYZ and WR-135
3. *Streptococcus pneumoniae* Omniserum

If patient is <6 months of age, add:

4. Group B streptococci
5. *Escherichia coli* K1

An *H. influenzae* type B-positive control is used and consists of *H. influenzae* purified polysaccharide vaccine diluted 1 : 1000. If the vaccine cannot be obtained, then an autoclaved, filtered suspension of *H. influenzae* grown for 18 to 24 hr in Levinthal's broth will suffice.

D. Concentration of Samples

If the volume is sufficient (5 ml), samples such as urine are concentrated 25–50 times before processing using a disposable multiple ultrafilter, the Minicon B-15 (Amicon Corp., Lexington, Massachusetts).

Each of eight isolated sample chambers of the Minicon B-15 holds up to 5 ml of sample. The chambers may be used for different samples or for the aliquots of the same sample. The inner surface of the chambers is a membrane of selective permeability backed, in turn, by absorbent pads that wick away water and permeating molecular species. Retained constituents are progressively concentrated in the chamber as sample volume is diminished. Graduation lines at 5, 10, 25, 50, and 100× indicate concentration ratios. At about 100×, the membrane has been treated to retard further concentration and inadvertent reduction to dryness. For CIE applications, an insufficient volume remains after concentration >50× but, because the capacity of the absorbent is about three times the total sample volumes, chambers may be refilled to process samples exceeding chamber capacity or to achieve higher concentrations.

Spinal fluid and other very dilute specimens may occasionally be concentrated beyond 100×. To ensure recovery of such irreplaceable samples, the concentration level should be checked periodically and the concentrate removed once it has reached 100×. In the event that the sample accidentally goes to dryness, most of the concentrate can be recovered by adding a small amount (50 μl) of normal saline.

If the initial volume of the specimen is small (less than 5 ml), a cold ethanol precipitation of capsular polysaccharide may be used as follows:

1. Centrifuge the specimen to free it of particulate residue.
2. Add 25 ml of 95% ethanol to 5.0 ml of urine (or proportionate amounts) in a conical centrifuge tube.
3. Refrigerate at 4°C for 1 hr.
4. Discard the supernatant and dry the tube at a 45°C angle for 30 min at room temperature. Mix with 0.25 ml of normal saline and resuspend the pellet. Centrifuge and test the supernatant for the polysaccharide.

E. Liquefaction of Sputum for CIE

Sputum may be tested for microbial polysaccharides but only after liquefaction.

1. Remove 1 ml of concentrated Sputolysin (*N*-acetylcysteine; Sigma Chemical Company) with a syringe, being careful not to introduce air into the vial. Dilute 1 : 10 with sterile distilled water. Unopened vials of Sputolysin are stable at room temperature. Reconstituted solutions should be used

immediately. If kept sterile, the solution is stable for at least 48 hr stored between 2 and 8°C.

2. Overlay the sputum with an equal volume of diluted Sputolysin.
3. Vortex for 30 secs.
4. Allow the mixture to remain at room temperature for 15 mins.
5. Centrifuge for 5 mins at approximately 1500 rpm and use the supernatant as the antigen in CIE.

F. Staining of CIE Plates

1. Prepare a deproteinizing solution, as follows:
   1.88 g sodium phosphate dibasic heptahydrate
   8.0 g NaCl
   0.08 g sodium phosphate monobasic
   0.1 g sodium azide
   Dissolve to 1.0 liter distilled water
2. Place the agarose sheet in the deproteinizing wash solution overnite.
3. Change deproteinizing wash to distilled water after 18 hr and wash for 1 hr.
4. Place the sheet on a test tube rack and dry in a 300°F oven for 30 to 60 min (until agarose gel is dried).
5. Prepare Coomassie brilliant blue stain:

Coomassie BBR	6 g
Absolute methanol	1350 ml
Glacial acetic acid	300 ml
Distilled water	1350 ml

   and a destaining solution:
   50% reagent ethanol with 10% acetic acid
   900 ml 50% ethanol and 100 ml glacial acetic acid
6. Stain for 5 min with Coomassie blue stain.
7. Wash with distilled water. Place in destaining solution for at least 10 min. Rinse with distilled water. Blot dry.

V. CLINICAL APPLICATIONS OF CIE

Anhalt *et al.* (1978) list the following five applications of CIE in infectious disease:

1. Detection of antigen in body fluids
2. Determination of antibody titers (poor sensitivity)
3. Prognostic assessment
4. Identification and/or typing of clinical isolates
5. Elucidation of role of circulating antigens in disease pathogenesis

Counterimmunoelectrophoresis is used primarily for the detection of microbial antigens in body fluids and the direct immunologic identification of certain bacteria such as the beta hemolytic streptococci (Croix *et al.*, 1975; Dajani, 1973; Edwards and Lawson, 1973; Portas *et al.*, 1976). In some cases, however, it may be necessary first to extract the cell-associated antigen with either heat, acid, or enzymes. Edwards and Larson (1973) noted that on 400 strains of groups A, B, C, D, E, and F streptococci, CIE was a more sensitive and faster identification method than the capillary precipitin test.

REFERENCES

Anhalt, J. P., and Yu, P. K. W. (1975). *J. Clin. Microbiol.* **2,** 510.

Anhalt, J. P., Kenny, G. E., and Rytel, M. W. (1978). "Cumitech 8." (T. L. Gavan, ed.), Soc. Microbiol., Washington, D.C.

Bussard, A. (1959). *Biochim. Biophys. Acta.* **34,** 258.

Cohn, J., and Kahan, M. (1976). *J. Immunol. Methods.* **11,** 303.

Croix, J. C., Bajolle, F. and Dalle, M. (1975). *Ann. Biol. Clin.* **33,** 149.

Dajani, A. S. (1973). *J. Immunol.* **6,** 1702.

Dochez, A. R., and Avery, D. T. (1917). *J. Exp. Med.* **26,** 477.

Edwards, E. A. (1971). *J. Immunol.* **106,** 314.

Edwards, E. A., and Larson, G. L. (1973). *Appl. Microbiol.* **26,** 899.

Kelkar, S. S., and Niphadkar, K. B. (1974). *Lancet* **1,** 1394.

Moody, G. J. (1976). *Lab. Pract.* **25,** 575.

Portas, M. R., Hogan, N. A., and Hill, H. R. (1976). *J. Lab. Clin. Med.* **88,** 339.

Rytel, M. W. (1979). "Rapid Diagnosis in Infectious Disease." CRC Press, Boca Ratan, Florida.

Tilton, R. C. (1978). *CRC Crit. Rev. Clin. Lab. Sci.* December, pp. 347–365.

Vincent, M. H., and Bellot, M. (1901). *Bull. Acad. Natl. Med. (Paris)* **61,** 326.

6

*Application of Counterimmunoelectrophoresis to the Diagnosis of Meningitis**

RICHARD H. PARKER

Veterans Administration Medical Center
and
Howard University College of Medicine
Washington, D.C.

I.	Introduction	97
II.	Meningococcal Meningitis	98
III.	*Haemophilus influenzae* Meningitis	99
IV.	Pneumococcal Meningitis	100
V.	Group B Streptococcal Meningitis	101
VI.	*Escherichia coli* Meningitis	102
VII.	Conclusion	102
	References	102

I. INTRODUCTION

Optimal management of bacterial meningitis requires the rapid and specific determination of etiology. Since the early 1970s counterimmunoelectrophoresis (CIE) has been widely used for this purpose. Because of this experience, CIE is now generally accepted as a useful adjunct to the standard microbiological studies [i.e., Gram-stained smear and culture of cerebrospinal fluid (CSF)]. Because of its very objective end point, CIE may be more sensitive and appears to be as specific as microscopic examination of CSF, particularly when the microscopist has had minimal experience. In addition, because the presence of soluble antigen is not dependent on the presence of viable microorganisms, CIE may be positive

*Refer also to Chapters 12 and 14, this volume.

The Direct Detection of Microorganisms in Clinical Samples

ISBN 0-12-187780-9

when prior antimicrobial therapy received by the patient results in negative cultures of CSF. However, in spite of the advantages of CIE, it must be emphasized that a negative result does not rule out bacterial meningitis in any patient, and for some organisms (e.g., *Listeria monocytogenes* or group B meningococci) good antisera are not available. This chapter reviews some of the specific experience with CIE and the diagnosis of bacterial meningitis caused by *Neisseria meningitidis, Streptococcus pneumoniae, Haemophilus influenzae* type b, *Escherichia coli,* and group B streptococci.

II. MENINGOCOCCAL MENINGITIS

The first report of the use of CIE for diagnosing bacterial infection was that by Edwards (1971), who described its use for the rapid detection of circulating antigen in fulminant meningococcal disease.

Greenwood *et al.* (1971; Table 1) studied 68 patients with presumed meningococcal meningitis, and 47 had positive CIE tests for meningococcal group A antigen. This included 5 cases in which CIE was positive yet culture was negative. These results were generally confirmed by Higashi *et al.* (1974). They studied 285 patients with a clinical diagnosis of bacterial meningitis. Positive cultures of CSF were obtained from 137 patients, and CIE of CSF was positive for meningococcal antigen in 121 (88%) of these. All meningococcal isolates except two were serogroup A, and all except two CIE-positive samples were with group A antiserum. The non-group A isolates were a group C and a group W-135 meningococcus. The CSF specimens were positive, as determined by CIE, for antigen of the corresponding serogroups. In 20 of the remaining 148 patients

TABLE I

***Neisseria meningitidis* Meningitis: Cerebrospinal Fluid Findings**

Number of specimens	% Positive			Reference
	Gram's stain	Culture	CIE	
68	25	62	69	Greenwood *et al.* (1971)
69	—	51	38	Hoffman and Edwards (1972)
265	40	52	56	Higashi *et al.* (1974)
191	—	70	72	Whittle *et al.* (1975)
83	72	77	59	Colding and Lind (1977)
24	58	83	33	Kaplan and Feigin (1979)
700	45[a]	62	60	

[a] Based on 440 specimens.

cultures were positive for other microorganisms, and CIE tests for meningococcal antigen were negative. In 28 (22%) of the 128 culture-negative CSF specimens, the CIE test was positive for group A meningococcal antigen.

Results of other studies are summarized in Table I. It is possible to increase the sensitivity in diagnosing meningitis by performing tests for antigen with serum and urine as well as with CSF (Feigin *et al.*, 1976). In one study meningococcal antigen was detected by CIE in either CSF, urine, or serum in all cases of meningitis. However, the same investigators demonstrated in a prospective study that in only 14 (58%) of 24 cases of meningococcal meningitis was antigen detected in at least one of these fluids (Kaplan and Feigin, 1979).

One major reason for the low yield of positive CIE results in meningococcal disease is an inability to detect group B antigen. The group B polysaccharide is not a good immunogen, and the quality of most group B antisera does not approach that of the other meningococcal group antisera. Even with non-group B meningococcal meningitis, false negative results may occur because the quantity of antigen in the CSF may be below that necessary for a positive CIE test. This was supported by the observations of Greenwood and Whittle (1974), who showed that concentration of CSF by negative pressure dialysis with a microconcentrator resulted in CIE detection of antigen in 7 of 24 previously antigen-negative CSF specimens. The lowest concentration of group A polysaccharide that could be detected by their system was 50 ng/ml. These authors also suggested that negative CIE tests for free antigen may be related to the presence of HA antibody.

III. *HAEMOPHILUS INFLUENZAE* MENINGITIS

Infection with *Haemophilus influenzae* type b is associated with the release of polyribophosphate (PRP) capsular material into serum and other body fluids. Coonrod and Rytel (1972) demonstrated that 18 of 19 patients with *H. influenzae* type b meningitis could be rapidly diagnosed by the use of CIE for detecting this type-specific antigen in CSF. Since the early 1970s numerous reports have supported these initial observations and indicated that CIE of CSF for *H. influenzae* type b antigen was positive in 69 to 93% of cases (Table II).

As with meningococcal meningitis it has been shown that CIE of urine and serum in addition to CSF in suspected cases of meningitis will yield a positive diagnosis in nearly 100% of cases of *H. influenzae* type b. The failure to detect antigen in the CSF of all cases has been related to the observation that CIE is not sufficiently sensitive to detect the low levels of antigen, particularly in milder infections. The lower limit for the detection of PRP of *H. influenzae* type b by CIE has been 5–10 ng/ml. Ward *et al.* (1978) noted that the CSF of 4 of 24 with *H. influenzae* type b meningitis had a concentration of PRP or <5 ng/ml.

TABLE II

***Haemophilus influenzae* Type b Meningitis: Cerebrospinal Fluid Findings**

Number of specimens	% Positive Gram's stain	Culture	CIE	Reference
19	89	100	89	Coonrod and Rytel (1972)
18	—	94	83	Shackelford *et al.* (1974)
13	38	100	69	Sillanpää *et al.* (1975)
64	—	100	89	Feigin *et al.* (1976)
31	84	97	77	Colding and Lind (1977)
14	64	50	93	Denis *et al.* (1977)
26	—	—	85	Granoff *et al.* (1977)
24	—	—	83	Ward *et al.* (1978)
132	82	95	87	Kaplan and Feigin (1979)
31	65	97	77	Naiman and Albritton (1980)
372	77[a]	95[b]	85	

[a] Based on 240 specimens.
[b] Based on 322 specimens.

IV. PNEUMOCOCCAL MENINGITIS

The utility of CIE for detecting capsular antigen of the pneumococcus in the CSF was first reported by Coonrod and Rytel (1972). They observed a positive CIE test of CSF for pneumococcal antigen in all 7 patients with culture-proven pneumococcal meningitis. There were no positive (false positive) tests when the same procedure was done on CSF from 30 patients with meningitis caused by other microorganisms or CSF from 10 patients without any evidence of meningitis. The test was useful in that it provided specific information when the Gram-stained smear was either negative or confusing. Fossieck *et al.* (1973) made similar observations in 6 additional patients. Pneumococcal antigen was detected in the CSF of 2 patients in which Gram's stain was negative. In addition, they noted wide variation in both the quantity and persistence of antigen in the CSF. These findings have been confirmed by the experience of others in larger series of patients (Table III).

The overall results do not suggest that CIE has added much to the rapid diagnosis of pneumococcal meningitis when compared to the frequency with which examination of the Gram-stained smear of CSF is positive in this disease. However, individual analysis of cases indicates that approximately 10–20% of patients may have a negative examination of a Gram-stained smear with a positive CIE test for the detection of antigen. Conversely, the CIE test should not replace microscopic examination of the Gram-stained smear because approx-

TABLE III

***Streptococcus pneumoniae* Meningitis: Cerebrospinal Fluid Findings**

Number of specimens	% Positive			Reference
	Gram's stain	Culture	CIE	
7	86	100	100	Coonrod and Rytel (1972)
6	66	100	83	Fossieck *et al.* (1973)
55	80	73	98	Denis *et al.* (1977)
32	81	94	59	Kaplan and Feigin (1979)
100	80	83	85	

imately 10–20% of patients may have a negative CIE test with a positive microscopic examination. Colding and Lind (1977) found that, by the use of a combination of CIE and microscopy, the etiology of meningitis was specifically determined within 1 hr in 94% of 32 cases of CSF culture-positive pneumococcal meningitis. Furthermore, all 3 patients with CSF culture-negative pneumococcal meningitis had either a positive CIE or microscopic examination.

The reasons for a negative (false negative) CIE test of CSF in patients with pneumococcal meningitis include an insufficient quantity of antigen present for detection and infection with either serotype 7 or 14, which have neutral capsular antigens and do not migrate adequately in the routine CIE system. The antigen of these serotypes can be detected by CIE with an adjustment in the buffer system (Anhalt and Yu, 1975). The sensitivity of CIE for the detection of pneumococcal antigen usually ranges from 20 to 500 ng/ml.

V. GROUP B STREPTOCOCCAL MENINGITIS

Rapid diagnosis of group B streptococcal meningitis was established by Webb *et al.* (1980) in 20 of 23 infants (87%) by CIE using antisera prepared in their laboratory. This sensitivity of CIE for detecting group B streptococcal antigen in CSF was comparable to that reported by many others but at variance with the results of Thirumoorthi and Dajani (1979). The latter investigators were unable to detect group B antigen in CSF, serum, or urine of any of 10 patients with group B streptococcal meningitis. The cause of the difference in results is not known but is probably related to the fact that different antisera were used. The possibility that the antiserum used by Thirumoorthi and Dajani (1979) was poorly reactive is supported by the fact that they were unsuccessful in detecting group B antigen in the body fluids from the 10 patients even with latex agglutination and staphylococcal coagglutination. The latter tests are potentially even more sensitive than CIE for detecting group B antigen.

The specificity of CIE for detecting group B streptococcal antigen is very high. For example, Webb *et al.* (1980) did not obtain any false positive results when CSF from 31 infants with meningitis caused by other microorganisms and 25 infants without meningitis were tested using group B streptococcal antiserum.

VI. *ESCHERICHIA COLI* MENINGITIS

McCracken *et al.* (1974) demonstrated the utility of CIE in neonatal meningitis caused by *E. coli* K1 strains. Group B meningococcal and *E. coli* polysaccharides are closely related both chemically and immunogenically. Neonatal *E. coli* meningitis is most often caused by the K1 strain, whereas *E. coli* associated with bacteremia (without meningitis) in neonates (and adult infections less frequently) contains the capsular K1 antigen. The aforementioned investigators found *E. coli* K1 antigen in the CSF of 29 of 41 infants (71%) using meningococcal group B antiserum in the CIE tests. Unfortunately, there was no comparison of this with results of microscopic examination.

VII. CONCLUSION

Counterimmunoelectrophoresis for the detection of bacterial antigen can be an adjunct to the armamentarium of tests used for diagnosing bacterial meningitis. It provides results rapidly, usually within an hour after a lumbar puncture is done. The primary value of the test is that, when positive, it provides specific information regarding etiology, which can be useful in planning antimicrobial therapy.

Negative results rule out neither bacterial meningitis in general nor any specific cause of meningitis. Conversely, a positive result does not prove that a patient has meningitis. The test must be evaluated in conjunction with complete CSF analysis, including cell count, glucose, and Gram-stained smears. If other tests indicate that a patient has bacterial meningitis, then CIE can be useful in determining the etiology.

Because bacterial antigen may be circulating in the blood and excreted in urine, the sensitivity of CIE in diagnosing most infections is increased by the simultaneous testing of CSF, serum, and urine.

REFERENCES

Anhalt, J. P., and Yu, P. K. W. (1975). *J. Clin. Microbiol.* **2,** 510.
Colding, H., and Lind, I. (1977). *J. Clin. Microbiol.* **5,** 405–409.
Coonrod, J. D., and Rytel, M. W. (1972). *Lancet* **1,** 1154–1157.
Denis, F., Samb, A., and Chiron, J. P. (1977). *J. Am. Med. Assoc.* **238,** 1248–1249.

Edwards, E. A. (1971). *J. Immunol.* **106,** 314–317.
Feigin, R. D., Wong, M., Shackelford, P. G., Stechenberg, B. W., Dunkle, L. M., and Kaplan, S. (1976). *J. Pediatr.* (*St. Louis*) **89,** 773–775.
Fossieck, B., Jr., Craig, R., and Paterson, P. Y. (1973). *J. Infect. Dis.* **127,** 106–109.
Granoff, D. M., Congeni, B., Baker, R., Ogra, P., and Nankervis, G. A. (1977). *Am. J. Dis. Child.* **131,** 1357–1362.
Greenwood, B. M., and Whittle, H. C. (1974). *J. Infect. Dis.* **129,** 201–204.
Greenwood, B. M., Whittle, H. C., and Dominic-Rajkovic, O. (1971). *Lanret* **2,** 519–521.
Higashi, G. I., Sippel, J. E., Girgis, N. I., and Hassan, A. (1974). *Scand. J. Infect. Dis.* **6,** 233–235.
Hoffman, T. A., and Edwards, E. A. (1972). *J. Infect. Dis.* **126,** 636–644.
Kaplan, S. L., and Feigin, R. D. (1979). *In* "Rapid Diagnosis in Infectious Diseases" (M. W. Rytel, ed.), pp. 105–113. CRC Press, Boca Raton, Florida.
McCracken, G. H., Jr., Sarff, L. D., Glode, M. P., Mize, S. G., Schiffer, M. S., Robbins, J. B., Gotschlich, E. C., Orskov, I., and Orskov, F. (1974). *Lancet* **2,** 246–250.
Naiman, H. L., and Albritton, W. L. (1980). *J. Infect. Dis.* **142,** 524–531.
Shackelford, P. G., Campbell, J., and Feigin, R. D. (1974). *J. Pediatr.* (*St. Louis*) **85,** 478–481.
Sillanpää, M., Vaha-Eskeli, E., and Willman, K. (1975). *Scand J. Infect. Dis.* **7,** 113–115.
Thirumoorthi, M. C., and Dajani, A. S. (1979). *J. Clin. Microbiol.* **9,** 28–32.
Ward, J. I., Siber, G. P., Scheifele, D. W., and Smith, D. H. (1978). *J. Pediatr.* (*St. Louis*) **93,** 37–42.
Webb, B. J., Edwards, M. S., and Baker, C. J. (1980). *J. Clin. Microbiol.* **11,** 263–265.
Whittle, H. C., Greenwood, B. M., Davidson, N. M., Tomkins, A., Tugwell, P., Warrell, D. A., Zalin, A., Bryceson, A. D. M., Parry, E. H. O., Brueton, M., Duggan, M., Oomen, J. M. V., and Rajkowic, O. D. (1975). *Am. J. Med.* **58,** 823–828.

7

Counterimmunoelectrophoresis for the Diagnosis of Pneumococcal Respiratory and Other Infections

GEORGE E. KENNY

Department of Pathobiology
School of Public Health and Community Medicine
University of Washington
Seattle, Washington

I.	Introduction	105
II.	Summary of the Principles of Counterimmunoelectrophoresis	106
III.	Counterimmunoelectrophoresis Methods in Pneumococcal Infections	106
IV.	Detection of Antigen in Clinical Samples	107
	A. Detection of Antigen in Serum	108
	B. Detection of Antigen in Sputum	108
	C. Detection of Antigen in Other Body Fluids	109
V.	Perspective	110
	References	110

I. INTRODUCTION

The diagnosis of pneumococcal infections is complicated by the fact that pneumococci (*Streptococcus pneumoniae*) are found in throats of healthy individuals. Consequently, isolation of the organism is not absolute proof of infection with *Streptococcus pneumoniae*. However, the virulence of the organism depends on the presence of a polysaccharide capsule, which is present in large qualtities on the organism. So much capsular polysaccharide is present in the blood, urine, and body fluids that it can be readily detected, a fact that was recognized as early as 1917 by Dochez and Avery. One of the most convenient and rapid means of detecting polysaccharide is counterimmunoelectrophoresis (CIE), which is the topic of this chapter.

The Direct Detection of Microorganisms in Clinical Samples

ISBN 0-12-187780-9

II. SUMMARY OF THE PRINCIPLES OF COUNTERIMMUNOELECTROPHORESIS

At optimum proportions, polyvalent or multideterminate antigens combine with antibodies to form stable precipitates. Although precipitin reactions in fluid medium require careful attention to optimum proportions, precipitin lines readily form in single or double immunodiffusion reactions in agar or agarose (Crowle, 1973); optimum proportions are achieved by the diffusion of a gradient of antigen into either a constant antibody concentration or a gradient of antibodies. However, double immunodiffusion reactions require 24- to 48 hr incubation periods and are also quite insensitive. The reaction can be accelerated greatly by placing the reactants in an electrical field. At pH values of approximately 8.6, most antigens have a negative charge and migrate to the anode (positive pole). Immunoglobulins are essentially uncharged at pH 8.6 on the average. However, the heterogeneity of charge of the population of specific antibody molecules indicates that the most positively charged of the antibodies will migrate to the cathode relatively rapidly, whereas most antibodies will migrate hardly at all, and the most negatively charged antibody molecules will migrate anodically. Moreover, in agarose or agar gels a significant electroendoosmosis is observed with a flow of fluid to the cathode, which provides for the cathodal movement of uncharged molecules and will accelerate the movement of positively charged molecules. Consequently, a situation can be arranged in which antigen will move electrophoretically to the anode and meet antibody drifting electroendosmotically to the cathode. When antigen and antibodies pass over each other (eclipse; Kenny and Foy, 1975), a precipitin line is formed and is a monument to the antigen–antibody reaction. This technique is termed immunoelectroosmophoresis or counterimmunoelectrophoresis (Crowle, 1973). The fact that both antigen and antibody move toward each other in one dimension gives a 10- to 100-fold increase in sensitivity over diffusion methods in which antigens and antibodies diffuse in two dimensions (i.e., double immunodiffusion; Kenny and Foy, 1975).

III. COUNTERIMMUNOELECTROPHORESIS METHODS IN PNEUMOCOCCAL INFECTIONS

The typical methods involve the use of a thin agarose gel on a glass slide at pH 8.2–8.6. Opposing rows of holes (wells) are cut, being separated by 2 to 3 mm (Coonrod and Rytel, 1973a; Kenny *et al.*, 1972). Electrophoresis is carried out at 2 to 6 v/cm (measured directly on the agarose with a voltmeter) for 30 to 60 min. Serum or body fluids suspected of containing polysaccharides are placed in the cathodal wells so that the polysaccharide included will migrate electrophoretically toward the anodal wells, where specific antipneumococcal antisera have been placed. Depending on the balance of the relative quantities of antigen and antibody, the precipitin line may form between adjacent antibody wells, in the well (where it will be essentially invisible), or between the opposing rows of wells (the normal positive reaction). In practical terms, with the available Danish antisera (Lund, 1960) polysaccharides in most normal body fluids will form precipitin lines in between the wells, with the anode behind the antibody row. However, certain polysaccharides, namely, types 7 and 14, have a positive charge at pH 8.6 and thus will not form precipitin lines during CIE but will form precipitin lines in double immunodiffusion (Kenny *et al.*, 1972). Anhalt and Yu (1975) have shown that the detection of these polysaccharides can be greatly enhanced by the addition of phenylboronic acid to the buffer system to increase the negative charge of the polysaccharide by the formation of borate complexes. Szu and Oravec (1982) used reversed CIE (anode on the antigen side) at pH 4.75 and were able to detect approximately 1 μg/ml of type 7 and 14 polysaccharides. If the latter method were adopted, it would be essential to use both pH 8.6 with normal polarity and pH 4.75 at reversed polarity to detect all types of polysaccharides. Overall, the sensitivity for the detection of negatively charged polysaccharides is 0.1–1.0 μg/ml, depending to a certain extent on the amount of sample employed in the test (usually 2.0–10 μl) and whether Omniserum, grouping or monospecific sera are used.

Critical conditions for the test include the electroendosmotic flow induced by the agar support. Agaroses with $-m_r$ (Wieme, 1965) values of about 0.1 give sufficient electroendosmosis to provide for reasonable cathodal movement of the antibody without impairing the mobility of the antigen. The electrical force employed is defined in volts per centimeter as measured in the agarose: 2.5–6 V/cm have been employed for 40 to 60 min. Increasing the amount of time beyond that necessary to form precipitin lines does not increase the sensitivity of the test and may in fact decrease the sensitivity because precipitin lines may be reinforced by nearby antigen and antibody during washing in preparation for staining. An additional critical factor is the ionic strength of the buffer employed. Concentrations exceeding 0.05 *M* will cause heating because excessive amperage will be required to maintain the desired voltage. Positive reactions are indicated by a definite precipitin line between the antigen and antibody well. Upon staining, artifactual lines are frequently observed on the other side of the antibody well and are most likely due to the entrapment of serum lipoproteins.

IV. DETECTION OF ANTIGEN IN CLINICAL SAMPLES

Specimens typically employed are blood (serum), sputum, body fluids, and bacterial cultures.

A. Detection of Antigen in Serum

The most certain diagnosis of pneumococcal infection is the isolation of the organism from the bloodstream. Therefore, the best means of assessing a putative new method for the detection of infection is to compare the results with the demonstration of bacteremia. Theoretically, a single organism could give rise to a positive blood culture; accordingly, isolation from the bloodstream should be the most sensitive detection method. However, the ability to invade the bloodstream must indicate that the potential invader overwhelmed the host defenses, indicating that a relatively large number of organisms are involved. Furthermore, antibiotic therapy or eventual host response may eliminate the organism and give rise to false negative blood cultures. The efficiency of detecting polysaccharide is about 60% in patients from whom pneumococci have been isolated from the bloodstream, provided that it is known that an efficient test can be derived for the isolated serotypes (Coonrod and Rytel, 1973a; Kenny *et al.*, 1972). As might be expected a significant number of infections were detected in individuals who were culture-negative, with 20 to 25% additional positives found by antigen detection (Legrand *et al.*, 1979). Spencer and Savage (1976) found about 10% positive reactions in specimens from pneumonia patients (from whom pneumococci were not isolated). In these cases the patient had received prior antibiotic therapy. In pneumococcal meningitis the correlation between antigen detection and culture was excellent in the study by Denis (1979). The detection of polysaccharide in serum was closely correlated with a later increase in specific antibody as measured by enzyme-linked immunosorbent assay (ELISA) (Trollfors *et al.*, 1979). Although antigen was found to persist in one patient for as long as 200 days (Kenny and Foy, 1975), in most patients antigen persists for 10 to 50 days (Coonrod and Rytel, 1973b; Coonrod and Drennan, 1976; Kenny and Foy, 1975; Spencer and Savage, 1976). False positive reactions appear to be rare. No positive reactions were found in a survey of 245 pneumonia patients from a normal civilian population (Foy *et al.*, 1975).

B. Detection of Antigen in Sputum

The effectiveness of sputum culture for the diagnosis of pneumococcal infections is a controversial subject (Barret-Connor, 1971; Drew, 1977; Tempest *et al.*, 1974; Thorsteinsson *et al.*, 1975). Accordingly, a great deal of interest in CIE for the direct testing of sputum has arisen (El-Refaie and Dulake, 1975; Leach and Coonrod, 1977; Perlino and Shulman, 1976; Tugwell and Greenwood, 1975). Overall, the detection of polysaccharide in sputum has been highly successful and apparently distinguishes colonization from infection to a degree. Although little correlation between the number of colonies and antigen positivity was found in an early study (Verhoef and Jones, 1974), Downes and Ellner (1979) showed some correlation between "heavy" growth and positivity. However, patients treated with antibiotics may yield negative cultures with positive antigen detection (Congeni and Nankervis, 1978; El-Rafaie and Dulake, 1975; Trollfors *et al.*, 1979). Antigen persists in sputum for about 2 to 5 days or even longer depending on the type of infection (El-Refaie and Dulacke, 1975; Sands and Green, 1980; Trollfors *et al.*, 1979). Overall, antigen detection correlated well with clinical diagnosis and was a more sensitive measure for the detection of infection than culture from sputum (Downes and Ellner, 1979; El-Refaie and Dulake, 1975; Miller *et al.*, 1978; Leach and Coonrod, 1977; Sands and Green, 1980; Tugwell and Greenwood, 1975). Because the pneumococcal antiserum employed is polyvalent and contains antibodies against polysaccharides that are known to cross-react with a wide variety of materials (Kabat and Mayer, 1961), false positive reactions are more likely to occur with sputum than with blood. Furthermore, pneumococci are carried in the throat of normal individuals. Control tests of saliva or secretions from subjects without sputum or disease were essentially negative (Congeni and Nankervis, 1978; Miller *et al.*, 1978). Tests of sputum from subjects judged not to have pneumococcal pneumonia showed relatively high positivity rates [~25% (Leach and Coonrod, 1977; Sands and Green, 1980; Tugwell and Greenwood, 1975)]. However, Trollfors *et al.* (1979) showed that the detection of polysaccharide in the blood or sputum correlated very well with a positive antibody response by ELISA in nearly all patients tested. Additional antibody studies will be most important in establishing the sensitivity and specificity of antigen detection in sputum.

C. Detection of Antigen in Other Body Fluids

Antigen is readily detected in urine, a material that is readily available and that can be concentrated (Coonrod and Rytel, 1973a,b; Legrand *et al.*, 1979; Tugwell and Greenwood, 1975). For this reason the detection of antigen in urine was more effective than that in blood but less effective than that in purulent sputum (Tugwell and Greenwood, 1975). It is interesting that the polysacharide in urine is of smaller molecular weight than that found in serum, which is indicative of selective excretion of possibly degraded polysaccharide (Coonrod, 1974). Antigen is also found in pleural fluid (a specimen available only with appropriate clinical indications) with high rates of positivity (Legrand *et al.*, 1979; Tugwell and Greenwood, 1975). In pneumococcal meningitis, spinal fluid is more often positive (77%) than serum (51%), or urine (65%), as shown by Legrand *et al.* (1979). Antigen has even been found in synovial fluid (Dorff *et al.*, 1975). Counterimmunoelectrophoresis is also effective for identifying and detecting pneumococci in culture (Artman *et al.*, 1980; Sottile and Rytel, 1975).

V. PERSPECTIVE

Clearly, CIE is a highly useful method of detecting pneumococcal infections in a variety of disease syndromes. However, the techniques have yet to be optimized and standardized. From a practical standpoint the use of Omniserum is the best choice but, because the sensitivity of the test is a function of the power of the antibody, some sacrifice in sensitivity must be made (Kenny *et al.*, 1972). Nonetheless, the results reported herein suggest that the practical test using Omniserum works well. Only a limited number of purified polysaccharides are available from the many pneumococcal serotypes. Accordingly, the test is of unknown sensitivity for less common serotypes. Standardization of both the agar support to control electroendoosmosis and the methods of detecting positively charged polysaccharides will be required. The positively charged serotype 14 is particularly important because of its high prevalence in positive blood cultures (~20%) and its strong association with fatal infections (Shapera and Mattson, 1972). One of the shortcomings of the test is that it can detect neither untypable pneumococci nor those to which the antibodies happen to be weak. Thus, negative reactions in CIE have less significance than negative cultures from patients who have not had previous antibiotic therapy. Another problem is the demonstrated false positivity of sputum specimens because of possible cross-reactions with food materials and other microorganisms. Pneumococci might be studied for common antigenic components that might be detectable in sputum and that would verify the type-specific identification. The fact that pneumococcal polysaccharides persist in certain patients for an extremely long time is of considerable diagnostic importance because antigen detection in the blood or body fluids may represent an infection long past. For the future it will also be important to compare CIE with such techniques as antigen capture assay in ELISA (Drow and Manning, 1980) and coagglutination (Edwards and Coonrod, 1980). Agglutination of antigen-sensitized latex is apparently not a sensitive method (Coonrod and Rylko-Bauer, 1976). Further studies on the detection of antigen in immune complexes in serum (Coonrod and Leach, 1978) will be important for detecting bound antigen. Although initial studies had suggested a correlation between the amount of polysaccharide and the severity of disease and mortality (Kenny *et al.*, 1972), more quantitative studies failed to show a clear correlation (Kenny and Foy, 1975). Finally, correlation of antigen detection methods with antibody response is most important in order to provide a more stringent means of detecting pneumococcal disease than sputum culture, as shown by Trollfors *et al.* (1979).

REFERENCES

Anhalt, J. P., and Yu, P. K. W. (1975). *J. Clin. Microbiol.* **2,** 510–515.
Artman, M., Weiner, M., and Frankl, G. (1980). *J. Clin. Microbiol.* **12,** 614–616.

Barrett-Connor, E. (1971). *Am. Rev. Respir. Dis.* **103,** 845–848.
Congeni, B. L., and Nankervis, G. A. (1978). *Am. J. Dis Child.* **132,** 684–687.
Coonrod, J. D. (1974). *J. Immunol.* **112,** 2193–2201.
Coonrod, J. D., and Drennan, D. P. (1976). *Ann. Inter. Med.* **84,** 254–260.
Coonrod, J. D., and Leach, R. P. (1978). *J. Clin. Microbiol.* **8,** 257–259.
Coonrod, J. D., and Rylko-Bauer, B. (1976). *J. Clin. Microbiol.* **4,** 168–174.
Coonrod, J. D., and Rytel, M. W. (1973a). *J. Lab. Clin. Med.* **81,** 770–777.
Coonrod, J. D., and Rytel, M. W. (1973b). *J. Lab. Clin. Med.* **81,** 778–786.
Crowle, A. J. (1973). "Immunodiffusion." Academic Press, New York.
Denis, F. (1979). *Pathol. Biol.* **27,** 549–553.
Dochez, A. R., and Avery, O. T. (1917). *J. Exp. Med.* **26,** 477–493.
Dorff, G. J., Ziolkowski, J. S., and Rytel, M. W. (1975). *Arthritis Rheum.* **18,** 613–615.
Downes, B. A., and Ellner, P. D. (1979). *J. Clin. Microbiol.* **10,** 662–665.
Drew, W. L. (1977). *J. Clin. Microbiol.* **6,** 62–65.
Drow, D. L., and Manning, D. D. (1980). *J. Clin. Microbiol.* **11,** 641–645.
Edwards, E. A., and Coonrod, J. D. (1980). *J. Clin. Microbiol.* **ll,** 488–491.
El-Refaie, M., and Dulake, C. (1975). *J. Clin. Pathol.* **28,** 801–806.
Foy, H. M., Wentworth, B., Kenny, G. E., Kloeck, J. M., and Grayston, J. T. (1975). *Am. Rev. Respir. Dis.* **111,** 595–603.
Kabat, E. A., and Mayer, M. M. (1961). "Experimental Immunochemistry," pp. 838–850. Thomas, Springfield, Illinois.
Kenny, G. E., and Foy, H. M. (1975). *In* "Microbiology 1975" (D. Schlessinger, ed.), pp. 97–102. Am. Soc. Microbiol. Washington, D.C.
Kenny, G. E., Wentworth, B. B., Beasley, R. P., and Foy, H. M. (1972). *Infect. Immun.* **6,** 431–437.
Leach, R. P., and Coonrod, J. D. (1977). *Am. Rev. Respir. Dis.* **116,** 847–851.
Legrand, P., Lemoine, J. L., Squinazi, F., and Geslin, P. (1979). *Pathol. Biol.* **27,** 555–558.
Lund, E. (1960). *Bull. W.H.O.* **23,** 5–13.
Miller, J., Sande, M. A., Gwaltney, J. M., and Hendley, J. O. (1978). *J. Clin. Microbiol.* **7,** 459–462.
Perlino, C. A., and Shulman, J. A. (1976). *J. Lab. Clin. Med.* **87,** 496–502.
Sands, R. L., and Green, I. D. (1980). *J. Appl. Bacteriol.* **49,** 471–478.
Shapera, R. M., and Matsen, J. M. (1972). *Infect. Immun.* **5,** 132–136.
Sottile, M. I., and Rytel, M. W. (1975). *J. Clin. Microbiol.* **2,** 173–177.
Spencer, R. C., and Savage, M. A. (1976). *J. Clin. Pathol.* **29,** 187–190.
Szu, S. C., and Oravec, L. S. (1982). *J. Clin. Microbiol.* **15,** 1172–1175.
Tempest, B., Morgan, R., Davidson, M., Eberle, B., and Oseasohn, R. (1974). *Am. Rev. Respir. Dis.* **109,** 577–578.
Thorsteinsson, S. B., Musher, D. M., and Fagan, T. (1975). *J. Am. Med. Assoc.* **233,** 894–895.
Trollfors, B., Berntsson, E., Elgefors, B., and Kaijser, B. (1979). *Scand. J. Infect. Dis.* **11,** 31–34.
Tugwell, P., and Greenwood, B. M. (1975). *J. Clin. Pathol.* **28,** 118–123.
Verhoef, J., and Jones, D. M. (1974). *Lancet* May 4, p. 879.
Wieme, R. J. (1965). "Agar Gel Electrophoresis," pp. 110–113. Elsevier, Amsterdam.

8

Counterimmunoelectrophoresis for the Diagnosis of Intrapleural Empyema

H. DAVID WILSON

Department of Pediatrics
University of Kentucky Medical Center
Lexington, Kentucky

I. Introduction 113
II. Methods 114
References 116

I. INTRODUCTION

The testing of pleural fluid, serum, and urine by counterimmunoelectrophoresis (CIE) can add to the results of routine smears, cultures, and serology in making a definitive diagnosis of empyema in childhood. Most children referred to our hospital with empyema have received a variety of antimicrobial agents, and frequently all cultures are sterile. Under these circumstances establishing an etiologic diagnosis by CIE can significantly change therapy. Optimal treatment for empyema differs significantly depending on the etiologic agent, that is, *Staphylococcus aureus, Haemophilus influenzae, Streptococcus pneumoniae,* or *Streptococcus pyogenes.* Not only is the choice of antibiotic different, but the length of treatment may vary; *Staphylococcus* is treated for 4 to 6 weeks, whereas the others usually respond in 2 weeks.

When initial smears of pleural fluid are negative for organisms and when cultures later prove sterile, CIE may allow a specific diagnosis to be made, as illustrated by studies with empyemas caused by *Streptococcus pneumoniae, Haemophilus influenza* type b, *Staphylococcus aureus,* and *Klebsiella pneumoniae* (Coonrod and Wilson, 1976; Lampe *et al.*, 1976; Anhalt *et al.*, 1980; Holsclaw and Schaeffer, 1980). An example is demonstrated in Fig. 1, in which sterile empyema fluid obtained from a 6-year-old child previously treated with

The Direct Detection of Microorganisms in Clinical Samples

ISBN 0-12-187780-9

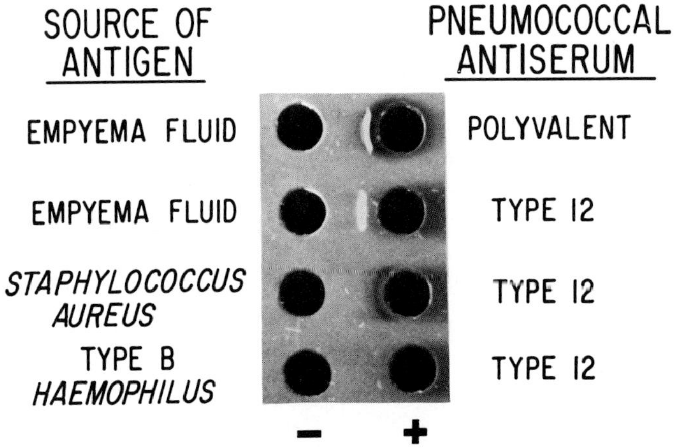

Fig. 1. Precipitin reactions in agarose gel obtained by CIE of empyema fluid. Precipitin bands are present between the two upper sets of wells, indicating that antigen in the pleural fluid reacted with polyvalent antipneumococcal serum (with antibody against 83 types) and with pneumococcal type 12 antiserum. The specificity of these reactions is indicated by the absence of precipitin bands between pneumococcal type 12 antiserum and an overnight culture of *Staphylococcus aureus* or purified *Haemophilus influenzae* type b polysaccharide (10 μg/ml). The dark band between the third set of wells is an artifact. (Reproduced with permission of J. D. Coonrod and *American Review of Respiratory Disease* published by American Lung Association; *Am. Rev. Respir. Dis.* **113,** 637, 1976.)

antibiotics had detectable pneumococcal antigen. Depending on the availability of good-quality antisera, antigens of other organisms, such as *Escherichia coli* (K1 serotype), *Pseudomonas aeruginosa, Neisseria meningitidis,* group B *Streptococcus,* and others, may also be found by CIE.

II. METHODS

Pleural or empyema fluids should be centrifuged at 3000 rpm for 5 min to remove debris, which can cause nonspecific deposits in agarose and make specific precipitin bands difficult to detect. The supernatant is tested immediately or can be stored for later testing. To conserve valuable antisera we do not immediately test fluids with visible organisms on smear. Serum and urine can also be stored or tested along with empyema fluid for optimal results (Coonrod and Rytel, 1973; Feigin *et al.,* 1976).

We routinely test the empyema supernatant undiluted and at 1 : 50 and 1 : 100 dilutions in the CIE buffer or sterile saline. In using this method we have avoided false negative reactions caused by the prozone phenomenon with excess antigen in some pleural fluids (Coonrod and Wilson, 1976).

Positive controls for *Haemophilus influenzae* and *Streptococcus pneumoniae* should be included in each test of empyema fluid. As with all materials tested by CIE, the limitations are the need for high-quality antisera, sufficient antigen in the specimen for detection, and cross-reactions with other microbial antigens. Another limitation is neutral-reacting capsular antigens, which do not migrate toward the anode in the usual buffer system. This limitation can be overcome for type 7 and 14 pneumococci by the use of a sulfonated phenylboronic acid or *m*-carboxyphenylboronic acid (CPB) in the buffer system (Anhalt *et al.*, 1980). Alternatively, if CPB or sulfonated phenylboronic acid is not available, one might consider a simple Ouchterlony gel diffusion with pleural fluid and antisera to type 7 and 14 pneumococci. We have found Burroughs Wellcome and Hyland Laboratories antisera for *Haemophilus,* Statens Seruminstitut antisera for pneumococci, and Burroughs Wellcome antisera for meningococci to be most satisfactory. Good-quality antisera for staphylococci and group B streptococci have usually been made in the laboratories of individual researchers.

When reporting the results of CIE to physicians, one should emphasize that a negative test does *not* rule out infection caused by the organisms tested for in CIE. Fortunately, pleural fluids usually have high concentrations of antigen (Coonrod and Wilson, 1976).

Cross-reactions can occur with antisera to *Haemophilus influenzae* type b and *Staphylococcus aureus, Escherichia coli,* and certain pneumococci (Ingram *et al.*, 1972: Lampe *et al.*, 1976). In practice, however, cross-reacting antigens have rarely caused difficulties with clinical specimens.

No comparative studies of CIE with enzyme-linked immunosorbent assay (ELISA) or latex agglutination for pleural fluid are available. All indications are that ELISA, and in some instances latex agglutination, would be more sensitive. ELISA would usually require more time to perform than CIE, whereas latex agglutination would be both sensitive and rapidly available.

Studies to date have not correlated the quantity of antigen present in pleural fluid with prognosis. It is possible that there would be a correlation between antigen concentration, and hence absolute number of infecting organisms, and residual lung or pleural damage. Only a careful study could answer such a question.

In summary, published studies (Coonrod and Wilson, 1976; Lampe *et al.*, 1976; Anhalt *et al.*, 1980; Holsclaw and Schaeffer, 1980) and our personal experience suggest that CIE is useful in making a specific and rapid etiologic diagnosis in cases of empyema when the pleural fluid smears and cultures are negative. The use of pleural fluid, serum, and urine results in the greatest number of positive results. The advantages of making a precise etiologic diagnosis are important to both patient and physician.

REFERENCES

Coonrod, J. D., and Wilson, H. D. (1976). *Am. Rev. Respir. Dis.* **113,** 637–641.

Lampe, R. M., Chottipitayasunodh, T., and Sunakorn, P. (1976). *J. Pediatr.* (*St. Louis*) **88,** 557–560.

Holsclaw, D. S. and Schaeffer, D. A. (1980). *Chest* **78,** 867–869.

Anhalt, J. P., Kenny, G. E., and Rytel, M. W. (1980). "Cumitech 8" (T. L. Gavan, ed.), Am. Soc. for Microbiol., Washington, D.C.

Feigin, R. D., Wong, M., Shackelford, P. G., *et al.* (1976). *J. Pediatr.* (*St. Louis*) **89**, 773–775.

Coonrod, J. D., and Rytel, M. W. (1973). *J. Lab. Clin. Med.* **81,** 770.

Ingram, D. L., Anderson, P., and Smith, D. H. (1972). *J. Pediatr.* (*St. Louis*) **81,** 1156–1159.

9

Problems with Precipitin Methods for Detecting Antigenemia in Bacterial Infections

J. DONALD COONROD

Division of Infectious Diseases
Department of Medicine
University of Kentucky
and
Veterans Administration Medical Center
Lexington, Kentucky

I. Introduction 117
II. Variables in the Detection of Antigenemia 118
III. Future Trends 123
References 123

I. INTRODUCTION

Early studies of the detection of microbial antigens in the circulation were carried out by Dochez and Avery (1917), who used capillary tube precipitation to measure pneumococcal capsular polysaccharide in body fluids of patients with pneumonia. Many of their patients lacked detectable circulating capsular polysaccharide, which they speculated was due to the specific antipneumococcal serum that some of the patients had received. Unfortunately, their results proved to be illustrative of the problems inherent in the detection of circulating antigens by simple immunologic techniques. Counterimmunoelectrophoresis (CIE), although much more sensitive than the capillary precipitin test, has proved to be a poor test for antigenemia.

The Direct Detection of Microorganisms in Clinical Samples

ISBN 0-12-187780-9

II. VARIABLES IN THE DETECTION OF ANTIGENEMIA

Factors that affect the detection of antigenemia by CIE include the chemical composition of the antigens, the amount of antigen produced, and patterns of antigen distribution and excretion. Most of the bacterial antigens that have been detected in the circulation of patients with bacterial infections have been polysaccharides. This may be because polysaccharides are major structural components of bacteria, but it is probably also because polysaccharides can resist degradation by mammalian enzymes. There is very little information on the rate of production of polysaccharides or proteins by bacteria during growth *in vivo*. We attempted to estimate the rate of production of capsular polysaccharide by studying mucoid type 3 pneumococci during log-phase growth in broth (Coonrod and Drennan, 1976). We found that a single pneumococcus in log phase secreted about 4×10^{-7} μg polysaccharide per hour or 1×10^{-5} μg polysaccharide per day. With only a few hundred viable pneumococci per milliliter of blood, as is commonly the case in pneumococcal pneumonia (Tilghman and Finland, 1937), we calculated that one could not develop detectable antigenemia if antigen in the circulation were derived solely from the bacteria proliferating intravascularly. However, there is fairly good evidence from clinical studies that antigenemia with bacterial polysaccharides is not usually detected by CIE (or techniques of similar sensitivity) unless there is bacteremia. Bacterial polysaccharides can apparently cross serous membranes readily despite their large mean molecular weight, and one might anticipate that extensive local infections would be associated with polysaccharide antigenemia in the absence of bacteremia. Nonetheless, in pneumococcal pneumonia in which there is a large number of pneumococci and up to several grams of polysaccharide in the lungs (Ney and Harris, 1937), many patients have negative tests for capsular polysaccharide in the serum. Most patients with positive tests are bacteremic. Thus, in one review (Coonrod, 1979a) the results of five studies of antigenemia in pneumococcal pneumonia showed that circulating capsular polysaccharide could be detected with CIE in 55% of 80 bacteremic patients but in only 9% of 77 nonbacteremic patients.

A careful study of the relationship between the number of viable bacteria in the circulation and the level of circulating *Haemophilus* type b capsular polysaccharide has been carried out by Granoff and Nankervis (1977). The studies of these investigators confirmed that capsular polysaccharide antigenemia, as detected by CIE, is highly associated with bacteremia and that antigen levels in the circulation can be correlated with both the duration and the level of bacteremia. Figure 1 (Granoff and Nankervis, 1977) shows the temporal course of *Haemophilus* capsular polysaccharide antigenemia and the levels of viable *Haemophilus* organisms in the blood of suckling albino rats infected by the intraperitoneal route. Capsular polysaccharide antigenemia was infrequent early in infection; only 20% of the animals had positive serum tests at 12 hr of infection, and only

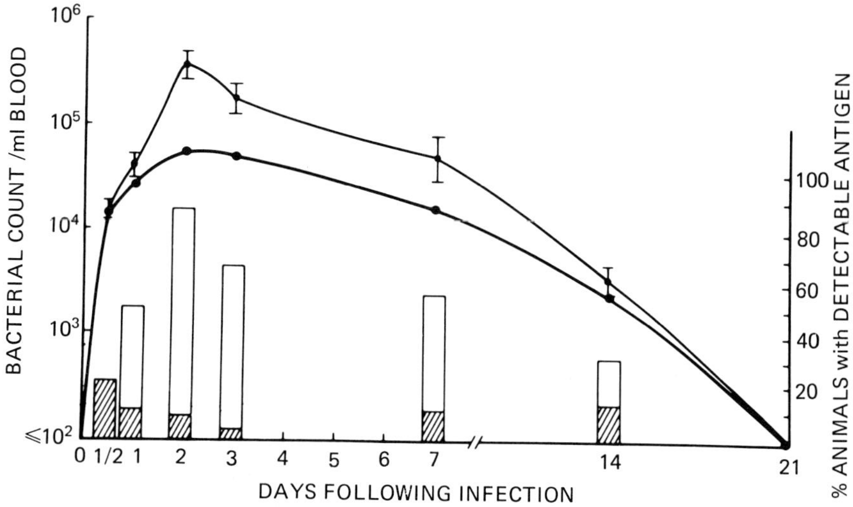

Fig. 1. Relation of quantitative bacterial count in blood (lines) to circulating capsular antigen (bars). Data from 96 rats infected ip at 12 to 14 days of age with 10^4 colony-forming units of *Haemophilus influenzae* type b; mean ± SE. Key: □, positive; ▨, trace; Ɨ, mean; •, median. Published with permission from the University of Chicago Press. D. M. Granoff and G. A. Nankervis, *J. Infect. Dis.* **136,** 292–296, 1977.)

50% were positive by 24 hr. Polysaccharide was detected in the circulation of 85% of the animals by 2 days of infection, corresponding to the time of the peak level of bacteremia (i.e., 10^5–10^6 bacteria per milliliter of blood). After 48 hr the number of bacteria in the blood and the level of circulating antigen declined concomitantly.

Even in the presence of bacteremia, many patients do not develop detectable antigenemia. If all of the antigen produced in the circulation remained there throughout the course of infection, antigenemia would undoubtedly be detected more frequently. However, there are mechanisms for the rapid clearance of bacterial polysaccharides from the circulation. Evidence suggests that specific antibodies greatly facilitate the elimination of polysaccharides, at least in detectable form. In many patients with pneumococcal pneumonia, antigenemia with capsular polysaccharide is brief (1–2 days), with an early exponential clearance of antigen from the blood. Our studies (Coonrod and Drennan, 1976) indicate that this exponential clearance of pneumococcal polysaccharide in pneumococcal pneumonia occurs concomitantly with the development of type-specific antibodies (Fig. 2). Evidence of immune complex formation in pneumococcal disease was obtained by subjecting sera from patients with pneumococcal infections to heat (56°C) and electrophoresis (Coonrod and Leach, 1977). With this technique we could detect capsular polysaccharide in the serum of some patients who

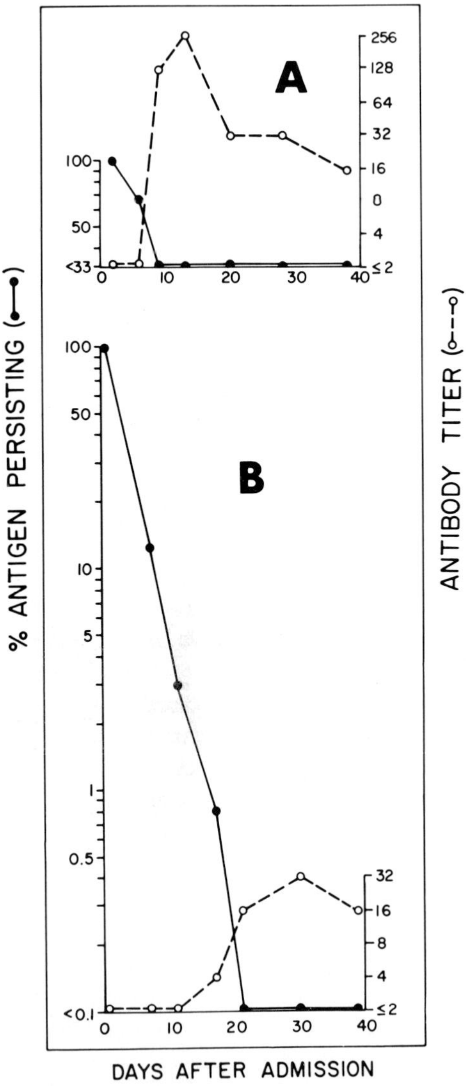

Fig. 2. Clearance of circulating pneumococcal capsular polysaccharide associated with the development of measurable type-specific hemagglutinating antibody. (A) Patient with type 12 pneumococcal pneumonia. The peak antigen level was 0.15 μg/ml. (B) Patient with type 7 pneumococcal pneumonia. The peak antigen level was 51.2 μg/ml. (Published with permission from the *Annals of Internal Medicine*. J. D. Coonrod and D. P. Drennan, *Ann. Intern. Med.* **84,** 254–260, 1976.)

had negative results in the regular CIE test. The technique probably has no practical importance for clinical diagnosis, however, because it is cumbersome and the number of additional positive reactions obtained is relatively small. Schiffman *et al.* (1974) have reported the detection of complexed capsular polysaccharide in the circulation in pneumococcal infections. These investigators used radioimmunoassay (RIA) with pepsin digestion of the serum to free antigen from complexes. No practical application of their observations to clinical diagnosis has been reported.

In a few patients with pneumococcal infections, pneumococcal polysaccharides persist in the circulation for a long period, similar to the prolonged antigenemia described with "polysaccharide paralysis" in animals that have been injected with large doses of polysaccharide (Fig. 3). Perhaps patients with persistent antigenemia are genetically unable to produce antibodies to certain bacterial polysaccharides and therefore cannot clear antigen as rapidly. This has never been investigated in man, but some animals appear to be genetically unresponsive to certain bacterial polysaccharides. As a practical point one must be aware that prolonged antigenemia can lead to confusion diagnostically, unless the patient's history and serologic results are known.

A detailed quantitative study of capsular polysaccharide antigenemia in *Haemophilus influenzae* type b meningitis has been reported by O'Reilly *et al.* (1975). These investigators used a sensitive RIA capable of detecting 0.0005 μg of *Haemophilus* type b polysaccharide per milliliter (whereas CIE has a sensitivity of only 0.01 to 0.05 μg of type b polysaccharide per milliliter. O'Reilly

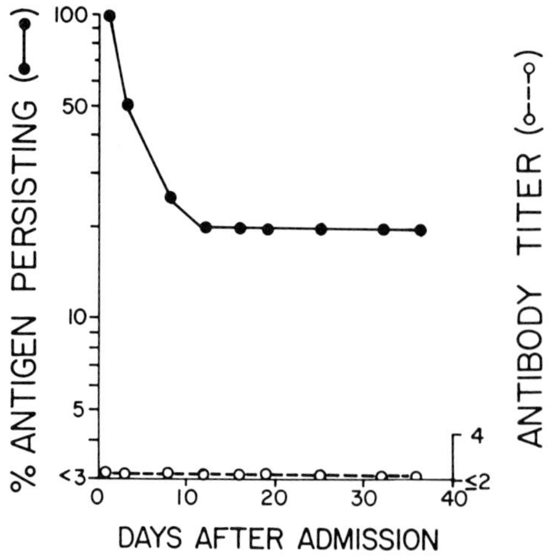

Fig. 3 Kinetics of capsular polysaccharide antigenemia in a patient with type 3 pneumococcal pneumonia who had prolonged persistence of circulating antigen. The peak antigen level was 1.6 μg/ml. (Published with permission from the *Annals of Internal Medicine*, same reference as in Fig. 2.)

et al. detected capsular polysaccharide in the serum of 34 of 38 patients with *Haemophilus* infection. Thirty-five percent of the antigenemic patients had <0.01 μg polysaccharide per milliliter of serum and would not, therefore, have had detectable antigenemia as determined by CIE. These findings indicate that false negative results with CIE are due partly to the relative insensitivity of CIE. However, O'Reilly and colleagues (1975) also obtained good evidence of immune complex formation. This finding was suggested by the shape of antigen clearance curves, which showed a rapid loss of circulating antigens in patients with a prompt antibody response. Several patients showed release of polysaccharide by pepsin digestion of their serum. In one patient 86% of the total capsular polysaccharide in the serum was released by pepsin digestion, strongly suggesting the presence of circulating immune complexes. Complexes were detected in the circulation for prolonged periods in some of the patients. Overall, immune complex formation appeared to play a significant role in masking circulating antigen in these studies.

Immune complex formation has proved to be a problem in detecting circulating antigens in staphylococcal infections. Wheat and colleagues (1978; see also Chapter 15, this volume) developed a solid-phase RIA that detected as little as 0.3 μg/ml of purified staphylococcal teichoic acid with the antigen dissolved in buffer; when the antigen was dissolved in normal rabbit or human sera the sensitivity was greatly decreased. Normal sera contained low titers of antistaphylococcal antibodies, which appeared to be responsible for the decreased sensitivity. Although these investigators could detect teichoic acid antigen in the serum of each of 12 nonimmune rabbits with experimental *Staphylococcus aureus* endocarditis (Wheat *et al.* 1978), only two of eight immune rabbits with *S. aureus* endocarditis and only one of nine patients with *S. aureus* bacteremia had positive tests (Wheat *et al.*, 1979). These results appeared to be due to complexing of teichoic acid antigen by antistaphylococcal antibodies in immune rabbits and humans.

Polysaccharides are very readily taken up by the reticuloendothelial system, even in the absence of immune complex formation. In the early studies of Kaplan *et al.* (1950) it was demonstrated that pneumococcal polysaccharides injected into mice are sequestered in reticuloendothelial tissues throughout the body, and similar sequestration seems to occur in pneumococcal infection (Kaplan *et al.*, 1950). It is likely that there is rapid clearance of polysaccharides from the circulation of some patients via the reticuloendothelial system even without antibody production. It also seems likely that much of the antigen produced in closed-space infections is sequestered locally by phagocytic cells and never enters the circulation.

Although pneumococcal polysaccharides and other bacterial polysaccharides resist degradation by mammalian enzymes, these antigens may nonetheless eventually be degraded by host tissues. We have found that type-specific pneumococcal capsular polysaccharide in the urine is physically and immunologically different from the type-specific polysaccharide present concomitantly in the circulation (Coonrod, 1974). The mean molecular size of polysaccharide in the urine is much smaller than that in the serum, and urine polysaccharide precipitates more slowly in agarose gels. When homogeneous, high molecular weight pneumococcal capsular polysaccharide is injected into rats, the polysaccharide appears to be altered *in vivo* and is excreted in a lower molecular weight form in the urine (Coonrod, 1979b). Thus, some proportion of the polysaccharide produced during infection appears to be broken down by mammalian tissues and excreted by the kidney. This process is slow, however, and degradation may not contribute substantially to the lowering of polysaccharide antigen levels in the blood during acute infection.

III. FUTURE TRENDS

Most investigators have found that CIE and other precipitin tests are too insensitive to be of value for the detection of antigenemia, although these techniques have added to our knowledge of the kinetics of polysaccharide antigenemia and the relation of antigenemia to bacteremia. Radioimmunoassays and enzyme-linked immunoassays are more sensitive and could give a higher number of positive results. However, these tests are not rapid. Moreover, there may be serious technical difficulties in adapting enzyme assays and radioimmunoassays to organisms such as pneumococci with multiple antigenic variants. Nor are enzyme assays and radioimmunoassays free from certain of the problems encountered with precipitin tests. Thus, immune complex formation *in vivo* may pose a major problem in antigen detection by these methods, as evidenced by the results of Wheat *et al.* (1979) in staphylococcal infections. Several laboratories are currently working on techniques that may overcome some of the problems caused by immune complex formation *in vivo*. In an ingenious approach, monoclonal antibodies are developed that can react at antigenic sites which are different from those that are covered by antibodies produced by the host. In theory, complexed antigen could be detected as easily as free antigen by these very specific monoclonal antibodies. However, it is possible that antigens that are complexed with native antibody may be rapidly removed from the circulation and, if so, they may not be available for detection by monoclonal antibodies.

REFERENCES

Coonrod, J. D. (1974). *J. Immunol.* **112,** 2193–2201.

Coonrod, J. D. (1979a). *In* "Handbook Series in Clinical Laboratory Science" (D. Seligson, ed.), Sect. F, Vol. 1, Part 2, p. 209. CRC Press, Boca Raton, Florida.

Coonrod, J. D. (1979b). *Proc. Soc. Exp. Biol. Med.* **162,** 249–253.
Coonrod, J. D., and Drennan, D. P. (1976). *Ann. Intern. Med.* **84,** 254–260.
Coonrod, J. D., and Leach, R. P. (1977). *J. Clin. Microbiol.* **8,** 257–259.
Dochez, A. R., and Avery, O. T. (1917). *J. Exp. Med.* **26,** 477–493.
Granoff, D. M., and Nankervis, G. A. (1977). *J. Infect. Dis.* **136,** 292–296.
Kaplan, M. H., Coons, A. H., and Deane, H. W. (1950). *J. Exp. Med.* **91,** 15–30.
Ney, R. N., and Harris, A. H. (1937). *Am. J. Pathol.* **13,** 749–767.
O'Reilly, R. J., Anderson, P., Ingram, D. L., Peter, G., and Smith, D. H. (1975). *J. Clin. Invest.* **56,** 1012–1022.
Schiffman, G., Summerville, J. E., and Costagna, R. (1974). *Fed. Proc., Fed. Am. Soc. Exp. Biol.* **23,** 758.
Tilghman, R. C., and Finland, M. (1937). *Arch. Intern. Med.* **59,** 602–619.
Wheat, J. L., Kohler, R. B., and White, A. (1978). *J. Infect. Dis.* **138,** 174–180.
Wheat, J. L., Kohler, R. B., Tabbarah, Z., and White, A. (1979). *J. Infect. Dis.* **140,** 54–61.

10

Evaluation of Counterimmunoelectrophoresis in the Diagnosis of Infectious Diseases

WILLIAM L. ALBRITTON

Department of Medical Microbiology
University of Manitoba
Winnipeg, Manitoba, Canada

I.	Introduction	125
II.	Counterimmunoelectrophoresis as the Test of Choice	126
	A. Test Procedure	126
	B. Predictive Value	126
III.	Use of Counterimmunoelectrophoresis in Early Diagnosis	129
IV.	Culture-Negative Diagnosis	130
V.	Identification of Pathogens in the Presence of Mixed Flora	130
VI.	Discussion	131
	References	133

I. INTRODUCTION

A number of immunologic methods are available for detecting soluble antigen, including flocculation, agglutination, and precipitation. Among the precipitin techniques, counterimmunoelectrophoresis (CIE) has been used widely in the diagnosis of infectious diseases. An evaluation of CIE in the diagnosis of infectious diseases should include a description of the performance characteristics of the test under standardized laboratory conditions and a comparison with other diagnostic tests, if any, used for the same purpose. It is equally important, however, to include a description of the comparative performance characteristics of the test under the clinical conditions in which it will be used, as well as an analysis of the test results for each clinical syndrome and, again, a comparison with other diagnostic tests used for the same syndrome, including an evaluation

The Direct Detection of Microorganisms in Clinical Samples

ISBN 0-12-187780-9

of turnaround time and cost. Such an evaluation has not been reported for CIE, and the introduction and widespread application of a test without complete evaluation is of some concern.

II. COUNTERIMMUNOELECTROPHORESIS AS THE TEST OF CHOICE

Many tests are available for the detection of soluble antigen, and the selection of CIE as the test of choice depends on many factors involving the performance of the test procedure as well as the applicability of the test to a specific clinical problem. Although it is possible to use CIE to detect antigen in asymptomatic patients and thus to determine the carrier state, most studies have used CIE to detect antigen in early symptomatic patients.

A. Test Procedure

The CIE methodology is discussed in Chapter 5 of this volume as well as in the Cumulative Techniques and Procedures in Clinical Microbiology series published by the American Society for Microbiology (Anhalt *et al.*, 1979). Factors related to the test procedure that favor CIE as the test of choice are the following. (*a*) The test detects preformed antigen and does not require growth of the organism, (*b*) the turnaround time for the test is faster than that for cultural methods, and (*c*) any antigen to which specific antibody has been prepared theoretically can be identified. Such factors have to be weighed against the cost and predictive value of the test when compared with other available tests. Most reported studies have evaluated only selected aspects of the test, and it is difficult, if not impossible, to evaluate the test in such a piecemeal fashion. Although evaluation and quality control of the procedure and reagents are prerequisites for any laboratory test, determination of the predictive value of the test in a clinical setting and its comparative evaluation will determine its clinical usefulness in the diagnosis of infectious disease.

B. Predictive Value

The predictive value is perhaps the most important determinant to be considered. The vast majority of clinical situations in which CIE will be used are related to patient management either in diagnosis or as an aid in prognosis. The decision to perform the test is based on the assumption that the result of the test will help determine the management of the patient. In most diagnostic situations the test parameter of concern is the sensitivity (number of true positive patients/total number of patients with disease). The sensitivity of CIE has been evaluated in several studies and depends on (*a*) the source of fluid for analysis, (*b*) the interval from the onset of symptoms to the time at which the sample is obtained, (*c*) the quantity of antigen present, and (*d*) the performance characteristics of the test, especially choice and source of antiserum reagents (Anhalt *et al.*, 1979; Finch and Wilkinson, 1979; Jacobs *et al.*, 1981). In general, CIE is said to be less sensitive than other immunologic tests such as latex agglutination (LA), immunofluorescence, and radioimmunoassay when sensitivity is determined by the lowest concentration of antigen detected by the test (Leinonen and Dayhty, 1978; Olcen, 1978; Scheifele *et al.*, 1981; Ward *et al.*, 1978), although a few studies have shown no difference in the sensitivities of LA and CIE in meningitis (Dirks-Go and Zanen, 1978). Such studies show considerable variability in the level of antigen or minimum number of organisms needed for a positive test and depend very much on the choice of test procedure and source of antiserum (Tobin and Jones, 1972; Leinonen and Dayhty, 1978; Fung and Wicher, 1981). Furthermore, the performance characteristics of the test determined in the laboratory may not necessarily determine the performance characteristics in the clinical setting, where the sensitivity will be the capacity of the test to detect patients with disease. The threshold of antigen detection by the test determined in the laboratory with prepared antigen may not be related to the threshold needed to detect patients with disease, and the capacity of the test to detect disease may depend on the particular disease to be diagnosed and the source of the test specimen. For example, CIE and LA appear to have similar sensitivities in clinical studies of meningitis, but LA appears to be more sensitive in non-meningitis *Haemophilus influenzae* infections.

Determining the test sensitivity in a clinical setting is not as straightforward as it might appear from performance tests in the laboratory. The probability that the test will be positive in patients with disease depends very much on the case definition. If one accepts only culture-positive patients to determine the sensitivity of a nonculture diagnostic test such as CIE, the selection has been biased because not all patients with disease are culture-positive. If we accept culture-negative patients in our determination of sensitivity, then we have to distinguish true positive from false positive results, and this is not always possible in infectious diseases in which different organisms cause similar syndromes. Nor has the specificity of the CIE test (specificity is the number of true negative patients/total number of patients without disease) been adequately determined in a clinical setting.

Reference to Table I may be helpful in this discussion of test performance characteristics as they apply to CIE, and the reader is referred to Griner *et al.* (1981) for a more detailed discussion of the selection and interpretation of diagnostic tests. Determination of the sensitivity only in a population of culture-positive patients is not entirely valid because the performance of the test may be quite different in clinical settings, where culture results may be negative, for

TABLE I

Performance Characteristics of Counterimmunoelectrophoresis[a,b]

	CIE test result	
Disease	Positive	Negative
Present	True positive (a)	False negative (b)
Absent	False positive (c)	True negative (d)

[a] Modified from Griner *et al.* (1981).
[b] Sensitivity = $a/(a + b)$; predictive value (positive) = $a/(a + c)$. Specificity = $d/(c + d)$; predictive value (negative) = $d/(b + d)$.

example, after antibiotic therapy or early in the course of disease when the number of organisms is low. The clinical evaluation of the specificity of the CIE test is usually limited to a discussion of cross-reactions of antisera with other known organisms in the laboratory, test artifacts, and the few observed false positive cases in which CIE results vary from positive culture results (Naiman and Albritton, 1980; Chan and Folds, 1981). Nevertheless, specificity is very important if clinical management decisions are to be made in culture-negative patients or before the culture results are available because intergeneric cross-reactions in CIE tests could lead to inappropriate antimicrobial therapy. Thus, the best evaluation of a test such as CIE includes the determination of the predictive value of a positive test and the predictive value of a negative test. These values indicate to the user the probability that a positive test represents disease and that a negative test rules out disease. Even with high predictive values it is necessary to determine the value of the test in a clinical setting and when compared to more traditional laboratory tests. A physician faced with a febrile 5-year-old child with meningitis who has a cerebrospinal fluid (CSF) with 2000 polymorphonuclear neutrophils, a glucose of 10 mg %, and gram-negative pleomorphic rods on Gram's stain of CSF rarely requires a CIE test with high predictive value for patient management. However, the same physician faced with a febrile 5-year-old child with meningitis who has been taking oral antibiotics for a urinary tract infection and who has a CSF with 200 white blood cells (70% polymorphonuclear neutrophils), a glucose of 30 mg %, and a negative Gram's stain of CSF would find a test with 100% predictive value for a positive or negative test very useful for patient management. Unfortunately, false negative results tend to limit the predictive value of a negative CIE test (Hoffman and

Edwards, 1972; Dirks-Go and Zanen, 1978). A careful evaluation of the test requires the determination of all four values in the 2×2 table for each infectious disease in which CIE is offered as a diagnostic test and the predictive value determined for both a positive and negative test.

III. USE OF COUNTERIMMUNOELECTROPHORESIS IN EARLY DIAGNOSIS

By far the most frequent use of CIE has been in early diagnosis (Greenwood *et al.*, 1971; Coonrod and Rytel, 1972; Granoff *et al.*, 1977; Estela and Heinricks, 1978; Storm and Lemburg, 1980). The rapid determination or exclusion of an etiologic agent in a clinical setting of suspected infectious disease would be of value. The value, however, depends on the specific clinical syndrome, the etiologic agent, and the results of other clinical and laboratory findings. For example, in patients with clinical and laboratory findings suggestive of meningitis there are no studies that show the CIE test to have a higher predictive value for a positive test than a Gram's stain; that is, given a positive Gram's stain of the CSF, it is unlikely that the CIE test will be of added value. Since the Gram's stain procedure is even faster and less expensive than the CIE test, a comparative evaluation would most likely favor the Gram's stain (Olcen, 1978). However, given a negative Gram's stain in the same clinical setting, any increased sensitivity of CIE and high predictive value for a positive test would establish the value of the CIE test (Whittle *et al.*, 1975). In the same clinical setting with a negative Gram's stain, a high predictive value for a negative test would also be valuable. It should be remembered, however, that initial antimicrobial therapy in most suspected bacterial infections depends on the established epidemiology of the disease. The choice of reagent antisera for CIE testing also is determined by the established epidemiology of the disease. Thus, the predictive value of a negative CIE test applies only to the organisms sought, and there is always the risk of a new or unusual organism. Therefore, sound clinical judgment as well as traditional culture techniques should always be used. Even if the Gram's stain and CIE test with high predictive value for a positive test are positive, the initial choice of antimicrobial therapy is based on established sensitivity patterns for the suspected organism. Because antimicrobial susceptibility cannot yet be determined by noncultural methods, the value of determining a specific etiologic agent from a group of agents is, at best, limited to those organisms with uniform susceptibility to a given antibiotic. Definitive antimicrobial therapy, in general, still requires isolation of the organism.

In our study of CIE in the diagnosis of acute infection, the test identified 45 of 65 patients with culture-proven meningitis, and the early diagnostic value defined by a positive CIE and negative Gram's stain was 25%. Nevertheless, in 13 of the 16 patients in whom a presumptive etiologic agent was identified by CIE before culture results were available, patient management was not altered from an epidemiologic approach because the positive test was for *H. influenzae* and antimicrobial susceptibility was unknown. Thus, a change in patient management could have been recommended less than 5% of the time, assuming that the test specificity for the other organisms was 100%, which it was not.

The clinical usefulness of such rapid diagnostic tests for the early diagnosis of infections due to common rapidly growing pathogens remains to be determined. The clinical usefulness of such tests for the early diagnosis of infections due to slowly growing pathogens such as *Mycobacterium tuberculosis* also remains to be determined, but it is likely to be much greater.

IV. CULTURE-NEGATIVE DIAGNOSIS

Although the sensitivity of culture methods and Gram's stain has been shown to be affected by prior antibiotic therapy, the sensitivity of the CIE test is, in general, less affected, and a positive test persists for a variable time after the institution of specific therapy (Hoffman and Edwards, 1972; Granoff *et al.*, 1977; Naiman and Albritton, 1980; Whittle *et al.*, 1975). As such, when the culture and Gram's stain are negative, the detection of antigen by methods such as CIE is frequently the only method for establishing a specific etiologic diagnosis. Determining the specific etiologic agent may be of value in certain diseases for prognosis, may be useful in establishing the need for prophylaxis in certain diseases, such as meningococcal meningitis, and may be useful in ongoing surveillance of diseases in remote areas lacking microbiology laboratory facilities (Pickering, 1976; Jacobs *et al.*, 1981; Whittle *et al.*, 1975).

It might be possible, for example, to determine the appropriateness of an epidemiologic approach to initial therapy for bacterial meningitis in remote areas by determining the incidence of various etiologic agents using CIE of CSF in a distant laboratory (Whittle *et al.*, 1975). However, there have been no comparative studies specifically evaluating CIE and other noncultural methods in such clinical settings.

V. IDENTIFICATION OF PATHOGENS IN THE PRESENCE OF MIXED FLORA

The predictive value of a positive Gram's stain of a normally sterile body fluid such as CSF or synovial fluid is very much higher than the predictive value of a positive Gram's stain of fluid from a mucosal surface. The problem of identifying etiologic agents in the presence of mixed flora is probably best represented by the problems in diagnosing pneumonia. Selective isolation media and quantitative techniques have been of value in establishing the etiology of gastroenteritis and urinary tract infections. Similar techniques, however, have been less successful in the diagnosis of pneumonia. Transtracheal and transthoracic aspirations have been used to avoid the problems encountered in culture and Gram's stain examination of expectorated sputum. Careful attention to the criteria for interpreting Gram's stain, however, significantly improves the accuracy of this test in identifying pneumococci in sputum (Rein *et al.*, 1978). However, CIE has the potential for detecting antigen in other body fluids such as urine and blood as well as detecting excess antigen in sputum. A high predictive value for a positive or negative test would be very useful for even a few of the many organisms epidemiologically associated with pneumonia, especially organisms commonly found in the respiratory tract flora. A limited evaluation of CIE in pneumococcal pneumonia in adults has been attempted, but more work is needed before its usefulness can be established.

Coonrod and Rytel (1973) found that 46.7% of 30 patients in all diagnostic categories for pneumococcal pneumonia had a positive urine CIE test. The urine CIE test was positive in 63.6% of patients with positive blood culture but in only 30% of patients with only a positive Gram's stain. Perlino and Shulman (1975), however, found that 8 of 8 patients with positive blood cultures for pneumococci had a positive sputum CIE test. Overall, 18 of 19 patients in the definite or probable diagnostic categories had a positive CIE test on sputum, compared with only 2 of 14 positive urine CIE tests and 5 of 18 positive sputum cultures. In our study (Naiman and Albritton, 1980), primarily in children, the CIE test on urine was also of value both in early diagnosis and in definitive diagnosis in patients with negative cultures. Schmid *et al.* (1979), however, found the Gram's stain of sputum to be as sensitive and as specific as CIE in pneumococcal pneumonia. A careful comparative study using diagnostic criteria for interpreting the Gram's stain and the CIE test is needed to determine the predictive value of a positive and negative test and its usefulness in the diagnosis of pneumonia.

VI. DISCUSSION

As can be seen from the previous discussion, the clinical usefulness of any laboratory test depends very much on its performance characteristics in the clinical setting. Most clinical diseases, including infectious diseases, fall into various system syndromes such as meningitis, endocarditis, and osteomyelitis, or more diffuse syndromes without organ localization such as "pyrexia of unknown origin" and sepsis. After a careful history has been taken and a physical examination carried out, a good clinician has developed a differential diagnosis that, for the purposes of our discussion, includes an infectious disease. As indicated earlier the use of CIE for determining the asymptomatic carrier state of various organisms is not considered. The clinician faced with a patient with a high clinical probability of an infectious disease based on history and physical findings now turns to the laboratory to (*a*) confirm the presence or absence of an infection and (*b*) if present, determine the etiology. Frequently, because of the nature of infectious diseases, empirical therapy is initiated before the laboratory results are available. Thus, in order to be of value in initial patient management the laboratory test of choice must be rapid. Although CIE has a relatively rapid turnaround time, for practical considerations in the laboratory it cannot approach the rapidity of the Gram's stain procedure. Furthermore, it is important to remember that we are asking two questions. First, is CIE the test of choice to confirm the presence or absence of disease? It probably is not, because for the two syndromes in which CIE has been most frequently reported, meningitis and pneumonia, traditional examination of the CSF for pleocytosis and hypoglycorrhachia and the chest X ray and sputum examination, respectively, are better indicators of the presence or absence of disease. The false negative rate for CIE is much higher than the false negative rate for traditional laboratory tests. The CIE test is a much better choice for the second question: What is the etiology? If the specific etiology could be determined before initial patient management, then presumably specific therapy, rather than empirical therapy based on epidemiology, could be instituted. Unfortunately, as indicated, this is true only for organisms with predictable antimicrobial susceptibility patterns.

At least one approach to the evaluation of a test such as CIE is presented in Table II. This approach applies to a test proposed for general use in clinical laboratories for the diagnosis of infectious diseases and does not necessarily apply to specific uses developed for CIE in research or reference laboratories. The first step is to determine the sensitivity, specificity, limitations, and reproducibility of the test under standardized laboratory conditions including refer-

TABLE II

Evaluation Scheme for CIE in the Diagnosis of Infectious Diseases

Step 1. Determine laboratory test parameters
- A. Sensitivity, specificity, reproducibility
- B. Availability of standardized reagents and reference cultures
- C. Standardized test procedure

Step 2. Determine predictive value in clinical setting
- A. Definition of patients with disease
- B. Separate determinations for every proposed system syndrome
- C. Source of specimen or combination of specimens

Step 3. Determine clinical usefulness
- A. Establish purpose of test
- B. Establish benefit/cost ratio

ence strains and antisera and clinical control strains. Monoclonal antibodies and the availability of a wider range of commercial reagents will undoubtedly improve the CIE test performance characteristics in the future. The limits of acceptability of these laboratory test parameters depend very much on the clinical setting in which the test will be used and how its performance compares with that of the established test of choice, if there is one. The second step is the most important and will determine the overall clinical usefulness of the test. A well-designed prospective study is necessary and should include a determination of the predictive value of a positive and negative test in every proposed syndrome for each source of specimen or combination of specimens. In addition, in the same study it is necessary to compare the results with similar determinations for existing tests under equally rigorous standardization of test procedures. Finally, it is necessary to undertake a cost/benefit analysis and to establish the purpose of the test. The fact that a test performs well does not necessarily mean that it is needed. As a rule of thumb, if the test result does not affect patient management it is not needed.

The laboratory test parameters for CIE have been reasonably well established, but the evaluation of CIE in the diagnosis of infectious diseases will remain incomplete until the comparative predictive value of the test has been determined in clinical settings and the cost/benefit analysis has been evaluated. Such definitive studies have not yet been reported for this procedure.

REFERENCES

Anhalt, J. P, Kenny, G. E., and Rytel, M. W. (1979). *In* "Cumitech 8" (T. L. Gavan, ed.), pp. 1–11. Am. Soc. Microbiol., Washington, D.C.

Chan, J. K., and Folds, J. D. (1981). *Clin. Microbiol.* **13,** 877–879.

Coonrod, J. D., and Rytel, M. W. (1972). *Lancet* **1,** 1154–1157.

Coonrod, J. D., and Rytel, M. W. (1973). *J. Lab. Clin. Med.* **81,** 778–786.

Dirks-Go, S. I. S., and Zanen, H. C. (1978). *Clin. Pathol.* **31,** 1167–1171.

Estela, L. A., and Heinrichs, T. F. (1978). *Clin. Pathol.* **70,** 239–243.

Finch, C. A., and Wilkinson, H. W. (1979). *Clin. Microbiol.* **10,** 519–524.

Fung, J. C., and Wicher, K. (1981). *Clin. Microbiol.* **13,** 681–687.

Granoff, D. M., Congeni, B., Baker, R., Ogra, P., and Nankervis, G. A. (1977). *Am. J. Dis. Child.* **131,** 1357–1362.

Greenwood, B. M., Whittle, H. C., and Dominic-Rajkovic, O. (1971). *Lancet* **2,** 519–521.

Griner, P. F., Mayewski, R. J., Mushlin, A. I., and Greenland, P. (1981). *Ann. Intern. Med.* **94,** 553–592.

Hoffman, T. A. and Edwards, E. A. (1972). *J. Infect. Dis.* **126,** 636–644.

Jacobs, R. F., Yamauchi, T., and Eisenach, K. D. (1981). *Am. J. Clin. Pathol.* **75,** 203–208.

Leinonen, M., and Dayhty, H. (1978). *Clin. Pathol.* **31,** 1172–1176.

Naiman, H. I., and Albritton, W. L. (1980). *J. Infect. Dis.* **142,** 524–531.

Olcen, P. (1978). *Scand. J. Infect. Dis.* **10,** 283–289.

Perlino, C. S., and Shulman, J. S. (1975). *J. Lab. Clin. Med.* **87,** 496–502.

Pickering, L. K. (1976). *J. Am. Med. Assoc.* **236,** 1882–1883.

Rein, M. F., Gwaltney, J. M., O'Brien, W. M., Jennings, R. H., and Mandell, G. L. (1978). *J. Am. Med. Assoc.* **239,** 2671–2673.

Scheifele, D. W., Ward, J. I., and Siber, G. R. (1981). *Pediatrics* **68,** 888–890.

Schmid, R. E., Anhalt, J. F., Wold, A. D., Keys, T. F., and Washington, J. A. (1979). *Am. Rev. Respir. Dis.* **119,** 345–348.

Storm, W., and Lemburg, P. (1980). *Eur. J. Pediatr.* **135,** 65–67.

Tobin, B. M., and Jones, D. M. (1972). *J. Clin. Pathol.* **25,** 583–585.

Ward, J. I., Siber, G. R., and Smith, D. H. (1978). *J. Pediatr. (St. Louis)* **93,** 37–42.

Whittle, H. C., Greenwood, B. M., Davidson, N., McTomkins, A., Tugwell, P., Warrell, D. A., Zalin, A., Bryceson, A. D. M., Parry, E. H. O., Brueton, M., Duggan, M., Oomen, J. M. V., & Rajkovic, A. D. (1975). *Am. J. Med.* **58,** 823–828.

11

Agglutination Techniques for the Detection of Microbial Antigens: Methodology and Overview

J. DONALD COONROD

Division of Infectious Diseases
Department of Medicine
University of Kentucky
and
Veterans Administration Medical Center
Lexington, Kentucky

I.	Principles of Agglutination	135
II.	Advantages and Disadvantages of Agglutination for Antigen Detection	139
III.	Future Prospects	141
	References	142

I. PRINCIPLES OF AGGLUTINATION

Agglutination techniques have proved useful for many years in blood banking, and they are currently being used regularly for the measurement of rheumatoid factor, antibacterial antibodies, and other medically important substances. Part of the appeal of these techniques in the rapid diagnosis of infectious diseases is their familiarity. In several areas of microbial antigen detection they may be the techniques of choice, provided that the quality of reagents available and the intrinsic value of the results to clinicians justify the effort.

The principles of agglutination have been reviewed by Nichols and Nakamura (1980), to whom the reader is referred for greater detail than can be presented here. In theory, agglutination can occur whenever antibodies and antigens combine with extensive cross-linking. Microscopic or gross aggregation develops as complexes lose solubility. Many particles, such as bacteria, bear a slight negative change, which causes them to repel each other under normal circumstances.

The Direct Detection of Microorganisms in Clinical Samples

ISBN 0-12-187780-9

Such factors as buffer pH and ionic strength and the concentration of proteins or synthetic polymers in the suspension can affect the electrical charge on particles and thereby their state of aggregation. In general, it is helpful to include some albumin or polymers such as dextran or polyvinylpyrrolidone in buffers used in agglutination work.

Agglutination is a "secondary" manifestation of antigen–antibody interaction, and antibodies differ tremendously in their capacity to agglutinate with antigens. Immunoglobulin M may be as much as 750 times more efficient that IgG in agglutination reactions (Nichols and Nakamura, 1980). The efficacy of a particular antiserum in agglutination will thus vary with the schedule used to raise it and the time at which it is collected. Reported differences in the sensitivity of techniques such as counterimmunoelectrophoresis (CIE) and agglutination for the detection of bacterial antigens may reflect characteristics of the antiserum used as much as inherent differences in the sensitivity of the methods (Coonrod and Rylko-Bauer, 1976).

Whole antisera purified immunoglobulins, or monoclonal antibodies can be used to sensitize particles for agglutination. Whole antisera are usually used. They are convenient to use, usually as efficacious as purified immunoglobulins, and contain IgM antibodies, which are lacking in immunoglobulins prepared by salt fractionation of serum.

When particulate antigen is clumped by antibody, the reaction is classified as direct agglutination. Indirect ("passive") assays entail the agglutination of soluble antigens bound or absorbed onto particles. The inhibition of agglutination is also used to detect antigens or antibodies. Erythrocytes have been widely used in agglutination work in the past, including an early application to the etiologic diagnosis of *Haemophilus* meningitis by the detection of the type b capsular polysaccharide in cerebrospinal fluid (CSF) (Warburton and Keogh, 1949). Although erythrocytes have been used successfully in some diagnostic studies (e.g., Greenwood and Whittle, 1974), erythrocytes are not as convenient to use as latex or staphylococcal particles. Erythrocytes have a limited shelf life and frequently require alteration or chemical bonding of antibody for optimal results. Heterophile reactions can also cause nonspecific agglutination of erythrocytes. In the development of commercial kits, shelf life and stability are major considerations, and particles other than erythrocytes are usually used.

Latex particles (approximately 0.8μm) are inexpensive and effective carriers for antigens or antibodies. When washed and stored at 4°C, they remain stable indefinitely. Proteins can be passively absorbed on latex; the optimal dilution of antiserum is ascertained by a checkerboard titration. The dilution of immunoglobulin giving the greatest sensitivity with the least spontaneous agglutination is then selected for routine use.

The preparation of staphylococci for use in coagglutination takes considerable effort, although unsensitized staphylococcal particles can be purchased, or in some cases sensitized staphylococci may be available. In the method of Kronvall (1973) an isolate of *Staphylococcus aureus* containing a high concentration of protein A is grown in broth, washed, fixed with 0.5% formaldehyde and heat-treated at 80°C. These particles are stable for many weeks. Antibody attaches by reaction of the Fc portion with protein A on the staphylococci, leaving the specific Fab portion to react with the desired antigen. As with latex an optimal dilution of antibody for sensitization of the particles must be established. A 1% suspension of sensitized particles is usually used in coagglutination. Staphylococcal particles, unlike latex, may cause problems by agglutinating with immunoglobulins or with specific antistaphylococcal antibodies in clinical samples. The former difficulty appears to be common (Thirumoorthi and Dajani, 1979) but can be eliminated in most cases by heating samples at 65 to 80°C for 5 to 10 min. Absorption with soluble or insoluble protein A is also effective but more cumbersome. Interference by antistaphylococcal antibodies in samples appears to be quite uncommon and can be eliminated by absorption with staphylococci.

Direct agglutination tests are very simple to perform. A drop of clinical sample is added to a drop of optimally sensitized particles on a glass slide, the drops are mixed by gentle rotation or with an applicator, and the state of aggregation is determined after a given interval. Fewer than 10 min are needed to develop most slide agglutination reactions. The interpretation of the degree of agglutination is very subjective, however, as anyone who has tried to quantify results with agglutination has discovered. In reality, the interpretation of agglutination reactions takes practice. Even with the use of absorption, heat, etc., to minimize background agglutination, many samples of normal serum, urine, CSF, or sputa show some increase in granularity over controls. Moreover, the longer the reactions continue, the more marked is the background granularity. Several systems for grading agglutination have been devised, one of which the present author has found useful (Fig. 1). Although objective criteria are helpful, the quantification of agglutination reactions must be considered a subjective process.

In the reading of agglutination reactions, results with specifically sensitized particles must be compared simultaneously with results obtained with control particles coated with nonimmune serum. Preimmune serum from the same animal or animals that provide the specific immune serum is the best source of nonimmune serum for coating control particles. This is particularly important with clinical samples such as sputa that can harbor numerous microbial agents other than the one specifically being sought. Reactions between naturally acquired antibodies in "nonimmune" sera and the mixed flora in such samples is definitely possible. In addition, heterophile reactions are possible, and these are best monitored by the use of preimmune and immune sera from the same ani-

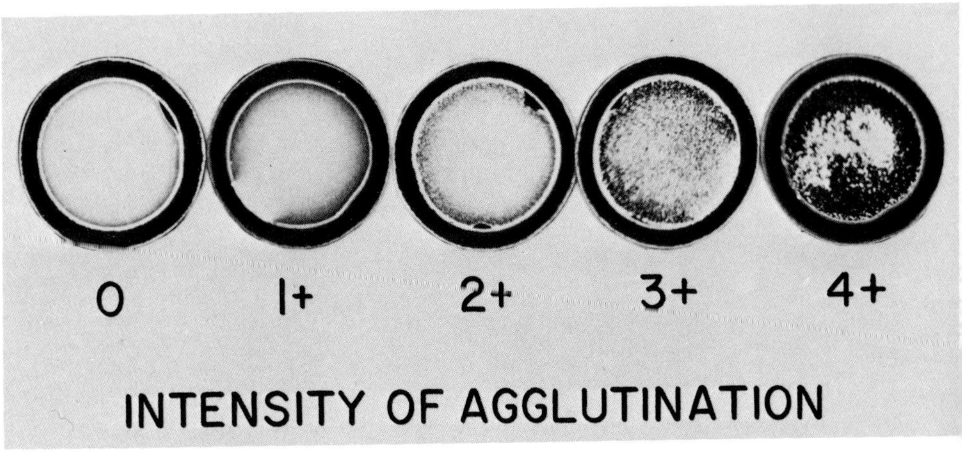

Fig. 1. Method for scoring the latex agglutination reaction. 1+, Fine granularity; 2+, coarse granularity with a cloudy background; 3+, coarse granularity with a clear background; 4+, complete agglutination and clearing. From Coonrod and Rylko-Bauer (1976). Published with permission.

mals. Only clear-cut differences in the degree of agglutination between control and specifically sensitized particles should be considered meaningful. The amount of time allotted for agglutination of the control and the specifically sensitized particles must be carefully monitored, because nonspecificity in agglutination reactions is time-dependent. Uncoated particles should be included as a control to help pinpoint the source of nonspecific reactions. Clinical samples must be freshly obtained or must have been properly stored (usually frozen) before use, and all samples should be clarified by centrifugation before use. Grossly bloody fluids or those that have an uncertain history (especially if they have been left at room temperature for several hours) are unsuitable for testing.

It is important to remember that agglutination tests can give erroneous results from ''zone'' phenomena. If there is a great excess of antigen or antibody in the reaction mixture, cross-linking among complexes (and consequently agglutination) may be impossible. This phenomenon can be understood by imagining a few antibody molecules, each completely surrounded by antigen. Antigen molecules on the perimeter would then have contact only with other antigen molecules, and cross-linking would be impossible. The problem of antigen excess in agglutination tests has been observed in the detection of bacterial polysaccharides in the CSF in meningitis by Severin (1972) and Newman *et al.* (1970). Negative samples are therefore best retested at several widely spaced dilutions to rule out excess antigen as the cause of the negative reaction.

II. ADVANTAGES AND DISADVANTAGES OF AGGLUTINATION FOR ANTIGEN DETECTION

Agglutination techniques are speedy and technically simple to perform. The materials required are inexpensive, except for the antiserum, and even antiserum goes much farther in agglutination than, for example, in precipitin tests. We (Edwards and Coonrod, 1980) calculated that 1 ml of antipneumococcal antiserum would make 100 ml of sensitized staphylococci, or enough for 4000 coagglutination tests. The same quantity of antiserum would yield only about 140 tests in CIE.

Although experience with latex tests is more extensive than that with coagglutination of staphylococcal particles, both systems appear to have excellent sensititivity when good antisera are used. As detailed elsewhere (Chapters 6, 9, and 12–14, this volume) latex agglutination regularly detects as little as 1 ng/ml of purified bacterial polysaccharide, and in a few cases (Ward *et al.*, 1978) as little as 0.2 ng/ml of polysaccharide has been detected. This is generally superior to results achieved with CIE. Sensitivity for antigen detection in clinical samples is generally severalfold lower than that achieved with the purified antigens with both precipitin and agglutinin techniques. In some clinical situations (e.g., *Haemophilus* type B nonmeningeal infections) agglutination techniques have proved superior to CIE for antigen detection. Isolated instances of the reverse situation are also encountered (Whittle *et al.*, 1974). Results with latex agglutination have been excellent in selected infections such as meningitis caused by *Haemophilus* type b or group B streptococci. Almost all patients with these infections have had detectable levels of type- or group-specific polysaccharides in the CSF, serum, or urine. When the antisera are not as good, either because of the variety of bacterial serotypes encountered (e.g., pneumococcal disease) or because of poor immunogenicity of the antigen (e.g., group B meningococci), the results are highly inferior.

Perhaps the most important question concerning the sensitivity of agglutination techniques is whether they are superior to the examination of Gram-stained smears for early presumptive diagnosis. This comparison has not been made adequately with agglutination techniques, but in one careful comparison of the results of Gram's stains and CIE in tests of the CSF in meningitis (Naiman and Albritton, 1980) up to 25% of culture-proven cases had a positive CIE but negative Gram's stain results. Undoubtedly, results will vary with the predominant bacterium encountered (because the antisera are better for, say, *Haemophilus* than for pneumococcus) and with the number of pretreated cases encountered. The care with which the Gram-stained smear is examined is also important. In my own studies of meningitis there were discrepancies between results obtained by house officers and the microbiology staff when smears of

CSF were prepared and read independently (J. D. Coonrod, unpublished data). The correct diagnosis in this situation might as easily be obtained by a well-performed Gram's stain in the laboratory as by an agglutination test for antigen. However, the latter test may still add certainty to the impressions obtained by the examination of a Gram-stained smear. In patients with community-acquired pneumonia, who do not have chronic bronchitis, there is a strong suggestion that the detection of pneumococcal antigen by CIE is superior to examination of Gram-stained smears of sputum for etiologic diagnosis. There are no data bearing on this point with agglutination techniques, but the sensitivity of these techniques is no less than that of CIE in this setting (Edwards and Coonrod, 1980). Negative tests for antigen in meningitis or pneumonia do not rule out infection with a particular pathogen. This is also true for the Gram's stain. In either test, negative results are indeterminate.

Agglutination techniques, like most laboratory methods, have a measurable level of nonspecificity. With latex agglutination for the detection of type or group bacterial polysaccharides, about 2% of results are "false" positive, if one includes the immunologic cross-reactions of similar antigens of different organisms. This is true even when standard precautions such as centrifugation and heating of clinical samples are taken to decrease nonspecific reactivity and even when the uninterpretable tests in which controls have also agglutinated are excluded from consideration. Coagglutination may give a slightly higher rate of misleading results than latex agglutination. On the surface this level of specificity appears to be excellent, but as pointed out by others (see, for example, Chapter 10, this volume) one must consider the predictive value of the results. Predictive value will vary with the clinical population studied if all other factors are held constant. Thus, if one is dealing with a rare disease (e.g., cryptococcal meningitis) one may achieve a low rate of false positivity in a diagnostic test (say 1%), yet, if only 1% of the samples tested are truly positive, the predictive value of a positive result can be no better than 50%. The bacterial meningitides and bacterial pneumonias are relatively common infections, and a test with only 1% false positivity would have correspondingly greater predictive value. The specificity of agglutination tests must be further studied before comprehensive appraisal of their predictive value is possible.

In Chapter 10 of this volume Albritton raises an extremely important critical point which is usually forgotten in the assessment of new clinical diagnostic tests. Given a sensitive and specific test, is decision making by the physician affected? Do the results of the test alter the treatment used? In the case of patients with clear-cut clinical evidence of bacterial meningitis or community-acquired pneumonia, certain kinds of antibiotics will most likely be given irrespective of the results of any single presumptive etiologic test. If the patient appears to respond to therapy, the antibiotics are likely to be continued even if cultures remain inconclusive. Yet the clinician may benefit in subtle ways if a presumptive etiologic diagnosis can be made. In meningitis, for example, physicians not only must treat the patient but must grapple with epidemiologic questions regarding family members and contacts (i.e., whether to give prophylactic vaccines or antibiotics). A presumptive etiologic diagnosis can give the clinician insight into the outcome of an infection on the basis of his own and published experiences with the different bacterial meningitides. In the bacterial pneumonias, clinicians are so accustomed to working without the benefit of a specific etiologic diagnosis that antibiotic "coverage" is usually initiated on the basis of guesses about statistical probabilities. Aids in etiologic diagnosis, such as Gram's stain or antigen detection, can give the clinician greater confidence that his guesses are correct and that his antibiotic choices are appropriate. In the case of severe infections with gram-negative rods, there has been renewed interest in the use of antiserum therapy (Ziegler *et al.*, 1982). If antiserum proves to be beneficial as an adjunct to antibiotics in the treatment of infections caused by high-grade pyogenic bacteria, detection and serotyping of microbial polysaccharides in clinical samples could become a primary factor in selecting therapy.

Whether or not clinical microbiology directors or physicians can benefit greatly from some of the current applications of agglutination techniques to presumptive etiologic diagnosis, the tests are rapidly becoming commercially available. Agglutination has long been widely used as an adjunct in the diagnosis of cryptococcal meningitis (e.g., the Crypto-LA kit, International Biological Laboratories). Now numerous kits are available for the detection of bacterial antigens as well, including Wellcogen Strep B (Wellcome Diagnostics), Phadebact (Pharmacia), Directogen (Becton Dickinson), and Bactogen (Wampole Laboratories). Other tests designed primarily for the rapid presumptive identification of bacteria in pure culture are Sero-Stat Staph Latex Slide Test (Scott Laboratories, Inc.) and Accu-Staph (Carr-Scarborough Microbiologicals, Inc.) for the identification of *Staphylococcus aureus*. Tests for Lancefield grouping of streptococci A–D, F, and G are also available (e.g., Streptex, Wellcome Diagnostics).

III. FUTURE PROSPECTS

There is a possibility that better agglutination tests for microbial antigens in clinical samples will eventually become available. Certainly, one might expect improved specificity with the use of selected monoclonal antibodies. Although specificity for the agglutination tests currently in use appears to be acceptable, even a low incidence of cross-reactivity or false positive results is worrisome in tests designed for use in life-threatening infections. The introduction of new, commercially available kits utilizing monoclonal antibodies might effectively eliminate many specificity problems and certainly would be very desirable.

REFERENCES

Coonrod, J. D., and Rylko-Bauer, B. (1976). *J. Clin. Microbiol.* **4,** 168–174.

Edwards, E. A., and Coonrod, J. D. (1980). *J. Clin. Microbiol.* **11,** 488–491.

Greenwood, B. M., and Whittle, H. C. (1974). *J. Infect. Dis.* **129,** 201–204.

Kronvall, G. (1973). *J. Med. Microbiol.* **6,** 187–190.

Naiman, H. L. and Albritton, W. L. (1980). *J. Infect. Dis.* **142,** 524–531.

Newman, R. B., Stevens, R. W., and Gaafar, H. A. (1970). *J. Lab. Clin. Med.* **76,** 107–113.

Nichols, W. S., and Nakamura, R. M. (1980). *In* "Manual of Clinical Immunology" (N. R. Rose and H. Friedman, eds.), Chapt. 3, pp. 15–34. Am. Soc. Microbiol., Washington, D.C.

Severin, W. P. J. (1972). *J. Clin. Pathol.* **25,** 1079–1082.

Thirumoorthi, M. C., and Danjani, A. S. (1979). *J. Clin. Microbiol.* **9,** 28–32.

Warburton, M. F., Keogh, E. V., and Williams, S. W. (1949). *Med. J Aust.* **1,** 135–137.

Ward, J. I., Siber, G. R., Scheifele, D. W., and Smith, D. H. (1978). *J. Pediatr. (St. Louis)* **93,** 37–42.

Whittle, H. C., Tugwell, P., Egler, L. J., and Greenwood, B. M. (1974). *Lancet* **2,** 619–621.

Ziegler, E. J., McCutchan, J. A., Fierer, J., Glauser, M. P., Sadoff, J. C., Douglas, H., and Braude, A. I. (1982). *N. Engl. J. Med.* **307,** 1225–1230.

12

Agglutination Tests for the Diagnosis of Meningitis

ADNAN S. DAJANI

Department of Pediatrics
Wayne State University School of Medicine
and
Division of Infectious Diseases
Children's Hospital of Michigan
Detroit, Michigan

I. Introduction 143
II. Antigen Determinants 144
A. *Haemophilus influenzae* Type B Antigens 144
B. Pneumococcal Antigens 145
C. Meningococcal Antigens 146
D. Group B Streptococcal Antigens 148
E. Cryptococcal Antigens 149
III. Antibody Determinants 150
IV. Specificity of Agglutination Tests 150
A. Latex Particle Agglutination 150
B. Staphylococcal Coagglutination 151
References 152

I. INTRODUCTION

Agglutination tests offer several distinct advantages for the rapid diagnosis of bacterial meningitis. The tests are extremely rapid, requiring minimal time after the cerebrospinal fluid (CSF) is obtained. Few or no technical skills are required for performing the tests, allowing for their bedside utilization. Because no major or specialized equipment is needed, the cost per test is very low. Finally, agglutination tests have been shown to be highly sensitive and superior to many other methods.

Two major agglutination tests have been used: the latex particle agglutination

The Direct Detection of Microorganisms in Clinical Samples

ISBN 0-12-187780-9

test (LPA) and the staphylococcal coagglutination test (SC). The basic principle of both tests is an antigen–antibody reaction. The rate of success of the tests is therefore dependent on both of these variable substances.

II. ANTIGEN DETERMINANTS

Antigens of a number of microorganisms have been detected in CSF specimens from patients with meningitis. These organisms include *Haemophilus influenzae* type B, *Streptococcus pneumoniae, Neisseria meningitidis,* group B streptococci, and *Cryptococcus neoformans.* Thus, most bacterial causes of meningitis in older children and adults can be determined by these methods; however, many bacteria that cause meningitis in newborn infants cannot be so detected. These include *Listeria monocytogenes, Escherichia coli, Klebsiella pneumoniae,* and other enteric gram-negative bacilli.

A. *Haemophilus influenzae,* Type B Antigens

The detection of *H. influenzae* type B antigen in CSF has been well documented. Table I compares LPA, SC, and counterimmunoelectrophoresis (CIE) in detecting *H. influenzae* in CSF. The most sensitive test appears to be LPA. In the various cited studies, 290 specimens were tested, and 260 (90%) were positive by the LPA test. The SC test detected 143 of 191 specimens (75%) and was comparable to CIE, which detected 159 of 211 specimens (75%). Only specimens that were confirmed by positive cultures are included in this analysis. All tests have reportedly detected *H. influenzae* antigen in CSF of patients with meningitis that was not confirmed bacteriologically.. It is quite possible that many of these instances represent examples of partially treated meningitis; however, the remote possibility of false positive reactions cannot be ruled out completely.

In a number of studies (see also Chapters 6 and 14, this volume) quantitative measurement of *H. influenzae* polyribophosphate (PRP) was performed, and a comparison of the various tests was made. Ward *et al.* (1978), using antiserum from the Bureau of Biologics, found the LPA assay to be five times more sensitive than CIE in detecting *H. influenzae* antigen in CSF. The lower limit of sensitivity for LPA was 0.2 ng/ml of PRP. Using the same source of antiserum, Daum *et al.* (1982) reported the *in vitro* sensitivity of LPA to be 1.0 ng/ml of PRP. A commercially prepared reagent (Bactogen) detected 0.5 ng/ml. In a primate model of *H. influenzae* meningitis (Scheifele *et al.*, 1979), PRP was detected in the CSF of 19 of 19 animals using LPA and in 19 of 20 using CIE. In 50% of the animals, LPA was positive before CIE, and in no case was CIE positive before LPA. Furthermore, LPA detected antigen in the CSF of all 12

TABLE I

Detection of *Haemophilus influenzae* Type B in CSF

Source of antiserum	No. tested	% positive LPA	SC	CIE	Reference
Author's laboratory	30	93	—	—	Newman *et al.* (1970)
Hyland	16	94	—	88	Whittle *et al.* (1974)
Hyland	40	95	—	—	Kaldor *et al.* (1977)
Hyland	14	93	—	33	Leinonen and Herva (1977)
Bureau of Biologics	25	100	—	83	Ward *et al.* (1978)
Bureau of Biologics	24	96	—	82	Scheifele *et al.* (1981)
Bactogen	16	100	—	—	Daum *et al.* (1982)
Bureau of Biologics	16	100	—	—	
Bureau of Biologics Statens Seruminstitut Burroughs Wellcome	60	77	57	38, 75, 92[a]	Dirks-Go and Zanen (1978)
Hyland	36	—	86	72	Suksanong and Dajani (1977)
Hyland	65	86	89	75	Thirumoorthi and Dajani (1979)
Author's laboratory	30	—	67	80	Olcen (1978)

[a] 38% positive with Burroughs Wellcome antiserum, 75% positive with Seruminstitut antiserum, and 92% positive with Bureau of Biologics antiserum.

animals with $\geq 10^3$ colony-forming units (cfu) per milliliter and in 5 of 7 with 10^2 cfu/ml. Counterimmunoelectrophoresis detected antigen in 6 of 7 animals with $\geq 10^4$ cfu/ml, and in only 4 of 13 with fewer bacteria. In this study CIE detected 1.0 ng/ml of PRP in CSF; LPA detected 0.5 ng/ml. In another study (Suksanong and Dajani, 1977) SC detected 2.0 ng/ml of PRP, whereas CIE detected 4.0 ng/ml.

B. Pneumococcal Antigens

The detection of *Streptococcus pneumoniae* antigen in CSF has also been reported (Table II). In general, LPA, SC, and CIE appear to be equally sensitive in detecting pneumococcal antigen (mean positive results are 79, 73, and 77%, respectively).

Dirks-Go and Zanen (1978) reported that the detection of *S. pneumoniae* types 7 and 14 was possible using LPA or SC but not CIE and attributed that to the fact that these polysaccharides are neutral. In another report (Burdash and West, 1982) the levels of pneumococcal capsular polysaccharide for various serotypes detected by CIE and SC (Phadebact) were compared. For pneumococcal serotypes 1, 3, 6A, 6B, and 8, SC detected lower levels (25–50 ng/ml) of

TABLE II

Detection of *Streptococcus pneumoniae* in CSF

Source of antiserum[a]	No. tested	% positive LPA	SC	CIE	Reference
Seruminstitut	87	82	—	98	Whittle *et al.* (1974)
Seruminstitut	10	80	—	—	Kaldor *et al.* (1977)
Seruminstitut	5	50	60	20	Thirumoorthi and Dajani (1979)
Seruminstitut	14	—	71	57	Olcen (1978)
Seruminstitut	73	78	74	64	Dirks-Go and Zanen (1978)
Seruminstitut	10	—	70	40	Burdash and West (1982)
Phadebact[b]					

[a] Seruminstitut refers to Statens Seruminstitut, Copenhagen, Denmark.
[b] Used for SC test only.

polysaccharide than did CIE (50–100 ng/ml). Serotypes 7 and 14 were detectable by CIE at polysaccharide levels of 10,000 ng/ml, whereas SC was positive at 50 ng/ml.

C. Meningococcal Antigens

Table III summarizes the results of studies assessing the detection of meningococcal antigens in CSF. There are marked variations among the reports with regard to the capacity of any test to detect such antigens. Careful analysis of these reports indicates that apparent discrepancies are related to the antisera used. Meningococci belong to nine recognized groups (A, B, C, D, X, Y, Z, 29E, and 135), and antisera against all these groups were either not available or not assessed in these studies.

A more appropriate assessment of the various methods can be made when data obtained from the use of homologous antisera are compared (Table IV). Serogroup A was detected equally well by LPA or CIE; SC was inferior to the other two methods. Most investigators achieved poor results in detecting group B meningococcal antigens in CSF regardless of the method employed. Detection of group C meningococci was comparable to that of group A.

Quantitative detection of meningococcal antigens was made in some studies, and marked variations were also noted. Leinonen and Herva (1977) were able to detect 50 ng/ml of group A polysaccharide using LPA. Whittle *et al.* (1974) detected pure group A meningococcal polysaccharide by LPA only at concentrations of 50,000 ng/ml or greater; CIE was more sensitive, detecting antigen at a concentration of 90 ng/ml. Olcen (1978) detected 31 ng/ml using CIE. The SC

TABLE III

Detection of *Neisseria meningitidis* in CSF

Source of antiserum	Antiserum against group	Organism		% positive			Reference
		Serogroup	No. tested	LPA	SC	CIE	
Author's laboratory	A	A	26	92	—	74	Severin (1972)
Author's laboratory	C	C	11	100	—	87	
Difco	A	A	126	88	—	89	Whittle *et al.* (1974)
Difco	Pool (A, C, D)	Unknown	9[a]	11	—	—	Kaldor *et al.* (1977)
Difco	A	A	57	79	—	60	Leinonen and Herva (1977)
Burroughs Wellcome	B	B	6	0	—	0	
Difco	C	C	4	100	—	50	
Author's laboratory	A	A	17	94	65	94	Dirks-Go and Zanen (1978)
Author's laboratory	B	B	33	52	27	61	
Author's laboratory	C	C	16	81	69	75	
Burroughs Wellcome	Pool (A, B, C, D, X, Y, Z)	Unknown	6	20	33	17	Thirumoorthi and Dajani (1979)
Author's laboratory	A, C	A, C	16	—	38	38	Olcen (1978)
Author's laboratory	B	B	9	—	44	33	

[a] Five of 9 isolates were group B, against which no specific antiserum was available.

TABLE IV

Detection of *Neisseria meningitidis* in CSF Using Homologous Antisera[a]

Serogroup	LPA	SC	CIE
A	196/226 (87%)	11/17 (65%)	150/174 (86%)
B	17/39 (44%)	13/42 (31%)	23/48 (48%)
C	28/31 (90%)	11/16 (69)%	23/31 (77%)

[a] Values indicate the number of positive reactions per the number tested.

test was more sensitive than CIE in detecting group A polysaccharide (8 versus 31 ng/ml), but both tests were comparable for the detection of group C polysaccharide (12 ng/ml each) (Olcen, 1978).

D. Group B Streptococcal Antigens

Only a few reports describe the detection of group B streptococcal antigens in CSF by agglutination reactions (Table V). The number of specimens tested was small, and there are marked variations among the reports. In all instances LPA was more sensitive than CIE in detecting group B streptococci. The SC test was

TABLE V

Detection of Group B Streptococci in CSF

Source of antiserum	No. tested	% positive LPA	% positive SC	% positive CIE	Reference
Author's laboratory	12 (early)	100	—	92	Edwards *et al.* (1979)
	26 (late)	54	—	42	
Center for Disease Control	6	100	—	33	Kumar and Nankervis (1980)
Burroughs Wellcome	6	50	—	33	
Center for Disease Control	7	43	—	43	Bromberger *et al.* (1980)
Phadebact (SC)	23	—	83[a]	87	Webb *et al.* (1980)
Author's laboratory (CIE) Burroughs Wellcome	10	0	0	0	Thirumoorthi and Dajani (1979)

[a] False positive reactions in 5 of 23 using group A, C, or G reagents.

comparable to CIE in the one study in which antigens were detected using these two methods (Webb *et al.*, 1980).

Edwards and associates (1979) found that both LPA and CIE were substantially more sensitive when used to test specimens obtained early in the course of meningitis. This suggests a relationship between the amount of group B streptococcal antigen present and the degree of positivity of these tests. Beyond 13 days of treatment none of 12 specimens had detectable antigen. In this report the quantity of antigen detected by CIE (0.86–110.0 μg/ml) was comparable to that detected by LPA (0.86–110.0 μg/ml); $r = 0.77$.

All isolates in the report of Edwards *et al.* (1979) were serotype III. The types in the other reports were not specified. In one study (Bromberger *et al.*, 1980) CIE was more sensitive than LPA in measuring type-specific antigen. Whether the nature of the antigen, the antibody concentration, or other factors are important variables should be explored.

E. Cryptococcal Antigens*

Bloomfield and associates (1963) first described the detection of *Cryptococcus neoformans* polysaccharide in body fluids by LPA. The LPA test was reported to be superior to India ink preparations or cultures in the diagnosis of cryptococcal meningitis (Goodman *et al.*, 1971). The authors detected antigen in CSF of three patients with chronic cryptococcal meningitis who had negative cultures on several occasions. In a study at the Center for Disease Control quoted by the authors, 36 of 39 (92%) patients with culture-proven cryptococcal meningitis had positive LPA tests. In another report, 11 of 14 (79%) tests were positive (Gordon and Vedder, 1966). No false positive results were reported.

The titer of cryptococcal antigen in CSF or serum is of prognostic value. Diamond and Bennett (1974) demonstrated that patients who died of cryptococcal meningitis had significantly higher antigen levels in CSF and serum than did survivors ($p < 0.001$). A high titer at the end of therapy or a failure of the titer to decrease was associated with treatment failure.

In all of the above instances, except for *N. meningitidis* serogroup B, the antigenic determinants were polysaccharides. The detection of a particular antigen in CSF may be influenced, at least in part, by the degree of solubility of the antigen. A higher concentration of soluble antigen in CSF will result in a greater percentage of positive results regardless of the method used. In a few studies attempts were made to increase the amount of antigen in body fluids. Ethanol precipitation at subzero temperature with the addition of albumin as an antigen coprecipitant resulted in a 5- to 20-fold concentration of CSF specimens (Doskeland and Berdal, 1980). The degree of concentration depended on the volume of

*See also Chapter 14, this volume.

available CSF (0.5–2.0 ml). The application of this procedure to clinical specimens may increase the yield of positive results and deserves exploration.

Heating CSF specimens was also assessed in some studies. Whittle *et al.* (1974) could not detect serogroup A meningococcal antigen by LPA after CSF had been heated to 100°C for 5 min. Such treatment did not influence the results of CIE testing. The antigen was not affected by heating to 65°C. The detection of pneumococcal and *H. influenzae* antigens was not affected by heating CSF specimens (Whittle *et al.*, 1974; Dirks-Go and Zanen, 1978; Doskeland and Berdal, 1980).

III. ANTIBODY DETERMINANTS

There is little doubt that the sensitivity of agglutination reactions, as well as CIE, is directly related to the quality of antiserum used. Dirks-Go and Zanen (1978) compared three sources of *H. influenzae* type B antisera using CIE. Whereas the commercial antiserum (Burroughs Wellcome) showed a positive reaction in 23 of 60 (38%) specimens and the serum from Statens Seruminstitut reacted in 45 of 60 (75%) instances, the best results (55 of 60, or 92% positive reactions) were obtained using an antiserum from the Bureau of Biologics. Kumar and Nankervis (1980) detected group B streptococcal antigens in 6 of 6 CSF specimens by LPA using antiserum obtained from the Center for Disease Control; however, only 3 of 6 specimens were positive using commerical antiserum (Burroughs Wellcome). Similar experiences have been reported by others (Webb *et al.*, 1980; Siegel and McCracken, 1978). Furthermore, the results obtained by the same investigator to assess LPA, SC, and CIE were not always comparable because different antisera were used (see Tables I, II, III, and V). Such findings underscore the need to use standardized reagents to ensure the proper interpretation of results obtained by different investigators.

IV. SPECIFICITY OF AGGLUTINATION TESTS

False positive reactions and cross-reactions have been reported with both LPA and SC. Because there are differences between the two tests, they are discussed separately.

A. Latex Particle Agglutination

Almost all reports indicate that false positive results are noted with LPA. The rate of such false results is quite variable, however. Comparison among the various reports is not possible because of the lack of standardization of antibodies, the use of different sources of latex particles, a lack of appropriate controls in some studies, and a failure to distinguish between false positive results and cross-reactions. The overall rate of false positive results, based on data obtained from several reports, is 4.9% (59 of 1200 specimens), with a range of 0 to 37%. This is generally a lower rate than has been reported for false positive reactions in sera.

Nonspecific reactions and some false positive reactions have been reduced or eliminated in some instances. Newman *et al.* (1970) found that LPA produced false positive reactions in 13 of 558 control CSF specimens and that heating the specimens at 100°C for 15 min eliminated the reactions in all instances. Heating true positive specimens did not decrease the degree of positivity. Dirks-Go and Zanen (1978) also reported that nonspecific reactions did not occur when CSF specimens were heated, although the details were not presented. Ward *et al.* (1978) and Daum *et al.* (1982) were able to eliminate or reduce nonspecific reactions in sera after the addition of dithiothreitol (0.0026 *M*). Such treatment of CSF specimens has not been assessed.

Cross-reactions have also been reported with LPA. Examples of such immunologic cross-reactivity are those between *S. pneumoniae* (in CSF) and group B streptococcal antibody (Edwards *et al.*, 1979); *Streptococcus faecalis* (in ascitic fluid) and pneumococcal Omniserum (Kaldor *et al.*, 1977); and *S. pneumoniae* (in CSF) and group B meningococcal antiserum (Dirks-Go and Zanen, 1978).

B. Staphylococcal Coagglutination

False positive results also occur with SC, and the reported rates vary markedly. Whereas some reports indicated no false positive results in a total of 92 specimens tested (Suksanong and Dajani, 1977; Olcen, 1978; Burdash and West, 1982), high rates of false positive results were reported by others. Thirumoorthi and Dajani (1979) noted nonspecific agglutinations with SC in 30 of 68 (44%) CSF specimens, most of which were either bloody or contained a large quantity of protein. The addition of protein A (600 μg/ml) to CSF specimens eliminated the nonspecific agglutinations in 15 of 21 (71%) specimens. Dirks-Go and Zanen (1978) stated that "most specimens showed nonspecific reactions" (no numbers were specified) that were eliminated after heating at 100°C for 1 to 2 min but not after treatment with pure protein A solution. Olcen *et al.* (1975) detected false positive reactions in 28 of 117 (24%) CSF specimens using antimeningococcal sera. The addition of protein A (1 mg/ml) eliminated the false positive reactions in all of 10 specimens tested, without affecting true positive reactions.

Cross-reactions with SC are also known to occur. Webb and associates (1980) described a cross-reaction between a CSF containing *Streptococcus bovis* and group B streptococcal antiserum. In the same study 5 of 23 CSF specimens containing group B streptococci reacted with Phadebact reagents for groups A, C, and G streptococci in addition to reacting with group B antiserum. A possible cross-reaction was also reported between *H. influenzae, S. pneumoniae,* and *E. coli* in CSF and meningococcal antiserum (Olscen *et al.,* 1975).

REFERENCES

Bloomfield, N., Gordon, M. A., and Elmendorf, D. F., Jr. (1963). *Proc. Soc. Exp. Biol. Med.* **114,** 64–67.

Bromberger, P. I., Chandler, B., Gezon, H., and Haddow, J. E. (1980). *J. Pediatr.* (*St. Louis*) **96,** 104–106.

Burdash, N. M., and West, M. E. (1982). *J. Clin. Microbiol.* **15,** 391–394.

Daum, R. S., Siber, G. R., Kamon, J. S., and Russell, R. R. (1982). *Pediatrics* **69,** 466–471.

Diamond, R. D., and Bennett, J. E. (1974). *Ann Intern. Med.* **80,** 176–181.

Dirks-Go, S. I. S., and Zanen, H. C. (1978). *J. Clin. Pathol.* **31,** 1167–1171.

Doskeland, S. O., and Berdal, B. P. (1980). *J. Clin. Microbiol.* **11,** 380–384.

Edwards, M. S., Kasper, D. L., and Baker, C. J. (1979). *J. Pediatr.* (*St. Louis*) **95,** 202–205.

Goodman, J. S., Kaufman, L., and Koenig, G. (1971). *N. Engl. J. Med.* **285,** 434–436.

Gordon, M. A., and Vedder, D. K. (1966). *JAMA, J. Am. Med. Assoc.* **197,** 961–967.

Kaldor, J., Asznowicz, R., and Buist, D. G. P. (1977). *Am. J. Clin. Pathol.* **68,** 284–289.

Kaplan, S. L., and Feign, R. D. (1980). *Pediatr. Clin. North Am.* **27,** 783–802.

Kumar, A., and Nankervis, G. A. (1980). *J. Pediatr.* (*St. Louis*) **97,** 328–336.

Leinonen, M., and Herva, E. (1977). *Scand. J. Infect. Dis.* **9,** 187–191.

Newman, R. B., Stevens, R. W., and Gaafar, H. A. (1970). *J. Lab. Clin. Med.* **76,** 107–113.

Olcen, P. (1978). *Scand. J. Infect. Dis.* **10,** 283–289.

Olcen, P., Danielsson, D., and Kjellander, J. (1975). *Acta. Pathol. Microbiol. Scand. Sect. B: Microbiol.* **83,** 387–396.

Scheifele, D. W., Daum, R. S., Syriopoulou, V. P., Siber, G. R., and Smith, A. L. (1979). *Infect. Immun.* **23,** 827–831.

Scheifele, D. W., Ward, J. I., and Siber, G. R. (1981). *Pediatrics* **68,** 888–891.

Severin, W. P. J. (1972). *J. Clin. Path.* **25,** 1079–1082.

Siegel, J. D., and McCracken, G. H. (1978). *J. Pediatr.* **93,** 491–492.

Suksanong, M., and Dajani, A. S. (1977). *J. Clin. Microbiol.* **5,** 81–85.

Thirumoorthi, M. C., and Dajani, A. S. (1979). *J. Clin. Microbiol.* **9,** 28–32.

Ward, J. I., Siber, G. R., Scheifele, D. W., and Smith, D. H. (1978). *J. Pediatr.* (*St. Louis*) **93,** 37–42.

Webb, B. J., Edwards, M. S., and Baker, C. J. (1980). *J. Clin. Microbiol.* **11,** 263–265.

Whittle, H. C., Tugwell, P., Egler, L. J., and Greenwood, B. M. (1974). *Lancet* **2,** 619–621.

13

Diagnosis of Pneumonia by Agglutination Techniques

J. DONALD COONROD

Division of Infectious Diseases
Department of Medicine
University of Kentucky
and
Veterans Administration Medical Center
Lexington, Kentucky

I. Introduction 153
II. Coagglutination in Pneumonia 154
III. Comparison of Coagglutination and Counterimmuno-electrophoresis 155
IV. Specificity 157
References 157

I. INTRODUCTION

A number of studies of antigen detection in bacterial pneumonia have been carried out with counterimmunoelectrophoresis (CIE) (Coonrod and Rytel, 1973; Perlino and Shulman, 1976; Leach and Coonrod, 1977; Congeni and Nankervis, 1978; Miller *et al.*, 1978; Schmid *et al.*, 1979; Trollfors *et al.*, 1979; Downes and Ellner, 1979). Those studies are reviewed in Chapter 7 of this volume. The results of the few studies on etiologic diagnosis of pneumonia with agglutination techniques are reviewed here. In general, agglutination techniques (including coagglutination) require much less antibody and are simpler to perform than precipitin methods. Because of the ease with which agglutination methods can be adapted to the clinical laboratory, clinical studies of the usefulness of these methods for diagnosis of bacterial pneumonia appear to be worthwhile.

The Direct Detection of Microorganisms in Clinical Samples

ISBN 0-12-187780-9

II. COAGGLUTINATION IN PNEUMONIA

Theoretically, agglutination techniques could be applied to the detection of a variety of bacterial antigens in essentially any body fluid. Thus far, studies of pneumonia have been limited largely to the detection of pneumococcal antigens in sputum and serum. The coagglutination test is the most promising method that has been evaluated. In the coagglutination test the desired antibody is attached via its Fc portion to the protein A component of the staphylococcal cell wall. The specific Fab portion is then available for reaction with antigen. A procedure for detecting pneumococcal antigens in human body fluids by coagglutination has been developed by Edwards *et al.* (1980). A general problem with coagglutination is that sensitized or unsensitized staphylococci are occasionally agglutinated by normal human serum or sputum. This reaction is thought to be due to nonspecific reactivity of the organisms with IgG in body fluids. Edwards *et al.* (1980) showed that it could be eliminated by heating serum or sputum at 65°C for 5 mins. Absorption of body fluids with unsensitized staphylococci is also effective in preventing nonspecific reactions, but it is much more cumbersome than heating. Because capsular polysaccharides are heat-stable there is no loss of pneumococcal antigen reactivity during heating. Guidelines for preparing clinical samples for the detection of pneumococcal polysaccharides by coagglutination are listed in Fig. 1.

To prepare staphylococci that are coated with antibodies to pneumococci, polyvalent, rabbit antipneumococcal serum (Omniserum, Copenhagen, Den-

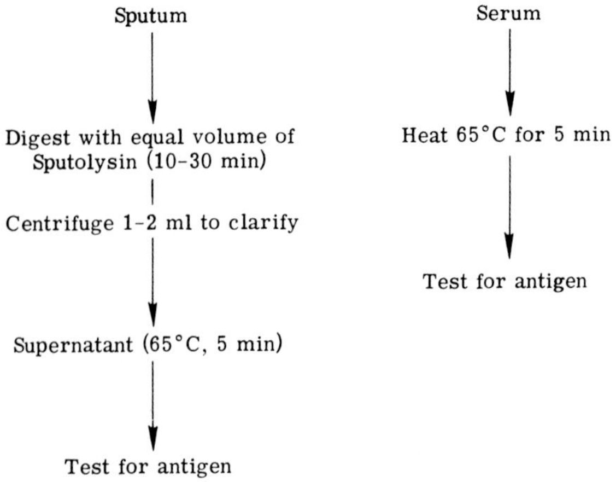

Fig. 1. Preparation of sputum and serum for the detection of pneumococcal antigen. (Published with permission from *Military Medicine*. E. A. Edwards, M. E. Kilpatrick, and D. Hooper, *Mil. Med.* **145,** 256–258, 1980.)

mark) is incubated with killed, washed *Staphylococcus aureus* organisms that have been prepared by standard techniques (Kronvall, 1973; Edwards *et al.*, 1980). The antigen test is performed by mixing a drop (0.025 ml) of sensitized or unsensitized staphylococci (as a control) with a drop of the clinical sample. The reactants are mixed for 5 min at room temperature before reading, although most positive reactions are evident by 2 min. The control detects samples that have interfering antistaphylococcal antibodies or residual Fc reactivity. Uncommonly, specific antibodies to *Staphylococcus aureus* are present in sufficient quantity to require absorption of the sample with unsensitized staphylococci before coagglutination can be performed successfully.

III. COMPARISON OF COAGGLUTINATION AND COUNTERIMMUNOELECTROPHORESIS

Coagglutination for the detection of pneumococcal capsular polysaccharides in sputum or serum has given results that are very comparable to the results observed with CIE. With serum, neither CIE nor coagglutination has proved to be sufficiently sensitive for routine use in the diagnosis of pneumococcal pneumonia (see Chapter 9, this volume, for a discussion of the problems of precipitin methods in detecting antigen in serum). However, the results with sputum have been more promising. Edwards and Coonrod (1980) carried out clinical studies of the efficacy of coagglutination with sputum samples from patients with proven (bacteremic) pneumococcal pneumonia (10 cases) or putative pneumococcal pneumonia (predominance of gram-positive cocci on sputum smears and a clinical response within 72 hr to narrow-spectrum antibiotic therapy; 34 cases). Pneumococcal antigen was detected by coagglutination in the sputum of 37 of the 44 patients (84%) when sputum was obtained early in the course of hospitalization (i.e., $<$12 hr of antibiotic therapy). Sputa were tested by CIE for comparison, and positive results were also obtained in 37 patients. Coagglutination and CIE gave concordant results in 35 of the 37 positive patients; 2 additional patients were positive only with CIE and 2 only with coagglutination. Thus, CIE and coagglutination had an identical, high level of sensitivity; the results of corresponding sputum cultures were not available for comparison with the antigen data in the study. Most investigators have found that sputum cultures are less sensitive and/or less specific than CIE for the diagnosis of pneumococcal pneumonia (Perlino and Shulman, 1976; Leach and Coonrod, 1977; Congeni and Nakervis, 1978; Downes and Ellner, 1979), but the frequency of isolation of pneumococci has varied widely in different studies, depending on culture conditions. In at least one study standard techniques of Gram's stain and culture were equivalent to CIE for detecting pneumococci in sputum (Schmid *et al.*, 1979). Certainly, one potential advantage of antigen detection methods over culture is that bacterial antigens can persist in respiratory secretions of patients who have received antibiotic therapy (Leach and Coonrod, 1977), whereas even a small amount of antibiotic therapy can render sputum cultures uninterpretable (Spencer and Philp, 1973). In the study of Edwards and Coonrod (1980) coagglutination appeared to be superior to CIE for the detection of pneumococcal antigen in the sputum of patients receiving antibiotic therapy (Fig. 2). Only 6 of 110 samples of sputum obtained after 1 to 3 days of antibiotic treatment were positive by CIE alone, whereas 23 were positive by coagglutination alone ($p < 0.01$). Differences between the efficacy of coagglutination and CIE did not appear to be related to the serotypes of pneumococci encountered in the study. There was only one patient with type 7 and none with type 14, which would have been missed with CIE.

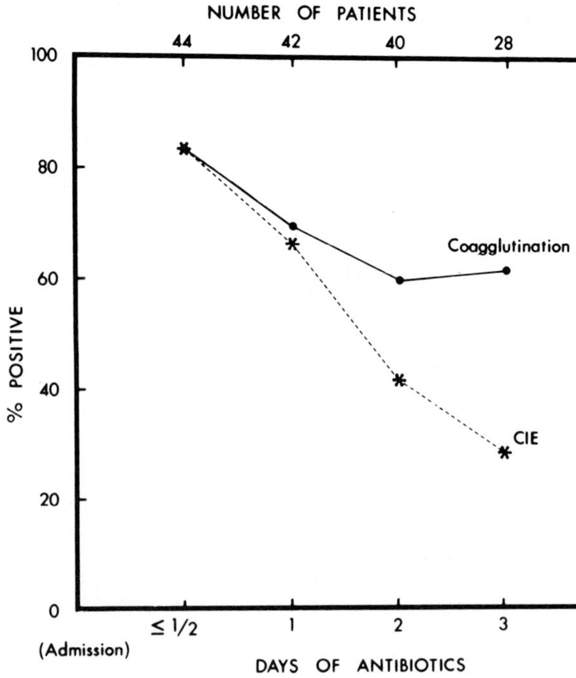

Fig. 2. Detection of pneumococcal antigens in sputum of patients with proven or probable pneumococcal pneumonia during antibiotic therapy. The difference between the results of coagglutination and of CIE was statistically significant at day 3 ($p < 0.05$). (Published with permission of the *Journal of Clinical Microbiology*, from E. A. Edwards and J. D. Coonrod, *J. Clin. Microbiol.* **11,** 488–491, 1980.)

IV. SPECIFICITY

A concern about coagglutination for the detection of pneumococcal antigens in sputum is the frequency with which nonspecific reactions occur. Edwards and Coonrod (1980) noted that sputum from 2 of 11 patients (18%) with proven nonpneumococcal pneumonia gave a positive coagglutination test for pneumococcal antigen. The specificity of coagglutination for pneumococcal antigens in the sputum of patients with other kinds of bacterial pneumonia should obviously be evaluated in a larger clinical study. This has been a difficult point to study because of the relative scarcity of cases of proven (bacteremic) nonpneumococcal pneumonia.

There is no question that coagglutination, like CIE, gives positive tests for pneumococcal antigen in patients who have culturable pneumococci in the sputum with chronic bronchitis and no pneumonia. We observed that 6 of 8 patients (75%) without pneumonia but with chronic bronchitis and pneumococci isolated from the sputum gave a positive coagglutination test (Edwards and Coonrod, 1980). Bronchitic patients in whom pneumococci cannot be cultured from the sputum give positive coagglutination tests much less frequently (3 of 15, or 20%, in our studies). Hence a positive coagglutination test for pneumococcal antigen in the sputum correlates with culture results and does not in itself prove lower respiratory infection with pneumococci; the same problem occurs with CIE of sputum (Leach and Coonrod, 1977). Because patients with chronic bronchitis often develop bacterial pneumonias, improved diagnostic tests are especially important, and neither CIE nor coagglutination, as currently performed, appears to be suitable. However, the use of a more quantitative approach with titration of pneumococcal antigens in sputum might help to resolve the problems of etiologic diagnosis of pneumonia in patients with bronchitis.

REFERENCES

Congeni, B. L., and Nankervis, G. A. (1978). *Am. J. Dis. Child.* **132,** 684–687.

Coonrod, J. D., and Rytel, M. W. (1973). *J. Lab. Clin. Med.* **81,** 778–786.

Downes, B. A., and Ellner, P. D. (1979). *J. Clin. Microbiol.* **10,** 662–665.

Edwards, E. A., and Coonrod, J. D. (1980). *J. Clin. Microbiol.* **11,** 488–491.

Edwards, E. A., Kilpatrick, M. E., and Hooper, D. (1980). *Mil. Med.* **145,** 256–258.

Kronvall, G. (1973). *J. Med. Microbiol.* **6,** 187–190.

Leach, R. F., and Coonrod, J. D. (1977). *Am. Rev. Respir. Dis.* **116,** 847–851.

Miller, J., Sande, M. A., Gwaltney, J. M., Jr., and Hendley, J. O. (1978). *J. Clin. Microbiol.* **7,** 459–462.

Perlino, C. A., and Shulman, J. A. (1976). *J. Lab. Clin. Med.* **87,** 496–502.

Schmid, R. E., Anhalt, J. P., Wold, A. D., Keys, T. F. and Washington, J. A., II (1979). *Am. Rev. Respir. Dis.* **119,** 345–348.

Spencer, R. C., and Philp, J. R. (1973). *Lancet* **2,** 349–350.

Trollfors, B., Berntsson, E., Elgefors, B., and Kaijser, B. (1979). *Scand. J. Infect. Dis.* **11,** 31–34.

14

Immunoassays in Meningitis

LORRY G. RUBIN AND E. RICHARD MOXON

Department of Pediatrics
The Johns Hopkins University School of Medicine
Baltimore, Maryland

I. Introduction 159
II. Problems in the Diagnosis of Meningitis 160
III. Detection of Bacterial Antigens in Meningitis 161
A. Background 161
B. Counterimmunoelectrophoresis 163
C. Latex Agglutination 163
D. Staphlococcal Coagglutination 164
E. Enzyme-Linked Immunosorbent Assay 166
IV. Fungal Meningitis: *Cryptococcus neoformans* 168
A. Background 168
B. Sensitivity of Latex Agglutination 169
C. Specificity 169
D. Assessment 170
V. Conclusions and Future Prospects 170
References 172

I. INTRODUCTION*

Meningitis is one of the most serious infections encountered in clinical practice. It may be fatal or cause permanent central nervous system damage even when optimal diagnostic and therapeutic modalities are used. Although virtually any bacterium or fungus has the potential to cause meningitis, relatively few account for the majority (Table I). Although primary prevention through immunization would be the ideal strategy for preventing the morbidity and mortality caused by meningitis and has proved successful against, for example, tuber-

*See also Chapters 6 and 12, this volume.

The Direct Detection of Microorganisms in Clinical Samples

ISBN 0-12-187780-9

TABLE I

Common Etiologic Agents of Bacterial and Fungal Meningitis

Bacterium	Fungus
Haemophilus influenzae type b	*Cryptococcus neoformans*
Streptococcus agalactiae (group B *Streptococcus*)	
Neisseria meningiditis	
Streptococcus pneumoniae	
Escherischia coli K1	
Listeria monocytogenes	

culosis and *Neisseria meningitidis* serogroup A and C meningitis, at present most cases of bacterial or fungal meningitis cannot be prevented by vaccination.

Secondary prevention would involve identification and intervention on behalf of individuals at high risk for the development of meningitis. The pathogenesis of meningitis involves either bacteremic or contiguous spread of organisms to the meninges. Thus, techniques for identifying patients with bacteremic or with focal infections contiguous with the central nervous system may allow early treatment of these patients and prevention of meningitis. For example, 6-month- to 2-year-old febrile children with leukocytosis are statistically at high risk for bacteremia with encapsulated bacteria such as *Streptococcus pneumoniae, Haemophilus influenzae,* or *Neisseria meningitidis,* and these bacteremic children have an increased risk of developing meningitis. The next best strategy is for the clinician to identify individuals at the earliest possible time in the evolution of meningeal infection. Common sense and clinical experience argue that a favorable outcome is determined largely by the speed with which appropriate treatment is initiated, and this in turn is influenced by the rapidity and accuracy with which the microbial agent is identified.

II. PROBLEMS IN THE DIAGNOSIS OF MENINGITIS

It may be helpful to summarize some of the problems that face the clinician who encounters individuals with suspected or established meningitis so that the overall contribution of pertinent laboratory tests can be placed in some perspective. Once a lumbar puncture has been performed and cerebrospinal fluid (CSF) obtained, the total leukocyte count and differential provides immediate and important information. Smears (e.g., Gram's stain, methylene blue stain, India ink preparations), glucose, and protein concentrations should also be obtained routinely. Thus, within a short time, meningitis (CSF pleocytosis and/or the presence of visible microorganisms) is found to be either present or absent. Only rarely does a microorganism grow from otherwise normal CSF. If there is evidence of meningitis, an immediate problem for the clinician is deciding whether to treat with an antimicrobial agent and, if so, which one. In cases in which Gram's stain unambiguously reveals bacterial forms and polymorphonuclear leukocytes in large numbers (>1000/mm^3), the diagnosis of bacterial meningitis is virtually certain. However, there is an overlap between the range of CSF findings in bacterial and nonbacterial meningitis, so this distinction is sometimes difficult. In these instances, a diagnostic limitation is imposed when standard methods of culturing microorganisms are used because many hours or days may be required for the growth and identification of the causative agent. Prior antibiotic treatment may often result in a negative Gram's stain and a sterile culture, making it difficult to distinguish between bacterial and nonbacterial meningitis. Furthermore, bacterial cultures may be sterile in 11% of cases of purulent meningitis (Geiseler *et al.*, 1980). Thus, some important characteristics of ancillary laboratory tests that make them maximally beneficial in the diagnosis of meningitis are as follows:

1. The test must be rapid; ideally, results should be available within minutes.
2. The test should be as sensitive as possible so that an etiologic diagnosis can be made despite prior antibiotic therapy.
3. Specificity is important so that a clear-cut distinction between bacterial and nonbacterial meningitis can be made and, if appropriate, specific antibiotics can be administered. The ancillary test(s) in conjunction with routine examinations of CSF should ideally allow precise diagnosis in close to 100% of cases.
4. The test should be economical, should not require specialized equipment, and should be commercially available.

An additional feature of these tests is that they may allow monitoring of the progress of treatment, predicting relapse and permitting judgments to be made about the optimal duration of treatment. Finally, such tests may singly or in combination permit the identification of patients who are at risk for acute or long-term complications (Feldman, 1977).

III. DETECTION OF BACTERIAL ANTIGENS IN MENINGITIS

A. Background

The observation that forms the basis for antigen detection in body fluids is that bacteria and fungi may elaborate microbial antigens, which may be detected by specific antibody. The most commonly used antigen has been the capsular polysaccharide of the microorganism. It is important to know the lowest concentration of antigen that can be detected (sensitivity) and the extent to which this antigen can be differentiated from other antigens that may also be present in the CSF (specificity).

Sensitivity is dependent on both the antibody used and the assay methodology. Antibodies, whether monoclonal or polyclonal, vary in their affinity as well as in the microbial epitope or determinants they recognize. However, for any particular assay method a limit exists below which no increased sensitivity can be achieved by varying antibody concentrations or other assay variables. This lower limit is substantially different for the various assays. Thus, in detecting the capsule of *Haemophilus influenzae* type b (polyribosylribitol phosphate) the sensitivities are as follows:

Assay method	Maximum sensitivity (ng/ml)
Precipitin	100
Counterimmunoelectrophoresis	2.5
Latex agglutination	0.2
Radioimmunoassay (inhibition assay)	0.5
Enzyme immunoassay	0.1

All of the assays currently available are insufficiently sensitive to detect antigen in all cases (sensitivity less than 100%). Therefore, it is important to use the body fluid with the greatest concentration of antigen. Although it is often CSF in the case of patients with meningitis, it is not invariably so, because polysaccharide antigens are excreted in the urine and a large volume of urine may be collected and concentrated severalfold.

Another point to be emphasized is that the amount of antigen may vary at different stages in the evolution of the infection, and therefore samples taken too early or too late may have a nondetectable concentration of antigen. Ideally, one should test sequential samples. It is also important to recognize that the concentration of bacterial antigen correlates only roughly with the concentration of bacteria. Thus, it has been well documented that organisms may be visible on Gram's or methylene blue stains of CSF, although antigen may not be detected in the same sample of CSF (Olcen, 1978). An important lesson from such observations is that the older, well-established routine tests should not be abandoned for these newer, ancillary assays.

Problems in using conventional antisera are that they are not highly specific and there may be lot-to-lot variation, resulting in poor reproducibility. The molecular similarity of several antigens has resulted in cross-reactivity in several studies (Lampe *et al.*, 1976; Ribner and Keusch, 1975; Shackelford *et al.*, 1974). Examples include reactions between the capsular antigens of *Haemophilus influenzae* type b and *Streptococcus pneumoniae* types 6, 15A, 20, and 35B, *Staphylococcus aureus, Staphylococcus epidermidis,* group A *Strep-*

tococcus, certain strains of *Escherichia coli, Streptococcus vividans, Bacillus* and diphtheroids. Polyvalent antiserum for use in *Streptococcus pneumoniae* detection has been found to cross-react with the capsular antigens of *E. coli, Proteus, Aerobacter aerogenes, Klebsiella pneumoniae, H. influenzae* type b, *Salmonella*, and *Staphylococcus aureus*. Antibodies for the detection of *Neisseria meningitidis* group B may cross-react with a capsule of *E. coli* K1. These problems may eventually be overcome, in part or in whole, through the use of monoclonal antibodies. Monoclonal antibodies would provide unlimited supplies of standardized rcagents and have the potential to lead to an increase in the sensitivity and specificity of enzyme-linked immunosorbent assays (ELISA). However, the lower affinity of many of these immunized animals and the single antigen combining site may limit the utility of these reagents for both agglutination assays and ELISAs (Yolken, 1982).

B. Counterimmunoelectrophoresis

Several problems have emerged from the decade or so of experience in the clinical application of counterimmunoelectrophoresis (CIE) to the diagnosis of bacterial meningitis. In a review summarizing several years of experience in the clinical use of CIE for the diagnosis of bacterial meningitis, only 13% of patients diagnosed by CIE had a negative Gram's stain. Furthermore, the residual diagnostic value after the culture report was received was only 9%. These authors concluded that CIE did help in culture-negative patients but that it was of limited sensitivity (Naiman and Albritton, 1980). Because other techniques (e.g., latex agglutination) have proved to be more sensitive with little, if any, loss of specificity, it would seem that CIE is no longer the assay of choice for the rapid diagnosis of bacterial meningitis, and its use for this purpose in microbiology laboratories is likely to decline. A summary of the sensitivity of various methods for detecting microbial antigen in meningitis is given in Tables II–V.

C. Latex Agglutination

Latex agglutination has proved to be a sensitive and highly specific technique for detecting antigen in the etiologic diagnosis of bacterial meningitis when measures are taken to eliminate nonspecific agglutination. Although more sensitive than CIE (Table III), the utility of the test in partially treated bacterial meningitis has not been evaluated. Definitive advantages of this technique are that the test does not require any special equipment, and kits are commercially available for the detection of group b *Streptococcus* (Wellcogen Strep B, Wellcome Diagnostics), *S. pneumoniae* (Directigen, Hynson, Westcott and Durning), *N. meningitidis* (groups A, B, C, Y—Bactogen, Wampole Laboratories; groups A, C—Directigen), and *H. influenzae* type b (Bactogen, Directigen). The use of monoclonal antibodies may provide standardized reagents with equivalent sensitivity and perhaps improved specificity.

D. Staphlococcal Coagglutination

Staphylococcal coagglutination is also a rapid assay that does not require special equipment and is more sensitive than CIE (Table IV). The commercial availability of coagglutination kits (Phadebact ®—*S. pneumoniae, H. influenzae* type b, *N. meningitidis* A, B, C, Y, W135) for antigen detection will allow comparative testing in many institutions using the same reagents and will allow small-volume laboratories to make antigen detection available. False positive reactions are rare, but care must be taken to run appropriate controls in order to avoid misinterpretation of nonspecific agglutination as a positive result. For practical purposes CSF specimens must be pretreated (heated and centrifuged

TABLE II

Sensitivity of CIE in Culture-Proven Bacterial Meningitis

Microorganism	Body fluid[a]	% antigen detected	References
Haemophilus influenzae	CSF, S, CU	76–100	Coonrod and Rytel (1972); Feigin *et al.* (1976); Ingram *et al.* (1972); Ward *et al.* (1978); Whittle *et al.* (1974)
Neisseria meningitidis A	CSF	69	Greenwood and Whittle (1974); Whittle *et al.* (1975)
N. meningitidis	CSF, S, CU	50–100	Coonrod and Rytel (1972); Hoffman and Edwards (1972); Shackelford *et al.* (1974); Spencer and Savage (1975); Whittle *et al.* (1974)
Streptococcus pneumoniae	CSF	40–100	Coonrod and Rytel (1972); Fossieck *et al.* (1973)
Escherichia coli K1	CSF	71	McCracken *et al.* (1974); Robbins *et al.* (1974)
Streptococcus agalactiae	CSF	70–100	Hill *et al.* (1975); Shackleford and Stechenberg (1977); Ingram *et al.* (1978, 1980)

[a] Abbreviations: CSF, cerebrospinal fluid; S, serum; CU, concentrated urine.

TABLE III

Sensitivity of Latex Agglutination for the Detection of Bacterial Antigen in Culture-Proven Bacterial Meningitis

Etiologic agent	Body fluid	% antigen detected	References
Haemophius influenzae type b	CSF	81, 90–100	Daum *et al.* (1982)
	Urine	100	Ingram *et al.* (1975)
	Serum	88	Newman *et al.* (1970); Shaw *et al.* (1982); Ward *et al.* (1978)
Streptococcus pneumoniae	CSF	82	Whittle *et al.* (1974)
Neisseria meningitidis			
Unspecified group	CSF	88	Whittle *et al.* (1974)
Group A	CSF	91	Severin (1972)
Group C	CSF	100	Severin (1972)
Streptococcus agalactiae (group B *Streptococcus*)	CSF	79–100	Bromberger *et al.* (1980)
	Urine	93–100	Webb *et al.* (1980)
	Serum	25	

TABLE IV

Sensitivity of Coagglutination for the Detection of Bacterial Antigen in CSF in Culture-Proven Bacterial Meningitis

Etiologic agent	Sensitivity (%)	Reference
Haemophilus influenzae type b	88	Thirumoorthi and Dajani (1979); Olcen (1978); Welch and Hensel (1982) Suksanong and Dajani (1977)
Streptococcus pneumoniae	66	Thirumoorthi and Dajani (1979); Burdash and West (1982); Olcen (1978)
Nelsseria meningitidis	40	Olcen (1978); Thirumoorthi and Dajani (1979); Olcen *et al.* (1975)
Streptococcus agalactiae (group B *Streptococcus*)	0, 83	Olcen (1978); Webb and Baker (1980)

TABLE V

Sensitivity of ELISA for the Detection of Bacterial Antigen in Culture-Proven Bacterial Meningitis

Etiologic agent	Body fluid	% antigen detected	References
Haemophilus influenzae type b	CSF	96	Pepple *et al.* (1980); Drow *et al.* (1979); Wetherall *et al.* (1980); Beuvary *et al.* (1979)
	Serum	100	Pepple *et al.* (1980); Drow *et al.* (1979)
	Urine	100	Pepple *et al.* (1980); Drow *et al.* (1979)
Streptococcus pneumoniae	CSF	100	Beuvary *et al.* (1979)
Neisseria meningitidis	CSF	88	Beuvary *et al.* (1979)
Streptococcus agalactiae	CSF	100	Polin and Kennett (1980)

and/or incubated with soluble staphylococcal protein A) before assaying, and some CSF specimens will nevertheless not be testable due to nonspecific agglutination.

E. Enzyme-Linked Immunosorbent Assay

Enzyme-linked immunosorbent assay is a sensitive solid-phase immunoassay system that can be used to detect and quantify antigen. It is similar in design and sensitivity to radioimmunoassay except that an enzyme is used in place of a radioactive isotope as the immunoglobulin marker. The clinical applications of ELISA to antigen detection have been largely for protein hormones, drugs, and certain viral infections including hepatitis B virus and rotavirus. Only relatively recently has this technology been applied to the detection of bacterial antigens.

Of the various ELISA methods (competition, double-antibody sandwich, inhibition), that used most commonly for the detection of bacterial antigens is the double-antibody sandwich method. In this assay antibody is bound to a solid phase such as a microtiter plate. Following the removal of unbound antibody, the solid phase can be stored until use. The clinical specimen is then mixed in buffer and added to the solid phase. After incubation and washing, an enzyme-labeled antibody is added (direct test) followed by another incubation and washing, after which the amount of bound enzyme-labeled antibody is determined by the addition of an enzyme substrate. The ensuing substrate–enzyme reaction is measured by quantification of substrate products, which is measured by spectrophotometry or visualization, and the amount is proportional to the amount of enzyme bound to the solid phase in the above reaction. The quantity of antigen is determined by comparing the reactivity of the test specimen with that of positive and negative control specimens.

1. *Sensitivity*

ELISA systems have been devised and tested for the detection of the polysaccharide capsular antigens of *Haemophilus influenzae* type b, *Streptotococcus pneumoniae, Neisseria meningitidis,* and *Streptococcus agalactiae* (group b *Streptococcus*) in body fluids in patients with meningitis. The maximum sensitivity, assessed using purified polysaccharide capsular antigens, ranges from 0.1 to 2 ng/ml, which is 1–2 logs more sensitive than CIE. As seen in Table V, ELISA performed on CSF from culture-proven cases of bacterial meningitis were positive in 96% of cases of *H. influenzae* type b meningitis, 88% of cases of *N. meningitidis,* and 100% of cases of *S. pneumoniae* and *S. agalactiae* types III tested. In cases of *H. influenzae* type b meningitis, ELISA has detected antigen in CSF specimens that were negative by latex agglutination and coagglutination. Pepple *et al.* (1980) detected *H. influenzae* type b antigen in 11 of 15 CSF specimens obtained 1–14 days following the institution of antibiotic therapy for documented *H. influenzae* type b meningitis. Thus, this assay may have utility in establishing an etiologic diagnosis in cases of partially treated meningitis when CSF cultures are often negative. Body fluids other than CSF may be helpful in the diagnosis of meningitis. In culture-proven *H. influenzae* type b meningitis, *H. influenzae* type b capsular antigen was detected in 100% of serum and urine specimens tested (Table V).

2. *Specificity*

The specificity of ELISA in the studies listed in Table V has been excellent. Assays were negative in 151 CSF specimens, 21 serum specimens, and 44 urine specimens from normal children and children infected with organisms other than those being tested. In several instances even the presence of cross-reacting antigens, for example, *Escherichia coli* K100 in the assay for *H. influenzae* type b antigen, failed to give positive reactions in the double-antibody sandwich assay (Drow *et al.*, 1979). Wetherall *et al.* (1980) found excessively high background and nonspecific readings when testing serum for *H. influenzae* type b antigen but did not use diluents such as fetal calf serum or *N*-acetylcysteine, which serve to minimize nonspecific reactions (Pepple *et al.*, 1980). The use of techniques to minimize the background and nonspecific reactions is critical for the maintenance of specificity without sacrificing the sensitivity of the assay. To avoid misinterpretation of nonspecific reactions, each specimen should be incubated with control wells (those containing nonimmune serum) in addition to the test wells containing hyperimmune serum.

3. Role in Diagonsis of Meningitis

Because ELISA has greater sensitivity than other antigen detection techniques (CIE, latex agglutination, and coagglutination), this technique has the greatest potential for establishing a diagnosis in ambiguous cases, such as partially treated meningitis, when cultures are negative and concentrations of antigen may be low. ELISA techniques can be adapted for the assay of single specimens (using polystyrene balls as the solid phase) or multiple specimens for the simultaneous detection of one or more antigens (using microtiter plates as the solid phase).

The major disadvantage of this technique is the relatively long time required (minimum of 4 hr) to complete the assay. For the detection of very low antigen concentrations, visualization of a color change is not sufficient, and small increments in optimal density must be assessed using a spectrophotometer. Yolken and Leister (1981) developed a rapid (30 min) assay system for the detection of *H influenzae* type b antigen by the use of short incubation times and/or simultaneous addition of test antigen and conjugate, but the system was 10-fold less sensitive than an assay run in the standard fashion.

Future prospects include the development of "homogeneous" ELISA systems, in which unreacted antibody does not have to be separated from antigen–antibody complex (Yolken, 1982). This assay system would have the advantages of rapidity (less than 30 min, because no washing steps are required) and simplicity because it could be performed without equipment other than test tubes and droppers.

At present, ELISAs for the diagnosis of meningitis are reserved for culture-negative cases in which CIE, latex agglutination, or coagglutination gives negative or nonspecific results.

IV. FUNGAL MENINGITIS: *CRYPTOCOCCUS NEOFORMANS**

A. Background

Immunoassays for antigen detection are of value in the diagnosis of fungal meningitis, specifically cryptococcal meningitis. Although the strict criterion for diagnosing cryptococcal meningitis is the culture of *Cryptococcus neoformans* from the CSF, cultivation may be slow or even negative in the face of documented infection (Goodman *et al.*, 1971; Snow and Dismukes, 1975). Although the CSF often has a modest lymphocyte-predominant pleocytosis and an elevated protein concentration and may have a depressed glucose concentration, these findings are nonspecific. The examination of India ink preparations of CSF and

*See also Chapter 12, this volume.

assays for serum antibody may be helpful but are insufficiently sensitive and specific to either verify or rule out the diagnosis.

Assays for antigen detection in serum and CSF have been a major advance in the diagnosis and management of patients with cryptococcal meningitis. Similar to the encapsulated bacteria described earlier in the chapter, *C. neoformans* contains a polysaccharide capsular antigen that elicits high-titer antibody in rabbits and may be present in soluble form in the serum and CSF of infected patients. Although CIE and ELISA systems for antigen detection have been described (Maccani, 1977; Scott *et al.*, 1980), the oldest and most widely employed technique is latex agglutination.

B. Sensitivity of Latex Agglutination

Latex agglutination is capable of detecting purified cryptococcal polysaccharide antigen at a concentration of 35 ng/ml (Scott *et al.*, 1980). It is of greater sensitivity than India ink preparations for CSF. The sensitivity is similar to the complement fixation assay for antigen detection but is easier to perform and can be performed on specimens with anticomplementary activity. The sensitivity of this test when performed on CSF of culture-proven cases has been consistently >90% and often higher (Kauffman *et al.*, 1981; Snow and Dismukes, 1975). The test has been positive in several culture-negative cases that were eventually proven to be cryptococcal meningitis (Snow and Dismukes, 1975; Goodman *et al.*, 1971). False negative reactions have occasionally been reported on CSF specimens in association with a prozone phenomenon (negative reaction using undiluted specimens but positive reaction following dilution in buffer) (Bloomfield *et al.*, 1963; Stamm and Polt, 1980).

C. Specificity

Although an occasional false positive reaction that was not due to nonspecific agglutination has been reported (MacKinnon *et al.*, 1978), most false positive reactions are due to nonspecific agglutination, and many of these are due to the presence of rheumatoid factor in the patient's serum. Dolan (1972) reported that nonspecific false positive reactions comprised 15% of the positive results in CSF and 85% of the positive results in sera. This problem was overcome by heating all specimens and assaying dilutions of positive specimens against both specific antibody-sensitized latex particles and normal serum-coated, control latex particles. In cases in which there is nonspecific reaction with the control latex particles, the agglutination titer must be at least fourfold higher against the anticryptococcal antibody-sensitized latex particles than the control latex particles for the test to be considered positive. Dithiothreitol pretreatment of serum is also helpful in eliminating rheumatoid factor while exerting a minimal effect on the sensitivity of the assay (Gordon and Lepa, 1974).

D. Assessment

The latex agglutination assay is highly sensitive and specific and useful both for diagnosis and following response to therapy when performed as described by Prevost and Newell (1978). Precautions include (*a*) testing the sensitivity of the reagents with cryptococcal polysaccharide antigen each time the test is performed, (*b*) inactivating the specimen by heat, (*c*) testing the specimen against both the anticryptococcal latex and the control latex, (*d*) interpreting weak agglutination reactions as negative, (*e*) testing serial dilutions of positive specimens, and (*f*) diluting and retesting negative specimens when the clinical suspicion is high. Although an earlier kit was found to be unreliable (Kaufman *et al.*, 1974), a commercial kit for the detection of cryptococcal antigen (Crypto-LA kit, International Biological Laboratories, Rockville, Maryland) is available and useful for the diagnosis of cryptococcal meningitis.

V. CONCLUSIONS AND FUTURE PROSPECTS

Since the early 1970s there has been an explosion of interest in antigen detection for the etiologic diagnosis of meningitis. These tests are based on the detection of soluble antigens, which are generally polysaccharide components of the capsule. Conveniently, these antigens are highly stable, making diagnosis possible even when samples may have been suboptimally handled. For the diagnosis of meningitis, concentrated urine is a valuable body fluid to assay in addition to CSF and is probably of greater utility than serum for antigen detection. Of the available systems, ELISA is clearly the most sensitive and specific. From a practical standpoint latex agglutination and coagglutination are the most convenient for the detection of *Haemophilus influenzae* type b and *Streptococcus agalactiae* (group b) antigen because they are faster and more sensitive than CIE. The utility of latex agglutination and coagglutination for the diagnosis of *Streptococcus pneumoniae* and *Neisseria meningitidis* will require further study.

Future developments may substantially reduce the time required for ELISA, and this may become the antigen detection method of choice. A rapidly developing area that is now in its infancy is the detection of microbial DNA by DNA hybridization techniques. This sensitive assay may be of particular utility for the detection of microorganisms such as *Streptococcus pneumoniae,* in which a large number of serotypes makes antigen detection difficult, or *Mycobacterium tuberculosis,* in which suitable antigens have not been identified. The detection of activity of bacterial enzymes that are not produced by host cells, for example, β-lactamase activity, is another nonantigenic test for the detection of bacteria. This technique may also provide information regarding antibiotic susceptibility (i.e., penicillin susceptibility) (Yolken and Hughes, 1980) and help the clinician in his choice of antibiotics.

There are major problems in the use of the current literature on antigen detection tests to evaluate the actual utility of antigen detection tests in clinical decisions. In order to ensure the validity of these studies, only patients who satisfied strict criteria for meningitis (positive CSF cultures) were admitted to the study. This tends to include cases that were obviously bacterial in origin and represents those cases in which the clinician did not require assistance in the diagnosis of bacterial meningitis. Similarly, the criteria for study eliminate the difficult cases in which there were equivocal CSF findings and in which patients may have received prior antibiotic treatments, which represent the cases in which ancillary techniques such as antigen detection were most needed. We now need prospective studies on the utility of antigen detection in clinical decision making with emphasis on cases in which the differentiation of bacterial and nonbacterial etiologies is difficult. Also, there have been no major studies using optimal antigen detection techniques aimed at the identification of patients statistically at high risk for the development of meningitis.

When the differentation between a bacterial and nonbacterial meningitis is the critical clinical issue, the large number of bacterial species that may be implicated precludes resolving the issue by a single assay that is specific for a particular bacterial species. An alternative approach is the use of a highly sensitive assay with reactivity directed against determinants from all bacterial causes of meningitis or the detection of a host response unique to bacteria or fungi (e.g., in the manner in which the *Limulus* lysate test detects almost all cases of gram-negative meningitis) (Ross *et al.*, 1975; Berman *et al.*, 1976). However, even this test is insufficiently broad in this reactivity because antibiotics cannot safely be withheld when the tests are negative due to the possibility of meningitis caused by gram-positive bacteria. C-Reactive protein determination is the prototype for tests that might recognize a host response unique to bacterial meningitis. In a prospective study, C-reactive protein was detected in the CSF from 24 of 24 patients with bacterial meningitis and 2 of 32 patients with nonbacterial meningitis (Corrall *et al.*, 1981).

The best approach to differentiating bacterial from nonbacterial meningitis may involve the use of a battery of tests that taken together predict virtually every case of bacterial meningitis. Using both standard tests on CSF (polymorphonuclear leukocyte, glucose, and protein concentrations) and ancillary tests on CSF (C-reactive protein, *Limulus* lysate reactivity, lactate, lactate dehydrogenase, and SGOT concentrations), C. J. Corrall, J. M. Pepple, J. R. Levin, W. T. Hughes, and E. R. Moxon (unpublished data) developed a scoring system that predicted the presence of culture-proven bacterial meningitis in

100% of cases and had a specificity of 62%. Thus, all patients with bacterial meningitis would have been started on antibiotic therapy, and the use of antibiotics would have been prevented in 62% of the patients with nonbacterial meningitis. Using CSF obtained 48–72 hr after antimicrobial therapy was instituted against bacterial meningitis as a model for partial treatment, they found that 100% of patients were placed in the bacterial group as a result of the battery of tests on CSF. This type of approach deserves further study.

REFERENCES

Berman, N. S., Siegel, S. E., Nachum, R., Lipsey, A., and Leedom, J. (1976). *J. Pediatr.* (*St. Louis*) **88,** 553–556.

Beuvary, E. C., vanRossum, F., Lauwers, S., and Coignac, H. (1979). *Lancet* **1,** 208.

Bloomfield, N., Gordon, M. A., and Elmendorf, D. F. (1963). *Proc. Soc. Exp. Biol.* **114,** 64–67.

Bromberger, P. I., Chandler, G., Gezon, H., and Haddow, J. E. (1980). *J. Pediatr.* (*St. Louis*) **96,** 104–106.

Burdash, N. M., and West, M. E. (1982). *J. Clin. Microbiol.* **15,** 391–394.

Coonrod, J. D., and Rytel, M. W. (1972). *Lancet* **1,** 1154–1157.

Corrall, C. J., Pepple, J. M., Moxon, E. R., and Hughes, W. T. (1981). *J. Pediatr.* (*St. Louis*) **99,** 365–369.

Daum, R. S., Siber, G. R., Kamon, J. S., and Russell, R. R. (1982). *J. Pediatr.* (*St. Louis*) **69,** 466–471.

Dolan, C. T. (1972). *Am. J. Clin. Pathol.* **58,** 358–364.

Drow, D. L., Maki, D. G., and Manning, D. D. (1979). *J. Clin. Microbiol.* **10,** 442–450.

Feigin, R. D., Stechenberg, B. W., Chang, M. J., Dunkle, L. M., Wong, M. L., Palkes, H., Dodge, P. R., and Davis, H. (1976). *J. Pediatr.* (*St. Louis*) **88,** 542–548.

Feldman, W. E. (1977). *N. Engl. J. Med.* **296,** 433–435.

Fossieck, B., Jr., Craig, R., and Paterson, P. Y. (1973). *J. Infect. Dis.* **127,** 106–109.

Geiseler, P. J., Nelson, K. W., Levin, S., Reddi, K. T., and Moses, V. K. (1980). *Rev. Infect. Dis.* **2,** 725–745.

Goodman, J. S., Kaufman, L., and Koenig, M. G. (1971). *N. Engl. J. Med.* **285,** 434–436.

Gordon, M. A., and Lepa, F. W. (1974). *Am. J. Clin. Pathol.* **61,** 488–494.

Greenwood, B. M., and Whittle, H. C. (1974). *J. Infect. Dis.* **129,** 201–204.

Hill, H. R., Riter, M. E., Menge, S. K., Johnson, D. R., and Matsen, J. M. (1975). *J. Clin. Microbiol.* **1,** 188–191.

Hoffman, T. A., and Edwards, E. A. (1972). *J. Infect. Dis.* **126,** 636–644.

Ingram, D. L., Anderson, P., and Smith, D. H. (1972). *J. Pediatr.* (*St. Louis*) **81,** 1156–1158.

Ingram, D. L., O'Reilly, R. J., and Pond, P. J. (1975). *Pediatr. Res.* **9,** 341.

Ingram, D. L., Pendergrass, E. L., Thullen, J. D., and Yoder, C. D. (1978). *Pediatr. Res.* **12,** 494.

Ingram, D. L., Pendergrass, E. L., Bromberger, P. I., Thullen, J. D., Yoder, C. D., and Collier, A. M. (1980). *Am. J. Dis. Child.* **134,** 754–758.

Kauffman, C. A., Bergman, A. G., Severance, P. J., and McClatchey, K. D. (1981). *Am. J. Clin. Pathol.* **75,** 106–109.

Kaufman, L., Cowart, G., Blumer, S., Stine, A., and Ross, W. (1974). *Appl. Microbiol.* **27,** 620–621.

Kronvall, G. (1973). *J. Med. Microbiol.* **6,** 187–190.

Lampe, R., Chottipitayasunondh, T., and Sunakorn, P. (1976). *Pediatrics* **88,** 557–560.

Maccani, J. E. (1977). *Am. J. Clin. Pathol.* **68,** 39–44.
MacKinnon, S., Kane, J. G., and Parker, R. H. (1978). *J. Am. Med. Assoc.* **240,** 1982–1983.
McCracken, G. H., Glodes, M. P., Sarff, L. D., Mize, S. G., Schiffer, M. S., Robbins, J. B., Gotschlich, E. C., Orskov, I., and Orskov, F. (1974). *Lancet* **2,** 246–250.
Naiman, H. L., and Albritton, W. L. (1980). *J. Infect. Dis.* **142,** 524–531.
Newman, R. B., Stevens, R. W., and Gaafar, H. A. (1970). *J. Lab. Clin. Med.* **76,** 107–113.
Olcen, P. (1978). *Scand. J. Infect. Dis.* **10,** 285–289.
Olcen, P., Danielsson, D., and Kjellander, J. (1975). *Acta Pathol. Microbiol. Scand. Sect. B: Microbiol.* **83,** 387–396.
Pepple, J., Moxon, E. R., and Yolken, R. H. (1980). *J. Pediatr.* (*St. Louis*) **97,** 233–237.
Polin, R. A., and Kennett, (1980). *J. Pediatr.* (*St. Louis*) **97,** 540–544.
Prevost, E., and Newell, R. (1978). *J. Clin. Microbiol.* **8,** 529–533.
Ribner, B., and Keusch, G. T. (1975). *Ann. Intern. Med.* **83,** 370–371.
Robbins, J. B., McCracken, G. H., Gotschlich, E. C., Orskov, F., Orskow, I., and Hanson, L. A. (1974). *N. Engl. J. Med.* **290,** 1216–1220.
Ross, S., Rodriguez, W., Controni, G., Korengold, G., Watson, S., and Khan, W. (1975). *J. Am. Med. Assoc.* **233,** 1366–1369.
Scott, E. N., Muchmore, H. G., and Felton, F. G. (1980). *Am. J. Clin. Pathol.* **73,** 798–794.
Severin, W. P. J. (1972). *J. Clin. Pathol.* **25,** 1079–1082.
Shackleford, P. G., and Stechenberg, B. W. (1977). *Pediatr. Res.* **11,** 505.
Shackleford, P. G., Campbell, J., and Feigin, R. D. (1974). *J. Pediatr.* (*St. Louis*) **85,** 478–481.
Shaw, E. D., Darker, R. J., Feldmann, W. E., Gray, B. M., Pifer, L. L., and Scott, G. B. (1982). *J. Clin. Microbiol.* **15,** 1153–1156.
Siegel, J. D., and McCracken, G. J., Jr. (1978). *J. Pediatr.* **93,** 491–492.
Snow, R. M., and Dismukes, W. F. (1975). *Arch. Intern. Med.* **135,** 1155–1157.
Spencer, R. C., and Savage, M. A. (1975). *Lancet* **1,** 1253.
Stamm, A. M., and Polt, S. S. (1980). *J. Am. Med. Assoc.* **244,** 1359.
Suksanong, M., and Dajani, A. S. (1977). *J. Clin. Microbiol.* **5,** 81–85.
Thirumoorthi, M. C., and Dajani, A. S. (1979). *J. Clin. Microbiol.* **9,** 28–32.
Ward, J. I., Siber, G. R., Scheifele, D. W., and Smith, D. H. (1978). *J. Pediatr.* (*St. Louis*) **93,** 37–42.
Webb, B. J., and Baker, C. J. (1980). *J. Clin. Microbiol.* **12,** 442–444.
Webb, B. J., Edwards, M. S. and Baker, C. J. (1980). *J. Clin. Microbiol.* **11,** 263–265.
Welch, D. F., and Hensel, D. (1982). *J. Clin. Microbiol.* (in press).
Wetherall, B. L., Hallsworth, P. G., and McDonald, P. J. (1980). *J. Clin. Microbiol.* **11,** 573–580.
Whittle, H. C., Tugwell, P., Egler, L. J., and Greenwood, B. M. (1974). *Lancet* **2,** 619–621.
Whittle, H. C., Greenwood, B. M., Davidson, N. M., Tomkins, A., Tugwell, P., Warrell, D. A., Zalin, A., Bryceson, A. D. M., Parry, E. H. O., Brueton, M., Duggan, M., Oomen, J. M. V., and Rajkovic, A. D. (1975). *Am. J. Med.* **58,** 823–828.
Yolken, R. H. (1982). *Rev. Infect. Dis.* **4,** 35–68.
Yolken, R. H., and Hughes, W. T. (1980). *J. Pediatr.* (*St. Louis*) **97,** 715–720.
Yolken, R. H., and Leister, F. J. (1981). *J. Clin. Microbiol.* **13,** 738–741.

15

Use of Immunoassays in Bacteremia

L. JOSEPH WHEAT

Indiana University Medical School
and
Indianapolis Veterans Administration Hospital
Indianapolis, Indiana

I. Introduction 175
A. Rationale 175
B. Technical Aspects 176
II. Problems of Immunoassays for the Detection of Antigenemia 177
A. Antigen Diversity 177
B. Nonspecific Inhibitors 177
C. Specific Inhibitors (Preexisting Antibodies) 177
D. Rheumatoid Factor 180
E. Protein A Contamination 180
III. Applications to the Rapid Diagnosis of Bacteremia 181
A. Radioimmunoassay 181
B. Enzyme-Linked Immunosorbent Assay 183
References 184

I. INTRODUCTION

A. Rationale

Assays to detect microbial antigens would be very useful for the diagnosis of bacteremia. They would be particularly helpful in patients who were already taking antibiotics, because blood cultures may be falsely negative or slow to become positive. Rapid tests for detecting bacterial antigens would also be useful in patients with clinical features of acute endocarditis or septicemia or in immunosuppressed patients, in which case early initiation of appropriate antimicrobial therapy is essential. Finally, they would be useful for detecting anaerobic bacteremia because of the relatively slow growth of those organisms.

The Direct Detection of Microorganisms in Clinical Samples

ISBN 0-12-187780-9

Soluble microbial substances were first detected in the blood and urine of patients with pneumococcal pneumonia in 1917 (Dochez and Avery, 1917). Since that report immunoassays for microbial antigens have been reported for several infections. This chapter reviews the use of radioimmunoassays and enzyme-linked immunoassays for the rapid diagnosis of bacteremia.

B. Technical Aspects

Technical aspects of enzyme-linked immunosorbent assays (ELISA) and radioimmunoassays (RIA) have been discussed in greater detail elsewhere. Certain aspects that are relevant to the diagnosis of bacteremia will be reviewed briefly.

Solid-phase antigen binding assays can be divided into direct or indirect antigen binding assays (sandwich) and antigen inhibition assays. Most direct sandwich-type assays use solid-phase supports coated with antimicrobial antibodies. Antigens in the test sample adhere to the solid-phase antibodies and are then detected with enzyme or radiolabeled antibodies. Assays using antisera produced with complex microbial antigens consequently can detect multiple microbial antigens. This broad specificity may be useful if the precise identity of the circulating antigen is unknown; however, cross-reactions may occur with antigens of other microorganisms. If the circulating microbial antigen could be identified, assays using antisera to that antigen might be more specific and more sensitive. The major advantages of direct solid-phase assays are ease and speed of performance. Indirect solid-phase assays use antibodies from a second animal species to detect antigens adherent to the solid-phase antibody; this second antibody is then measured with enzyme or radiolabeled antibody to the second immunoglobulin. Indirect assays may be more sensitive than direct assays (Kozaki *et al.*, 1979; Yolken and Stopa, 1980), and the same labeled antiimmunoglobulin can be used for multiple purposes. The major disadvantages of the indirect method are the requirement for two different antimicrobial antibodies and for an additional step. Most solid-phase assays can be completed within 1 day.

Solid-phase inhibition assays have also been described. Antigens in test samples inhibit the binding of labeled antigens to solid-phase antibodies. Inhibition assays that use purified radiolabeled antigens are specific for that antigen despite broad reactivity of the solid-phase antibody, thus decreasing the problem of cross-reactivity. In another type of inhibition assay, antigens in the test sample inhibited binding of antibody to purified solid-phase antigens (Rissing *et al.*, 1981). A disadvantage of inhibition assays is the long incubation required for the inhibition step, which decreases its rapid diagnostic potential.

In liquid-phase inhibition assays, labeled and unlabeled antigens compete for specific antibodies in a solution. After antibody and antigen complexes have formed, they are removed by a variety of separation methods: precipitation with a second antibody, ammonium sulfate, or sodium sulfate or attachment to protein A-rich *Staphylococcus aureus*. Liquid-phase systems may be more sensitive than solid-phase assays, but they require more work and are slow.

II. PROBLEMS OF IMMUNOASSAYS FOR THE DETECTION OF ANTIGENEMIA

A. Antigen Diversity

Although the antigenic composition of different strains of certain organisms such as *S. aureus* may be similar, thus permitting the development of assays that detect antigens of most strains, type-specific antigens occur in many organisms. For example, assays for meningococcal (Colding and Lind, 1977), pneumococcal (Colding and Lind, 1977), and *Legionella* antigens (Kohler; see Chapter 16 of this volume) are serotype-specific; assays for one serotype may not detect antigens of another. Similar problems occur with *Bacteroides fragilis* (Rissing *et al.*, 1981). Thus, antigen assays may not be practical screening tests. If antigens common to multiple organisms, such as the peptidoglycan backbone of gram-positive cocci and the core lipopolysaccharide of gram-negative bacilli, could be detected in the blood or urine of infected patients, assays for those antigens might be useful screening tests.

B. Nonspecific Inhibitors

Nonspecific serum factors may inhibit antigen detection. Heat-labile serum inhibitors have been described (Robern *et al.*, 1975; Tabbarah *et al.*, 1979). These inhibitors can be inactivated by heating at 56°C, by heparin, and by xymosan, thus improving the sensitivity of certain assays (Tabbarah *et al.*, 1979). The mechanism for this inhibition has not been elucidated, but early complement components appear to be involved; C1q may attach to the solid-phase immunoglobulin or directly to the solid-phase itself, thus competing with and preventing attachment of antigens to their specific antibodies.

C. Specific Inhibitors (Preexisting Antibodies)

Preexisting antibodies bind to microbial antigens and form immune complexes, thus inhibiting antigen detection (Wheat *et al.*, 1979). Antibodies to a variety of staphylococcal antigens occur in normal individuals, presumably because of prior infection or colonization with staphylococci or organisms containing similar antigens (Wheat *et al.*, 1978a, 1981). Although *S. aureus* antigens could be detected in rabbits with *S. aureus* endocarditis lacking preexisting staphylococcal antibodies (Wheat *et al.*, 1978b), they could not be detected in previously immunized rabbits or in humans with staphylococcal endocarditis (Wheat *et al.*, 1979). Intravenous injection of staphylococcal antigen into rabbits at different times during immunization demonstrated that specific antibodies diminish the magnitude and duration of antigenemia. This inhibitory effect of serum was present in the isolated IgG fraction, was stable at 56°C for 30 min, but could be inactivated by heating at higher temperatures. Others have demonstrated similar inhibition in *Haemophilus influenzae* (O'Reilly *et al.*, 1975), pneumococcal (Coonrod and Leach, 1978), and fungal infections (Weiner and Coats-Stephen, 1979). In *H. influenzae* meningitis, antigen could be detected only after dissociation of immune complexes in 26% of cases. In pneumococcal pneumonia, the sensitivity for antigen detection in the blood increased from 38 to 57% after the dissociation of immune complexes. In experimentally induced disseminated candidiasis, antigen could not be detected without prior dissociation of immune complexes but was detected in all 17 rabbits following dissociation (Weiner and Coats-Stephen, 1979). Thus, preexisting antibodies impair the detection of circulating antigens in a variety of infections, limiting the usefulness of antigen assays for the rapid diagnosis of infection.

Antigens and antibodies are held together in these complexes by ionic and hydrophobic forces. These immune complexes will dissociate under certain conditions. Two simple methods for dissociating immune complexes involve heating at 56°C or higher and acidification to pH 3.5 or lower. However, to detect the freed antigens by immunoassay, the specific antibodies must be inactivated or removed from the test sample. Otherwise, when the physical conditions (temperature or pH) are corrected to allow the freed antigen to react in the immunoassay, those antigens will recombine with their antibodies. Thus, not only must the immune complexes be dissociated, but the freed antibodies must be removed or destroyed to permit the detection of circulating antigens in subjects with preexisting antibodies to those antigens.

A variety of dissociation methods have been described which permit the detection of circulating antigens despite preexisting antibodies. Coonrod and Leach (1978) successfully dissociated pneumococcal immune complexes and physically separated the antigens from their antibodies by immunoelectrophoresis at 56°C. That method would not be useful for RIA or ELISA. O'Reilly dissociated *H. influenzae* immune complexes by treatment with 0.1 *M* citric acid, pH 2.0, containing 1 mg pepsin per milliliter for 2 hr at 56°C followed by neutralization with 0.25 *M* Na_2HPO_4 for 2 hr at 20°C and cooling in an ice bath (O'Reilly *et al.*, 1975). Presumably, the complexes were dissociated by the acidic conditions, and pepsin solubilized the acidified sample. However, the method would not be useful for antigens that were degraded by pepsin (Tabbarah *et al.*, 1980). The method also required fourfold dilution of the test sample, thus decreasing the sensitivity. Furthermore, intact Fab regions of pepsin-digested immunoglobulins are known to retain immunologic activity and may recombine with the freed antigens when the sample is neutralized. Acidic pepsin digestion was not successful in our hands (Tabbarah *et al.*, 1980).

To detect circulating mannan in candidiasis, Weiner diluted 20% serum with an equal volume of 0.04 *M* citrate buffer, pH 2.7, and then heated the sample at 96°C for 20 mins (Weiner and Coats-Stephen, 1979). The heated, acidified sample was then clarified by centrifugation at 3900 *g* for 12 mins, and the supernatant was neutralized with an equal volume of 0.35 *M* phosphate buffer. Circulating mannan was detected, but the 20-fold dilution before testing decreased the sensitivity. Weiner also described a modified dissociation method that includes a 30 min pepsin digestion step after heating of the sample (Weiner, 1980). The purpose of the pepsin treatment was presumably to aid in solubilization (O'Reilly *et al.*, 1975).

Meckstroth incubated 5 volumes of serum with 1 volume of 3 *N* NaOH at 56°C for 2 hr to dissociate immune complexes and denature the freed antibodies (Meckstroth *et al.*, 1981). The samples were then dialyzed against phosphate-buffered saline overnight at 4°C and concentrated to the original volume by evaporation. The processing needed in this method would limit its usefulness for rapid diagnosis.

In an assay for *Aspergillus* antigenemia, Shaffer vortexed 1 volume of 50% perchloric acid and 9 volumes of serum (Shaffer *et al.*, 1979). Specimens were centrifuged, and the supernatants were neutralized with 10 *N* KOH and centrifuged again to remove insoluble potassium salts. The supernatants were then dialyzed at 4°C for 24 hr, and the retentates were lyophilized. The lyophilized samples were reconstituted to one-tenth of their original volume and tested for antigen. Data proving the effectiveness of this method were not presented.

These methods are all time-consuming, and simpler methods are needed for assays that are to be used for the rapid diagnosis of life-threatening infections.

Several techniques are based on thermodissociation of immune complexes (Sugerman and Hart, 1973). Heating above 70°C dissociates the complexes and progressively denatures the freed antibodies, thus preventing their reassociation with antigen (Tabbarah *et al.*, 1980). However, undiluted serum coagulates when heated above 65°C. We have found that heating can be performed without coagulum formation, although some clouding occurs, if the immunoglobulins and immune complexes are first precipitated in 50% ammonium sulfate for 15 mins at 4°C and then resuspended in distilled water (Tabbarah *et al.*, 1980). Whereas staphylococcal antigen could not be detected in human serum containing high levels of specific antibodies, it could be detected following thermodissociation. Thermodissociation is simple, requires no special equipment, does not require dilution of the sample, and can be performed in less than 1 hr. Lew described a thermodissociation method that overcame the blocking effect of antibody in an ELISA for circulating candidal mannan (Lew *et al.*, 1982). Sera were diluted fourfold in phosphate-buffered saline, pH 7.4, containing 0.05% Tween 20 and then mixed with 0.1 volume of 0.2 *M* Na_2EDTA, pH 7.4. The mixture was autoclaved at 121°C for 5 min and clarified by centrifugation at 1600 *g*. Candidal mannan could not be detected when diluted in immune sera, whereas the heat extraction procedure restored the detection limit to 10 ng/ml.

D. Rheumatoid Factor

Rheumatoid factor, or rheumatoid factor-like activity, may cause false positive results in sandwich ELISAs (Wolters *et al.*, 1977; Waart *et al.*, 1978; Yolken and Stopa, 1979; Araujo and Remington, 1980; Yolken, 1982). Solid-phase IgG and IgG–enzyme conjugates are apparently recognized as "aggregated." Rheumatoid factor can bridge the solid-phase and conjugated IgG, producing false positive results. We have experienced nonspecificity in an ELISA for *Legionella pneumophila* antigens, presumable due to rheumatoid factor-like activity, but not in an RIA using the same materials (Sathapatayavongs *et al.*, 1982). Thus, rheumatoid factor may not interfere in RIAs, as suggested in our *S. aureus* antigen assay (Tabbarah *et al.*, 1980). Such nonspecific binding can be overcome by using only the $F(ab)_2$ portion of the IgG molecules, because rheumatoid factor binds to the Fc portion (Araujo and Remington, 1980). Other methods for eliminating false positivity caused by rheumatoid factor include treatment with 2-mercaptoethanol (Yolken and Stopa, 1979; Araujo and Remington, 1980) or *N*-acetylcysteine (Yolken and Stopa, 1979), heating at 100°C for 3 min (Doskeland and Berdal, 1980), or absorption with IgG-coated latex particles (Araujo and Remington, 1980) or whole serum or globulin from various animal species (Halbert and Auken, 1977; Yolken and Stopa, 1979; Pepple *et al.*, 1980). Specific inhibition of positivity by premixing the test specimen with antigen-specific or nonimmune serum (Wolters *et al.*, 1977; Yolken *et al.*, 1977; Waart *et al.*, 1978; Kacaki *et al.*, 1978; Merson *et al.*, 1980; Wetherall *et al.*, 1980) or by comparing the effect of immune versus nonimmune serum added to assay wells between the antigen and conjugate steps (Kacaki *et al.*, 1977; Locarnini *et al.*, 1978; Mathiesen *et al.*, 1978; Chao *et al.*, 1979) may also distinguish true from false positive results.

E. Protein A Contamination

Contamination of specimens with protein A-containing staphylococci is another potential source of false positive reactions (Wetherall *et al.*, 1980). Such specimens should become negative when absorbed with nonspecific IgG.

III. APPLICATIONS TO THE RAPID DIAGNOSIS OF BACTEREMIA

A. Radioimmunoassay

O'Reilly *et al.* (1975) detected circulating polyribophosphate (PRP) in the serum or spinal fluid of children with *H. influenzae* meningitis and bacteremia at concentrations ranging from 1 to 1100 ng/ml (Table I). Antigenemia occurred in 34 of 38 patients with a median concentration of 25 ng/ml and lasted up to 30 days. The magnitude and duration of antigenemia were correlated with clinical severity: severe illness, 54 ng/ml and 7.9 days; moderate or mild illness, 10 ng/ml and 2.0 days. Antigen could be detected only after the dissociation of immune complexes in 12 of 45 (27%) samples from 36 patients. This liquid-phase inhibition assay was relatively difficult to perform and required several centrifugation steps and two 2-hr incubations on the first day, followed by an overnight incubation at 4°C. Thus, it would not be a practical method for routine use. However, it demonstrates the potential of immunoassays for the diagnosis of severe infections.

Using a solid-phase RIA, Kohler *et al.* (1980) documented antigenemia in granulocytopenic rabbits with *Pseudomonas* bacteremia. Inactivating heat-labile serum factors by incubation at 56°C for 30 min before testing improved the sensitivity (Tabbarah *et al.*, 1979). Antigenemia was correlated with the magnitude of bacteremia; none of 15 rabbits with less than 10^3 colony-forming units of *P. aeruginosa* per milliliter of blood were antigenemic compared with 4 of 5 with higher colony counts. The effect of preexisting antibodies and dissociation of immune complexes was not studied, nor was the circulating antigen characterized. The IgG fraction of antisera produced in rabbits immunized over a 12-week period with serotype 6 organisms was used as the solid-phase and radiolabeled antibodies. Even though this assay was developed for serotype 6 *P. aeruginosa,* other serotypes could be detected (Kohler *et al.*, 1979).

Circulating staphylococcal antigens were detected in rabbits with *S. aureus* endocarditis (Wheat *et al.*, 1978b). Antigen levels did not correlate with the magnitude of bacteremia. Preexisting antibodies in human sera strongly impaired the detection of this antigen, presumably explaining the poor sensitivity for antigen detection in humans (1 of 20) with staphylococcal bacteremia (Wheat *et al.*, 1979). Thermodissociation of immune complexes permitted antigen detection despite preexisting antibodies, and antigenemia was documented in 4 of 26 patients with *S. aureus* bacteremia, including 3 of 5 with endocarditis (Tabbarah *et al.*, 1980). Antigen has not been detected in the urine of patients with staphylococcal bacteremia or in the synovial fluid, pleural fluid, or cerebrospinal fluid of a few patients with bacteremia complicated by localization to those areas

TABLE I

Antigen Detection in Bacteremia by Immunoassay

Bacterial antigen	Assay	Detection limit	Infection	Test sample	Special treatment	Clinical sensitivity	False positive reactions	Reference
Haemophilus influenzae polyribophosphate	Liquid-phase RIA, inhibition	0.5 ng/ml	Meningitis, humans	Serum Cerebrospinal fluid	Citric acid–pepsin Citric acid–pepsin	34/38 11/12	Not stated Not stated	O'Reilly *et al.* (1975)
Pseudomonas aeruginosa antigen	Solid-phase RIA, direct	Buffer, 0.5 μg/ml; serum, 10 μg/ml	Bacteremia, granulocytopenic rabbits	Serum	Heat, 56°C × 30 min	4/20	0/38	Kohler *et al.* (1980)
Staphylococcus aureus antigen	Solid-phase RIA, direct	Buffer, 0.31 μg/ml; rabbit serum, 1.25 μg/ml	Endocarditis, rabbits	Serum	No	12/12	0/54	Wheat *et al.* (1978)
S. aureus antigens	Solid-phase RIA, direct	Buffer, 0.31 μg/ml; human serum, 10 μg/ml	Bacteremia, humans	Serum	Ammonium sulfate precipitation, 90°C × 15 min	4/26	0/93	Tabbarah *et al.* (1980)
H. influenzae antigen	ELISA, indirect	1–5 ng/ml	Meningitis, bacteremia, humans	Serum Urine	56°C × 20 min 0.1 *M* NaOH	2/2 5/5	0/17 0/15	Drow *et al.* (1979)
H. influenzae antigen	ELISA, indirect	0.1 ng/ml	Meningitis, bacteremia, humans	Serum	1% *N*-acetylcysteine	21/21	0/26	Pepple *et al.* (1980)
Streptococcus pneumoniae antigen	ELISA, direct	1–3 ng/ml	Meningitis, bacteremia, rabbits	Serum	56°C × 30 min; diluted 1 : 2 phosphate-buffered saline– Tween	11/20	0/16	Harding *et al.* (1979)
Bacteroides fragilis antigen	ELISA, inhibition	20–50 ng/ml	Bacteremia, rats	Serum	22% polyethylene glycol	Not stated	Not stated	Rissing *et al.* (1981)

(L. J. Wheat, unpublished data). These assays used the IgG fraction of rabbit antisera following prolonged immunization with a sonically disrupted suspension of the Lafferty strain of *S. aureus* mixed with complete Freund's adjuvant. The clinical usefullness of this assay might be improved if the assay's sensitivity were greater.

B. Enzyme-Linked Immunosorbent Assay

Using an indirect ELISA, *Haemophilus* antigen was detected in each specimen tested from 11 patients with *H. influenzae* meningitis at concentrations of 10 to 4000 ng/ml (Drow *et al.*, 1979). Although serum did not inhibit the detection of purified type b *H. influenzae* polysaccharide, sera were heated at 56°C for 20 min before testing to inactivate intrinsic alkaline phosphatase. Urine samples were first adjusted to pH 8 with 0.1 *N* NaOH. The assay was technically simple to perform and could be easily completed within 1 day. The burro antiserum to PRP used to coat the solid phase was obtained from Dr. John Robbins, National Institutes of Health, and the rabbit anti-*H. influenzae* type b sera used as the second antibody was commercially available (Hyland Laboratories, Costa Mesa, California). The addition of 10% fetal calf serum after the solid phase was coated with the antibody minimized nonspecific binding. The use of polystyrene balls as the solid phase was thought to have contributed to the sensitivity and specificity of the assay.

Also using an indirect ELISA, Pepple *et al.* (1980) detected *H. influenzae* antigen in the serum of all 17 patients with bacteremic *H. influenzae* infection. Sera were diluted in an equal volume of *N*-acetylcysteine to minimize non-specific binding. Burro PRP antiserum (John Robbins, National Institutes of Health, Bethesda, Maryland) was used to coat the solid phase, and rabbit anti-*H. influenzae* was used as the second antibody (Health Research Inc., Albany, New York). This assay used a 1% bovine serum albumin blocking step and polyvinyl U-bottom microtiter plates (Dynateck 220–24). This assay could also be completed within 1 day.

Using a direct ELISA employing commercial antisera and polystyrene microtiter plates, Wetherall *et al.* (1980) found serum to behave erratically in an assay for *H. influenzae* antigen. They used a blocking step in which the test serum was incubated with either normal or immune rabbit globulin. In theory, true positive results would be specifically inhibited by the immune globulin, whereas false positive results would be either uninhibited or inhibited by both. However, the blocking step did not work for sera containing high antigen concentrations; neither the immune globulin nor the normal globulin was inhibitory. Sera containing high levels of rheumatoid factor were also false positive. The cause of this erratic behavior of serum was not fully evaluated, but the type of polystyrene was suggested to be a factor, because uneven coating with immunoglobulin could cause erratic results. Another difference between this assay and the two that successfully detected *H. influenzae* antigenemia is that the former is a direct method of measuring the antigen, whereas the others are indirect (Drow *et al.*, 1979; Pepple *et al.*, 1980).

Using a direct ELISA, Harding *et al.* (1979) detected pneumococcal antigen in over half of rabbits with pneumococcal meningitis and bacteremia. Sera were heated at 56°C for 30 min to remove heat-labile inhibitors and diluted 1 : 2 with phosphate-buffered saline containing 0.05% Tween 20 before testing. Of the 20 rabbits with meningitis, 12 were bacteremic; 8 of these 12 (75%) were antigenemic, compared with 3 of 8 (38%) nonbacteremic rabbits. The sodium sulfate γ-globulin precipitate of pneumococcal Omniserum, which contains antibodies against 83 pneumococcal capsular types, was used to coat the wells of polyvinyl chloride microtiter plates (Cooke Laboratory Products, Alexander, Virginia) and as the horseradish peroxidase labeled antibody. That assay took at least 8 hr to perform and thus could not practically be completed within 1 day. However, that study demonstrates the potential usefulness of immunoassay for the diagnosis of serious pneumococcal infection. Its sensitivity was better than that of CIE *in vitro* (1 versus 25 ng/ml) and *in vivo* (85 versus 65% of specimens were positive, respectively).

Using an antigen inhibition ELISA, Rissing *et al.* (1981) detected a lipopolysaccharide (LPS) antigen of *Bacteroides fragilis* in rats with bacteremia or soft-tissue abscesses. Although the results were not designated as positive or negative, clear differences existed between animals infected with *B. fragilis* and those infected with *Escherichia coli*. However, false positive results were seen in some rats with abscesses caused by Enterobacteriaceae. Also, animals infected with certain strains of *B. fragilis* were negative, suggesting serospecificity and further limiting the applicability of the assay. *Bacteroides fragilis* antiserum used in this assay was prepared by immunizing rabbits with sonicated whole organisms and outer membranes for at least 6 weeks, the first injection (1 mg/ml) given im with complete Freund's adjuvant, followed by weekly iv injections. Undiluted human serum did not impair the detection of *Bacteroides* antigen in this inhibition assay. The assay used polystyrene microtiter plates (Immulon, Dynatech Laboratories, Inc., Alexandria, Virginia) coated with 2 μg/ml of the *B. fragilis* outer membrane or LPS preparations. The test sample was incubated overnight with diluted hyperimmune rabbit serum in a polyvinyl transfer plate. The transfer plate reactants were placed in the solid-phase wells. Rabbit antibody to the LPS was then detected with goat anti-rabbit IgG–alkaline phosphatase conjugate. Antigen in the test sample would inhibit binding of the rabbit antisera to the solid-phase antigen.

REFERENCES

Araujo, F. G., and Remington, J. S. (1980). *J. Infect. Dis.* **141,** 144–150.

Chao, R. K., Fishaut, M., Schwartzman, J. D., and McIntosh, K. (1979). *J. Infect. Dis.* **139,** 483–486.

Colding, H., and Lind, I. (1977). *J. Clin. Microbiol.* **5,** 405–409.
Coonrod, J. D., and Leach, R. P. (1978). *J. Clin. Microbiol.* **8,** 257–259.
Dochez, A. R., and Avery, O. T. (1917). *J. Exp. Med.* **26,** 447–493.
Doskeland, S. O., and Berdal, B. P. (1980). *J. Clin. Microbiol.* **11,** 380–384.
Drow, D. L., Maki, D. G., and Manning, D. D. (1979). *J. Clin. Microbiol.* **10,** 442–450.
Halbert, S. P., and Auken, M. (1977). *J. Infect. Dis.* **136,** S318–S323.
Harding, S. A., Scheld, W. M., McGowan, M. D., and Sande, M. A. (1979). *J. Clin. Microbiol.* **10,** 339–342.
Kacaki, J., Wolters, G., Kuijpers, W. L., and Schuurs, A. (1977). *J. Clin. Pathol.* **30,** 894–898.
Kacaki, J., Wolters, G., Kuijpers, L., and Stulemeyer, S. (1978). *Vox. Sang.* **35,** 65–74.
Kohler, R. B., Wheat, L. J., and White, A. (1979). *J. Clin. Microbiol.* **9,** 253–258.
Kozaki, S., Dufrenne, J., Hagenaars, A. M., and Notermans, S. (1979). *Jpn. J. Med. Sci. Biol.* **32,** 199–205.
Lew, M. A., Siber, G. R., Donahue, D. M., and Maiorca, F. (1982). *J. Infect. Dis.* **145,** 45–56.
Locarnini, S. A., Garland, S. M., Lehmann, N. I., Pringle, R. C., and Gust, I. D. (1978). *J. Clin. Microbiol.* **8,** 277–282.
Mathiesen, L. R., Feinstone, S. M., Wong, D. C., Skinhoej, P., and Purcell, R. H. (1978). *J. Clin. Microbiol.* **7,** 184–193.
Meckstroth, K. L., Reiss, E., Keller, J. W., and Kaufman, L. (1981). *J. Infect. Dis.* **144,** 24–32.
Merson, M. H., Yolken, R. H., Sack, R. B., Froehlich, J. L., Greenberg, H. B., Huq, I., and Black, R. W. (1980). *Infect. Immun.* **29,** 108–113.
O'Reilly, R. J., Anderson, P., Ingram, D. L., Peter, G., and Smith, D. H. (1975). *J. Clin. Invest.* **56,** 1012–1022.
Pepple, J., Moxon, E. R., and Yolken, R. H. (1980). *J. Pediatr. (St. Louis)* **97,** 233–237.
Rissing, J. P., Buxton, T. B., Talledo, R. A., and Sprinkle, T. J. (1981). *Infect. Immun.* **27,** 405–410.
Robern, H., Dighton, M., and Dickie, N. (1975). *Appl. Microbiol.* **30,** 525–529.
Sathapatayavongs, B., Kohler, R. B., Wheat, L. J., White, A., Winn, W. C., Jr., Girod, J. C., and Edelstein, P. H. (1982). *Am. J. Med.* **72,** 576–579.
Shaffer, P. J., Kobayashi, G. S., and Medoff, G. (1979). *Am. J. Med.* **67,** 627–630.
Sugerman, D., and Hart, H. E. (1973). *Bull. Math. Biol.* **35,** 219–235.
Tabbarah, Z. A., Kohler, R. B., Wheat, L. J., Griep, J. S., and White, A. (1979). *J. Infect. Dis.* **140,** 822–825.
Tabbarah, Z. A., Wheat, L. J., Kohler, R. B., and White, A. (1980). *J. Clin. Microbiol.* **11,** 703–709.
Waart, M., Snelting, A., Cichy, J., Wolter, G., and Schuurs, A. (1978). *J. Med. Virol.* **3,** 43–49.
Weiner, M. H. (1980). *Ann. Intern. Med.* **92,** 793–796.
Weiner, M. H., and Coats-Stephen, M. (1979). *J. Infect. Dis.* **140,** 989–993.
Wetherall, B. L., Hallsworth, P. G., and McDonald, P. J. (1980). *J. Clin. Microbiol.* **11,** 573–580.
Wheat, L. J., Kohler, R. B., and White, A. (1978a). *Ann. Intern. Med.* **89,** 467–472.
Wheat, L. J., Kohler, R. B., and White, A. (1978b). *J. Infect. Dis.* **138,** 174–180.
Wheat, L. J., Kohler, R. B. and White, A. (1979). *J. Infect. Dis.* **140,** 25–30.
Wheat, L. J., Kohler, R. B., Tabbarah, Z. A., and White, A. (1981). *J. Infect. Dis.* **144,** 307–311.
Wolters, G., Kuijpers, L., Kacaki, J., and Schuurs, A. (1977). *J. Infect. Dis.* **136,** S311–S317.
Yolken, R. H. (1982). *Rev. Infect. Dis.* **4,** 35–68.
Yolken, R. H., and Stopa, P. J. (1979). *J. Clin. Microbiol.* **10,** 703–707.
Yolken, R. H., and Stopa, P. J. (1980). *J. Clin. Microbiol.* **11,** 546–551.
Yolken, R. S., Wha Kim, H., Clem, T., Wyatt, R. G., Chanock, R. M., Kalica, A. R., and Kapikian, A. Z. (1977). *Lancet* **2,** 263–266.

16

Diagnosis of Legionnaires' Disease by Radioimmunoassay and Enzyme-Linked Immunosorbent Assay

RICHARD B. KOHLER

Wishard Memorial Hospital
and
Department of Medicine
Indiana University Medical School
Indianapolis, Indiana

I. Introduction 187
II. Conventional Diagnostic Tests for Legionnaires' Disease 188
A. Demonstration of an Antibody Response 188
B. Demonstration of Bacilli in Clinical Specimens 189
C. Culture 190
III. Radioimmunoassay and Enzyme-Linked Immunosorbent Assay for the Detection of Legionnaires' Antigen 191
A. Solid-Phase Radioimmunoassay for Detecting Serogroup 1 *Legionella pneumophila* Antigens 191
B. Enzyme-Linked Immunosorbent Assay for Detecting Serogroup 1 Antigens 201
IV. Summary 203
References 205

I. INTRODUCTION

Legionnaires' disease is a febrile illness caused by *Legionella pneumophila*, a weakly gram-negative, serologically diverse bacillus first clearly recognized as a human pathogen in 1977. Pneumonia is the most important *L. pneumophila* infection. Serogroup one *L. pneumophila* probably causes about 1% of all pneumonias (Foy *et al.*, 1979). Individuals at increased risk include renal dialysis or transplant patients, with relative risks over 300 times appropriate controls of

The Direct Detection of Microorganisms in Clinical Samples

ISBN 0-12-187780-9

developing nosocomial Legionnaires' diseases; patients who receive immunosuppressive medication, with a relative risk of 26; patients with cancer, with a relative risk of 11; and patients with chronic bronchitis or emphysema, with a relative risk of 3.7 (England and Fraser, 1981). About 1 in 25 nosocomial pneumonias (Cohen *et al.*, 1979) and a significant proportion of pneumonias occurring in travelers who have stayed in large housing facilities (Grist *et al.*, 1979) are caused by *L. pneumophila*. Initial clinical and laboratory features which should suggest that pneumonia is caused by *L. pneumophila* include watery diarrhea, unexplained encephalopathy, high fever with relative bradycardia, and mucoid sputum containing a small number of polymorphonuclear leukocytes but few or no bacteria (Meyer and Finegold, 1980). In the course of the illness, negative cultures of blood and respiratory specimens on routine culture media or, more importantly, a poor therapeutic response to penicillins, cephalosporins, or aminoglycosides should also suggest the diagnosis of Legionnaires' disease.

Unfortunately, in many cases these features are absent, or their presence is insufficiently specific to allow one to make the diagnosis on clinical grounds. Because Legionnaires' disease is a relatively infrequent cause of pneumonia, empirical use of erythromycin, the antibiotic of choice for Legionnaires' disease, is not warranted except in highly suspected cases. Erythromycin alone is ineffective against many pathogens encountered in nosocomial pneumonias or pneumonias in immunosuppressed patients. Phlebitis is a frequent complication with intravenous erythromycin use. Furthermore, although erythromycin effectively treats *Streptococcus pneumoniae* and *Mycoplasma pneumoniae* infections, optimal treatment of Legionnaires' disease appears to require a longer course of therapy than is necessary for pneumonia due to these other agents (Kirby *et al.*, 1980). For these reasons specific diagnostic tests for Legionnaires' disease capable of providing a diagnosis early in the illness are highly desirable.

II. CONVENTIONAL DIAGNOSTIC TESTS FOR LEGIONNAIRES' DISEASE

A. Demonstration of an Antibody Response

As in other infections, rising levels of antibodies to *L. pneumophila* occur in most patients with Legionnaires' disease. The reference method for detecting this antibody rise is the indirect fluorescent antibody (IFA) test (Wilkinson *et al.*, 1979). Bacterial cells that fluoresce indicate the presence of antibody. The titration end point typically used is 1+ fluorescence.

In the absence of a foolproof method of diagnosing or ruling out Legionnaires' disease in a given patient, the sensitivity and specificity of IFA testing are not known with certainty. A few patients with compatible illnesses and organisms visualized in respiratory secretions by direct fluorescent antibody (DFA) staining have failed to demonstrate seroconversion after 6 to 11 weeks of follow-up (Edelstein *et al.*, 1980b). The test, therefore, is apparently not 100% sensitive. However, most patients followed serologically for at least 6 weeks have seroconverted (Kirby *et al.*, 1980). Seroconversion in Legionnaires' disease is relatively slow. In one large series the proportion of seroconverters was about 8% after 1 week of illness, 35% at 2 weeks, 54% at 3 weeks, 60% at 4 weeks, 62% at 5 weeks, and 100% at 6 weeks (Kirby *et al.*, 1980). Thus, seroconversion occurs too slowly to aid in the selection of antibiotic in most patients with Legionnaires' disease. The use of a single high titer to diagnose Legionnaires' disease, often taken as 1 : 256 or greater, cannot be recommended unles the patient is known to come from a population with a low frequency of titers at this level. Several populations with a fairly high proportion of asymptomatic individuals with "high titers" have been described (Broome *et al.*, 1979b; Politi *et al.*, 1979).

The frequency of seroconversion in the *Legionella* IFA by patients with infections due to non-*Legionella* organisms with cross-reactive antigens appears to be quite low. A few patients with *Bacteroides fragilis* infections seroconverted in the serogroup 1 *L. pneumophila* IFA test, probably due to cross-reacting antigens (Edelstein *et al.*, 1980a). Asymptomatic seroconversion and single high titers in asymptomatic individuals are well documented (Haley *et al.*, 1979). Thus, seroconversion in the *Legionella* IFA test is not specific for symptomatic Legionnaires' disease. However, the concurrence of a compatible clinical illness and seroconversion by IFA is reasonably specific and accepted by most investigators as proof of Legionnaires' disease.

Other serologic tests for *L. pneumophila* antibodies have been described but are beyond the scope of this discussion (Blackmon *et al.*, 1981).

B. Demonstration of Bacilli in Clinical Specimens*

Antisera prepared against serogroup 1 *L. pneumophila* strains and labeled with fluorescein isothiocyanate rarely cross-react with other bacteria. Among the many strains tested only a few strains of *Pseudomonas fluorescens* (Cherry *et al.*, 1978) and *Pseudomonas alcaligenes* (Broome *et al.*, 1979a) and three strains of *B. fragilis* (Edelstein *et al.*, 1980a) have been reported to cross-react. Antisera prepared against the other *L. pneumophila* serogroups appear to be equally specific, although they have been tested less extensively (Blackmon *et al.*, 1981). Assessing the sensitivity and specificity of DFA examination of clinical materials for Legionnaires' disease suffers from the same problems as those encountered in assessing the IFA test: Currently, no test can be relied on to detect

*See also Chapter 3, this volume.

all cases of Legionnaires' disease or to rule it out with certainty. The best available data suggest that the DFA test is positive in 50 to 60% of patients with Legionnaires' disease if transtracheal aspiration and, occasionally, lung biopsies are performed to obtain lower respiratory secretions when necessary (Edelstein *et al.*, 1980b). Definite false positive results in clinical diagnostic work have not been described. Because of the difficulty of ruling out Legionnaires' disease with certainty, the demonstration of definite false positive results requires the isolation of a microorganism from the DFA-positive sample that cross-reacts with the DFA reagent plus failure to isolate *L. pneumophila* from the same specimen.

The major advantages of the DFA test include its specificity, relative simplicity, and capacity to diagnose early in the illness. Disadvantages include the need for invasive procedures such as transtracheal aspiration or lung biopsy in some patients to maximize sensitivity because a significant proportion of patients with Legionnaires' disease produce minimal or no expectorated sputum. A reliable commercial source of the test reagents was not available at the time of preparation of this manuscript.

C. Culture

Legionella pneumophila and other *Legionella* species will grow on special, relatively simple laboratory media. Because they grow more slowly than many other commensal and pathogenic bacteria and are inhibited by other members of the human upper airway flora (Flesher *et al.*, 1980), a key problem has been the development of specifically selective media that will allow *Legionella* organisms to grow from specimens containing other organisms. Transtracheal aspirates, blood, and pleural fluid do not present this problem; the recovery of *Legionella* organisms from such specimens on charcoal yeast extract agar, with or without antibiotics, has been reasonably successful. Charcoal yeast agar containing cefamandole, polymyxin B, anisomycin, an inorganic buffer, and α-ketoglutarate appears to be a reasonably effective semiselective medium for *L. pneumophila* (Edelstein, 1981). The sensitivity of existing culture techniques as diagnostic tools is unknown; 30–60% might be a reasonable estimate (Edelstein *et al.*, 1980b). The isolation of *L. pneumophila* from human specimens presumably represents proof of causation of the infection, although experience is inadequate to rule out the existence of carrier or transient colonization states.

The major advantage of culture is its presumed specificity; positive cultures are regarded as the best proof of etiology. Disadvantages include the average 5-day delay in obtained recognizable growth, the need for special media and processing in the microbiology laboratory, and the need for transtracheal punctures or lung biopsies to optimize diagnostic yields.

III. RADIOIMMUNOASSAY AND ENZYME-LINKED IMMUNOSORBENT ASSAY FOR THE DETECTION OF LEGIONNAIRES' ANTIGEN

Because of problems with conventional approaches to diagnosis, attempts to diagnose Legionnaires' disease by antigen detection were initiated in several laboratories, including our own. Berdal *et al.* 1979) and Tilton (1979), using antisera developed at the Centers for Disease Control in Atlanta, Georgia, demonstrated that patients with Legionnaires' disease may excrete antigen(s) in their urine or sputum, which can be detected by enzyme-linked immunosorbant assay. In both studies only a small number of patient samples were tested, and the control populations examined were small and generally from healthy hosts. Our own efforts focused initially on the technique of solid-phase radioimmunoassay (SPRIA).

A. Solid-phase Radioimmunoassay for Detecting Serogroup 1 *Legionella pneumophila* Antigens

SPRIA is a widely used method that has been commerically available for the diagnosis of hepatitis B since 1972. On the basis of its success in hepatitis diagnosis, we elected to use SPRIA rather than ELISA for our developmental work in Legionnaires' disease because we possessed the necessary equipment, particularly a gamma counter, which could automatically sample a large number of tubes unattended, and because we were quite familiar with the technique and had the expertise to work with ^{125}I.

Our earlier experiences indicated that antigen detection in serum by SPRIA is complicated by inhibitory activities in serum not found in urine or crystalloid buffers (Tabbarah *et al.*, 1979; Kohler *et al.*, 1980). Furthermore previous work by others demonstrated that patients with various systemic infections excrete a detectable quantity of bacterial antigens in their urine, which can be detected even by relatively insensitive immunologic techniques (Dochez and Avery, 1917; Neill *et al.*, 1951; Feigen *et al.*, 1976). Thus, we elected to use urine in our initial attempts to detect antigens.

1. Development of Radioimmunoassay

The antiserum used in our SPRIA was initially developed for DFA testing by Sarah Zimmerman and Morris French at the Indiana University Medical Center serology laboratory (Kohler *et al.*, 1981). The vaccine used to elicit this antiserum differed from that obtained by the original method in several ways. Formaldehyde treatment of the bacterial cells was avoided. A method very similar to that reported by Farshy *et al.* (1978) to obtain a micro-ELISA serology antigen was used to prepare the vaccine. Organisms were harvested from agar, autoclaved, and allowed to set at 4°C for 2 weeks to allow soluble cell wall antigens to enter the solution. The vaccine consisted of equal volumes of bacteria resuspended in the soluble antigen solution and complete Freund's adjuvant. Injections were given subcutaneously and intramuscularly at multiple sites. The immunoglobulin G used for the initial radioimmunoassay (RIA) was made from an antiserum pool from bleedings at 6 and 7 weeks after the initial injection. Booster injections were given at 2- to 4-monthly intervals. Antiserum obtained 9 months and later after the initial immunization had higher titers of antibody to the vaccine antigen, as determined by RIA, than the 6- and 7-week antiserum (Fig. 1). Radioimmunoassays developed with immunoglobulin G from the late antisera have been superior in sensitivity to the assay initially reported.

Because the methods initially used for vaccine and antiserum production were successful, other antigens and immunization schedules have not been tested. Whether our vaccine preparation and immunization schedule are optimal is therefore, unknown. Of the six rabbits initially immunized, one yielded antisera with markedly higher titers than the other five. Although only limited experiments have been performed, these other antisera appear to work less well or not at all for antigen detection by SPRIA. Whether other antisera with the desirable properties found in this antiserum can be produced in large quantities is unknown.

A direct, or single-antibody sandwich, system was chosen because it requires only one animal species source of antibody and less time to complete than indirect systems. An indirect, or double-antibody sandwich, system was to be tried if the direct system proved inadequate. Indirect systems are reportedly more sensitive than direct systems (Kozaki *et al.*, 1979; Yolken and Stopa, 1980) but require at least two animal species sources of antisera and require one additional assay step. These differences are illusrated schematically in Fig. 2.

Conditions adopted for the RIA were based on data available from earlier work (Wheat *et al.*, 1978; Kohler *et al.*, 1979). Polystyrene tubes were chosen as the solid phase because they are at least as effective as other solid phases and are relatively inexpensive. The inside surfaces of the tubes were coated with immunoglobulin G at a concentration of 20 μg/ml, which had proved to be optimal in our *Pseudomonas aeruginosa* and *Staphylococcus aureus* assays. Coating is essentially completed within 1 hr at 37°C.

Following coating with immunoglobulin G an additional coating step with 5% bovine serum albumin was used to inhibit nonspecific binding to the polystyrene in later assay steps. In earlier experiments we had found 5% bovine serum albumin to be slightly superior to 1% bovine serum albumin. The buffer used for both coating steps was 0.01 *M* Tris HCl at pH 7.0, which we had earlier found to be equal to or slightly better than 0.1 *M* Tris, 0.15 *M* NaCl, pH 8.0, or 0.1 *M* sodium carbonate, pH 9.6.

The RIA was then carried out in the coated tubes, requiring two additional

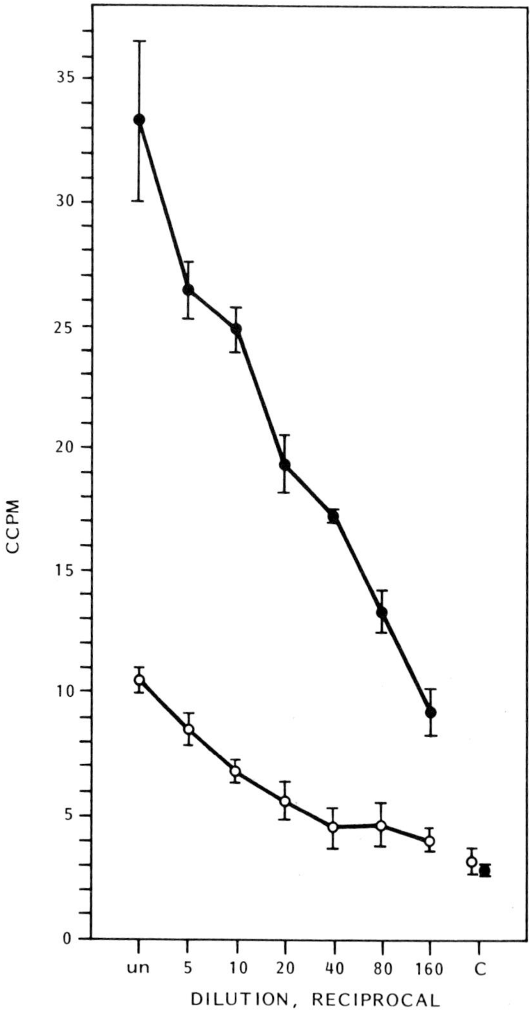

Fig. 1. Comparison of RIAs using antisera obtained after 6 weeks (old assay, ○) and 16 months (new assay, ●) of immunization from the same rabbit. An antigen-positive urine was serially diluted in normal urine. Triplicate tubes were tested for each dilution; the mean ± 1 SD is plotted for each triplicate set. Five antigen-negative control urines were tested (C). CCPM, Counts per minute (hundreds).

steps. The test urine or other test material was added to the coated tubes and incubated at 37°C for 1 hr. Again, our earlier work indicated that longer incubation periods produce no increase in sensitivity. The urine was aspirated and the tubes rinsed. Finally, radioiodinated immunoglobulin G was added at a con-

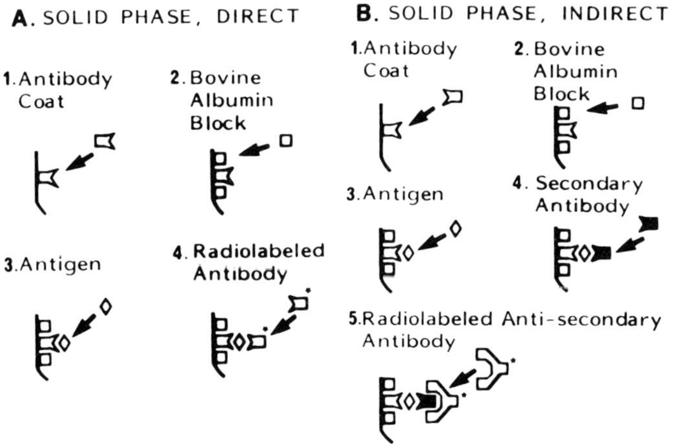

Fig. 2. Schematic comparison of direct and indirect SPRIAs.

centration of approximately 2 to 5 μg/ml. Within this concentration range, little change in sensitivity has been noted. After a 1-hr incubation at 37°C, the ^{125}I-labeled immunoglobulin G was aspirated, the tubes rinsed, and each of the tubes counted for 1 min in a gamma counter.

To produce the ^{125}I-immunoglobulin G, iodination was performed using a modified chloramine-T procedure (McConahey and Dixon, 1966). In our procedure 1 mg of DEAE-purified immunoglobulin G from anti-*Legionella* antisera was conjugated with 5 mCi of Na^{125}I in 0.05 *M* PO_4 buffer, pH 7.0, for 10 min. The protein-bound ^{125}I was separated from the unbound ^{125}I on Sephadex G-25M columns, purchased prepoured and discarded after each iodination. Specific activities of 0.4 to 1.0 mCi/mg generally result. The iodination procedure is very simple but must be performed in a hood that is specially equipped with a charcoal filtration system. Each radioiodinated immunoglobulin G preparation generally performs adequately for 4 to 6 weeks. The major factor limiting the useful life span of the preparations is the loss of radioactivity through radioactive decay of the ^{125}I, which has a half-life of about 60 days.

2. *Diagnosis of Serogroup 1 Infections with RIA*

For the *Legionella pneumophila* assay, the nature of the antigen(s) to be sought was unknown. Calculation of antigen concentration by comparison with a standard curve could therefore not be used to express results. Determining positivity thus required comparison of the test urine with urine from individuals known not to have Legionnaires' disease. Because means tend to vary less than standard deviations in day-to-day sampling from the same population, we elected to use mean counts per minute of control urines as the basic comparative unit.

We found that 5–10 control urine samples run each day give a reasonably consistent mean, which decreases gradually as the ^{125}I decays.

Initial studies with a small number of culture supernatants indicated that cross-reactions with organisms likely to be encountered frequently in urinary or lung infections were unlikely (Fig. 3). *Staphylococcus aureus* culture supernatants yielded significant reactivity, however. Whether this was due to protein A, which nonspecifically binds rabbit immunoglobulin G, or to other staphylococcal antigens is not known. The potential for false positive reactions in staphylococcal infections exists.

We then tested urine specimens from patients with urinary tract infections, pneumonias, other pulmonary conditions, other bacteremic infections, and miscellaneous other illnesses and compared the results with specimens from nine patients with Legionnaires' disease. As shown in Fig. 4 the counts per minute of the nine legionnaires' disease patients exceeded the normal control mean by a factor of 2.0 or more, whereas the counts per minute of the 241 patient controls were all below 1.6 times the control mean. Thus, the RIA appeared to be very specific. The sensitivity seemed reasonable, but nine patients were too few for definitive statements to be made.

As noted earlier the quality of the antiserum and resultant RIAs improved as rabbit immunization progressed (see Fig. 1). The initial group of 9 patients and 241 controls was studied with an assay that utilized antiserum from earlier bleeds. During the summer of 1980 an outbreak of Legionnaires' disease occurred in Burlington, Vermont. Drs. Washington Winn, Jr. and John Girod collected urine and other specimens prospectively from patients at the Medical Center Hospital of Vermont. We later tested the urine specimens blindly, interpreted them as positive or negative, and correlated the results with the findings of other laboratory tests (Kohler *et al.*, 1982). As shown in Table I 27 subjects, or 73%, of the 37 with evidence of Legionnaires' disease were positive in the RIA. In 21 patients the diagnosis was considered conclusive by other criteria; 15 were culture-positive for serogroup 1 *L. pneumophila,* and 6 others seroconverted in the IFA test and had respiratory secretions positive for *L. pneumophila* organisms by DFA examination. Ninety percent of this group had detectable urinary antigen. In 16 patients the evidence for infection consisted either of seroconversion by IFA only or of positive respiratory secretions by DFA examination; 50% of these patients had detectable antigen. A smaller total body bacterial load might account for the lower antigen detection rate in the latter 16 patients. Alternatively, either the DFA or IFA test may have occasionally represented the presence of asymptomatic *Legionella* carriage or asymptomatic infection. Our study did not enable us to distinguish between these possibilities. Nonetheless, the overall sensitivity of 73% suggests that the test can have great value in patient management.

Specimens from 25 other Vermont patients were studied. Five patients failed

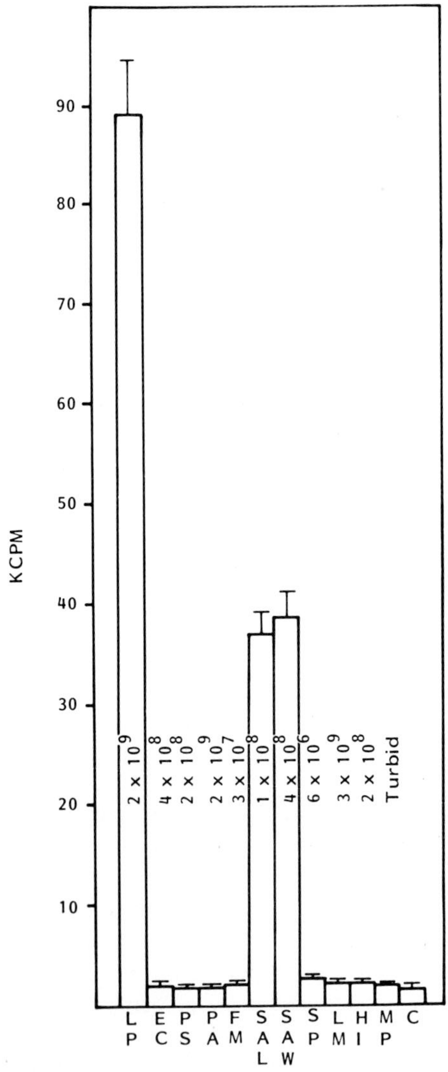

Fig. 3. Bacterial culture supernatants tested by RIA for serogroup 1 *Legionella pneumophila* antigens. Numbers within graph represent colony-forming units per milliliter. Abbreviations: LP, serogroup 1 *L. pneumophila;* EC, *Escherichia coli;* PS, *Providencia stuartii;* PA, *Pseudomonas aeruginosa;* FM, *Flavobacterium meningosepticum;* SAL, *Staphylococcus aureus,* Lafferty strain (produces protein A); SAW, *S. aureus,* Wood 46 strain (produces little or no protein A); SP, *Streptococcus pneumoniae;* LM, *Listeria monocytogenes;* HI, Haemophilus influenzae; MP, *Mycoplasma pneumoniae;* C, uninoculated beef heart infusion broth; KCPM, counts per minute (thousands).

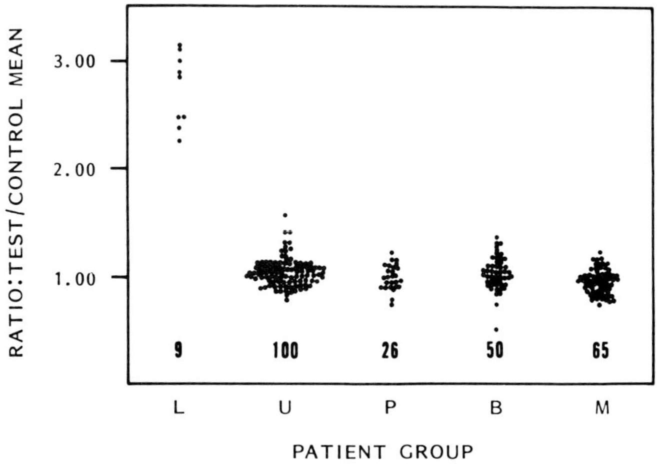

Fig. 4. Detection of *Legionella pneumophila* antigens in urine by RIA. Each day the assay was run, 10 negative control specimens were also tested and the mean counts per minute of the 10 controls were calculated. The test urine result was divided by the control group mean to obtain the ratio shown on the ordinate. All 9 Legionnaires' disease patients were clearly separated from the 241 control subjects. Abbreviations: L, Legionnaires' disease; U, urinary tract infection or contaminated urine; P, pulmonary disease; B, bacteremic infections; M, miscellaneous diseases. Numbers under test groups represent total number of specimens in that group. From Kohler *et al.* (1981). Published with permission of the Annals of Internal Medicine.

TABLE I

Detection of Serogroup 1 *Legionella pneumophila* Antigens by Radioimmunoassay in an Outbreak of Legionnaires' Disease

Patient status	No. of patients	No. positive
Legionnaires' disease		
Culture positive	15	14
Culture negative or not done, seropositive,[a] DFA[b] positive	6	5
Seropositive only	13	7
DFA positive only	3	1
Total	37	27
No Legionnaires' disease	5	0
Not classifiable	20	1

[a] Fourfold or greater rise in serum antibody levels to 1 : 128 or greater as determined by indirect fluorescent antibody testing with serogroup 1 *L. pneumophila* organisms.

[b] DFA, Direct fluorescent antibody examination of tissues or secretions for serogroup 1 *L. pneumophila* organisms.

to seroconvert over a period of 6 weeks or more and lacked other evidence of Legionnaires' disease. None of these were positive for antigen. Twenty patients could not be classified because of inadequate data, and 1 of these was positive for antigen. This patient had a few fluorescent bacilli in her sputum as found by DFA examination; her IFA titer was 1 : 128 on the fifth day and 1 : 64 on the thirty-third day after the onset of the illness. Urine specimens from 180 control patients from Indiana, different from the 241 tested earlier, were all negative in the SPRIA.

The assay appeared to be diagnostically useful even early in the patients' illnesses, with antigen being detectable within 3 days of the onset of symptoms nearly as often as later (Table II). Antigen remained detectable for variable periods after therapy was begun, disappearing as early as between days 1 and 4 in one patient and persisting into the fifth week in two others (Fig. 5). In patients followed at Indiana University, prolonged excretion for 8 weeks or more has frequently been seen. The longest period of detectable antigenuria was at least 283 days after initiation of therapy in a renal transplant recipient, at which time the patient remained antigen-positive.

The RIA and DFA examination of clinical materials complemented each other. One or both tests were positive in 31, or 84%, of the 37 patients (Table III). Of the 27 patients in whom both tests were performed, one or both were positive in 25, or 93%. Ten of 37 patients had no specimens examined by DFA, of whom 6 had detectable urinary antigen. However, 4 of the 10 patients who lacked detectable urinary antigen had specimens positive by DFA examination. Thus, both tests should be performed to optimize rapid diagnostic yields.

3. *RIA in Non-serogroup 1 Infection*

Our experience with urine specimens from patients with serogroup 2–6 Legionnaires' disease is quite limited. A single specimen, collected late in the

TABLE II

Detection of Serogroup 1 *Legionella pneumophila* Antigens by Radioimmunoassay: Correlation of Time of Urine Collection and Time of Onset of Symptoms

Time after onset of symptoms (days)	No. positive/ no. tested
0–3	7/11[a]
4–7	9/16[a]
>7	8/10[a]

[a] Differences not statistically significant.

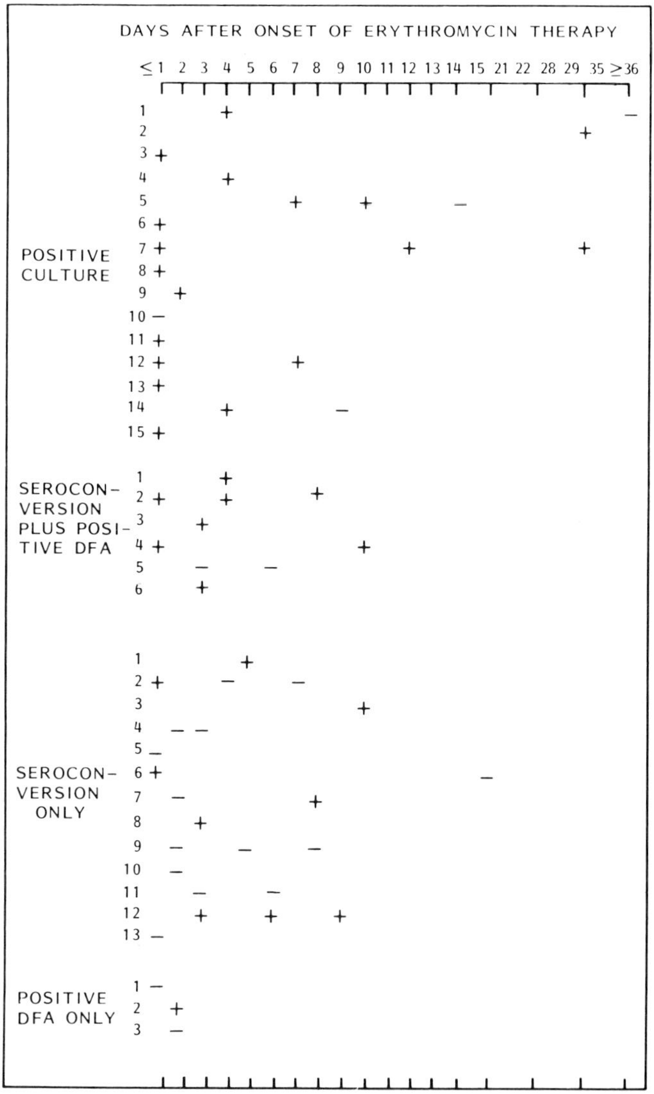

Fig. 5. Effect of erythromycin therapy on the detection of urinary *L. pneumophila* antigen. All patients shown here were diagnosed at the Medical Center Hospital of Vermont during an outbreak of Legionnaires' disease. Key: +, antigen present; −, antigen absent. Thirty-seven patients were studied.

TABLE III

Comparison of Direct Fluorescent Antibody Examination versus Radioimmunoassay for the Rapid Diagnosis of Legionnaires' Disease

Test		
DFA	RIA	No. of patients
Positive	Positive	16
Positive	Negative	4
Negative	Positive	5
Negative	Negative	2
Not done	Positive	6
Not done	Negative	4
		37

treatment course of serogroup 3 *L. pneumophila* infection by Paul Arnow, University of Chicago, was negative for antigen. Of four specimens from patients with serogroup 4 infections, two were positive; however, one of the two patients had concurrent serogroup 1 Legionnaires' disease. A single patient tested early in the treatment course of a serogroup 6 infection was also antigen-negative. Our experience with *Legionella* species other than *L. pneumophila* is even more limited. Urine collected acutely from a patient with *L. bozemanii* in lung tissue by DFA examination was antigen-negative, as was a specimen collected after completion of therapy from a patient with an *L. longbeachae* infection. It seems unlikely that non-serogroup 1 patients will have antigenuria detected frequently enough by our serogroup 1 system to be useful clinically.

4. Problems with RIA

Antisera for large-scale use were not available at the time of this writing. Commercial suppliers of diagnostic assays are currently considering the problem of large-scale production of antisera. A potential solution would be the development of hybridomas capable of producing suitable antibodies or the use of standard methods. It is anticipated that kits for diagnosing serogroup 1 Legionnaires' disease will be commercially available by 1984 or 1985.

The use of ^{125}I requires that precautions be taken, although many diagnostic laboratories already perform solid-phase RIAs with this isotope for other purposes. Because ^{125}I has a half-life of 60 days it does not pose a long-term radiation hazard, but for institutions that lack storage space or those producing very large quantities of waste ^{125}I, disposal poses special problems. Volatile ^{125}I-containing chemicals such as hydriodic acid may evaporate from sodium

^{125}I solutions. Inhaled or absorbed ^{125}I concentrates in the thyroid gland, posing potential hazards. Such wastes must therefore be packaged and handled carefully when stored or transported. At the time of this writing, only three radiation waste disposal sites existed in the United States. Transportation costs may be high for institutions located far from the states of Nevada, South Carolina, or Washington, in which these sites are located. Federal legislation requires that disposal sites be established within each of several geographic regions of the United States in anticipation of the demands for future disposal sites. Given widespread public fears about toxic and radiation waste hazards, prospects for the establishment of such sites remain equivocal.

The probable inability of the current successful assay to diagnose *Legionella* infections other than that caused by serogroup 1 *L. pneumophila* has been discussed. Whether antigens are excreted in detectable quantities in other legionelloses remains to be determined. If they are, a single assay capable of detecting multiple antigen types still remains a possibility. The sensitivity of less than 100% may be viewed as a problem, although improvements in antigen detection are possible in the future.

Occasionally, abrupt changes occur in the performance of the SPRIA, particularly with regard to high background activity and poor agreement among the tubes of triplicate sets. This may be caused by microbial contamination of the radioiodinated immunoglobulin G, and filter sterilization usually restores the utility of the preparation (although the protein content after filtration appears to decrease).

B. Enzyme-Linked Immunosorbent Assay for Detecting Serogroup 1 Antigens

ELISA eliminates the need for working with ^{125}I and appears to be equal in sensitivity to SPRIA in many systems (Wolters *et al.*, 1976, 1977; Halbert and Auken, 1977; Kacaki *et al.*, 1978). Work was performed by Boonmee Sathapatayavongs (Sathapatayavongs *et al.*, 1982) to establish an ELISA for Legionnaires' disease.

1. Development of ELISA

The first three steps of direct, single-antibody ELISA are identical to those of SPRIA. Many workers prefer to use Tween 20 in their buffer system in ELISA once tubes have been coated rather than to use bovine serum albumin as a secondary coat. Our experience suggests that coating with 5% bovine serum albumin gives better precision. We have therefore continued to use the bovine serum albumin and buffers without Tween 20. We have arbitrarily used polystyrene tubes rather than microtiter plates for our ELISA. Alkaline phosphatase-labeled immunoglobulin G is prepared by standard methods (Voller *et al.*, 1976).

2. *Comparison of ELISA and SPRIA*

The major differences between ELSIA and SPRIA for detecting serogroup 1 *L. pneumophila* antigens are illustrated in Fig. 6. The reaction time of the alkaline phosphatase-labeled immunoglobulin G step must be prolonged beyond that used for the ^{125}I-labeled immunoglobulin G to achieve the same sensitivity as with SPRIA. Furthermore, ELISA requires one more step than SPRIA: the development of a colored product from the enzyme substrate. Thus, for a typical laboratory performing immunoassays on the day shift only, an overnight incubation is required for the ELISA but not for SPRIA. By increasing the concentra-

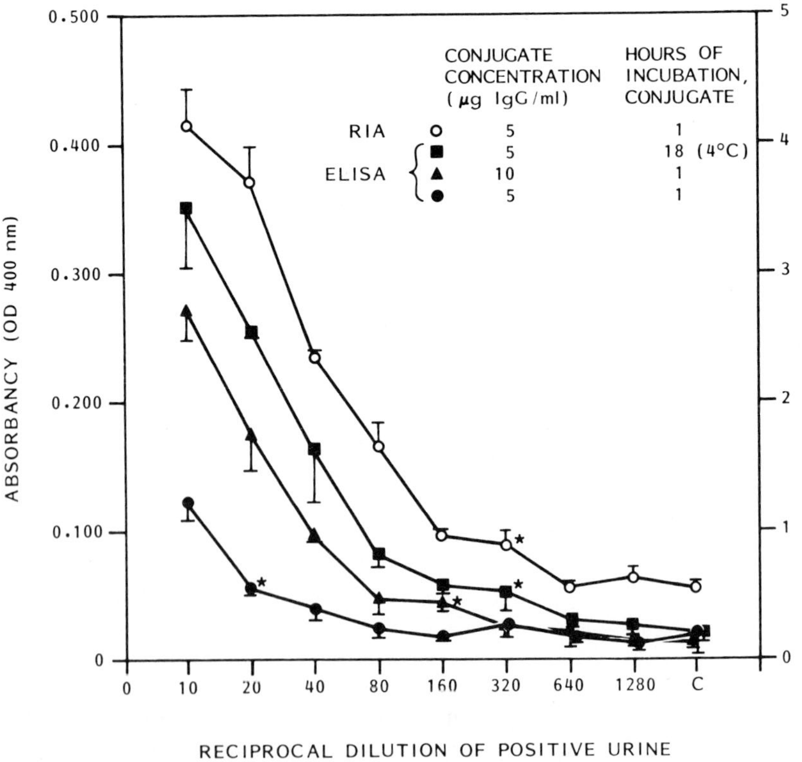

Fig. 6. Comparison of ELISA and RIA. Serial twofold dilutions of antigen-containing urine were tested in both assays. Except for the conjugate step, all assay steps were performed at 37°C with a 1-hr incubation time; conjugate step conditions were varied as shown. Sensitivity, defined as the highest dilution significantly different from the diluent (0.01 *M* Tris HCl, pH 7.0, labeled C), at *P* value 0.05 by *t*-testing, is denoted by the asterisk for each curve. Each symbol and bar represent the mean ± 1 SD for three tubes. KCPM, Counts per minute (thousands). From Sathapatayavongs *et al.* (1982). Published with permission of the American Journal of Medicine.

tion of alkaline phosphatase-labeled immunoglobulin G, the incubation time was decreased (Fig. 6). Under these circumstances ELISA was approximately as sensitive as SPRIA, although it required more antiserum. We have not determined why the alkaline phosphatase–immunoglobulin G conjugates react differently from the ^{125}I-immunoglobulin G conjugates. Ford *et al.* (1978) showed that, following coupling of alkaline phosphatase to immunoglobulin G with glutaraldehyde (the method we used), large polymers of immunoglobulin G moelcules are formed. Glutaraldehyde may therefore adversely affect the reactivity of the immunoglobulin G.

Another difference between ELISA and SPRIA is that occasional urine specimens are false positive by ELISA and negative by SPRIA. Of 178 control urine specimens, three were false positive by ELISA but not by SPRIA. However, these positive reactions were eliminated by heating at 100°C for 5 min. Heat treatment did not affect the sensitivity of the assay for true positive reactions.

Urine specimens from 47 patients with Legionnaires' disease were tested by both ELISA and SPRIA. Both tests were positive for 37 specimens and negative for 7. Two other specimens were positive by ELISA and negative by SPRIA, and one was positive by SPRIA and negative by ELISA. In all three cases in which the ELISA and SPRIA disagreed the test giving the positive result was barely positive. Thus, except for the need for longer performance time or increased immunoglobulin G consumption and the need for brief preheating of test specimens, ELISA produced results similar to those produced by SPRIA.

In addition to eliminating the need to work with ^{125}I, ELISA has an important advantage. Of the 39 ELISA-positive specimens, 35 were positive to the naked eye (Fig.7). Four specimens were equivocal by visual interpretation, and spectrophotometric analysis was needed to confirm their positivity. None of 178 preheated control specimens was positive by ELISA when read visually. Thus, the ELISA performed nearly as well as the SPRIA without special (and expensive) instruments. The supply of antiserum to meet the demands of large-scale use is a potential problem with ELISA, as with SPRIA.

IV. SUMMARY

The majority of patients with serogroup 1 *L. pneumophila* pneumonia excrete a detectable quantity of antigen in their urine. SPRIA and ELISA detect these antigens equally well. Each method has its own advantages and disadvantages. Barring unexpected difficulties in developing adequate quantities of suitable antisera, immunoassays to diagnose serogroup 1 *L. pneumophila* infections should become widely available in several years. Whether simpler immunoassay procedures will preempt SPRIA and ELISA is unclear. A simple reversed passive hemagglutination procedure for detecting *L. pneumophila* antigen has been

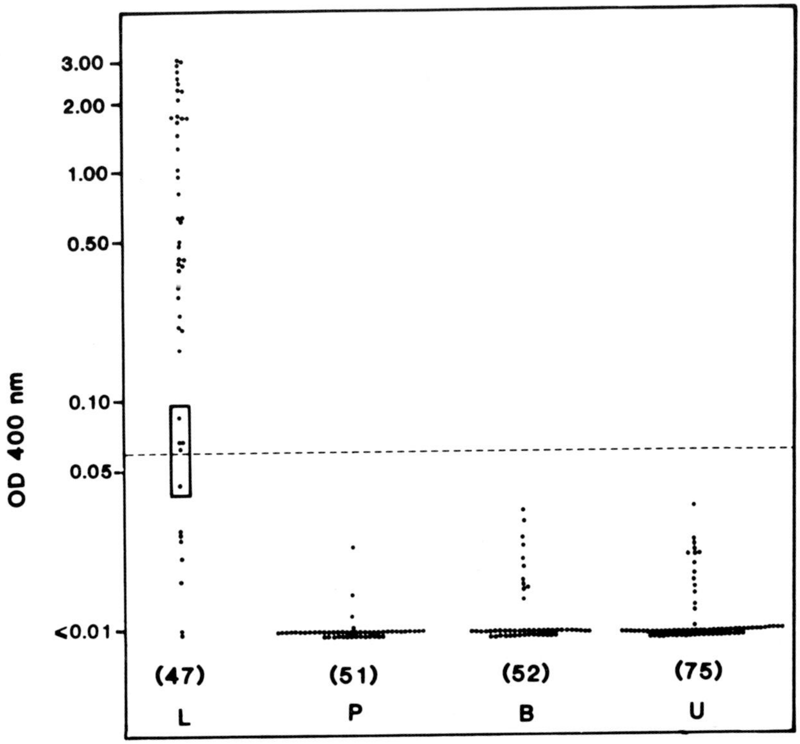

Fig. 7. Detection of *Legionella pneumophila* antigen in urine by ELISA. The absorbance at 400 nm (OD 400 nm), shown logarithmically on the ordinate, is the difference between the absorbance of the test specimen and the mean absorbance of seven negative controls run on the same day. A difference of 0.060 or greater was considered positive. By visual detection of color, urine specimens represented by dots in the rectangle were read as equivocal; the ones above and below the rectangle were read as definitely positive and definitely negative, respectively. The numbers in parentheses are the numbers of urine specimens tested in each group. Abbreviations: L, Legionnaires' disease; P, pulmonary disease; B, bacteremic infections; U, urinary tract infection. From Sathapatayavongs *et al.*, (1982). Published with permission of the American Journal of Medicine.

described but has not been tested adequately to determine its sensitivity and specificity in clinical situations (Mangiafico *et al.*, 1981). Our own work with latex agglutination suggests that it will detect about 80% of patients positive by SPRIA or ELISA, although it is 16-fold less sensitive when serial dilutions of antigen-containing urine are tested (Sathapatayavongs *et al.*, 1983). Current serogroup 1 antigen detection systems appear to have the capacity to simplify the management of pneumonia and thus to save lives through early specific diagnosis.

REFERENCES

Berdal, B. P., Farshy, C. E., and Feeley, J. C. (1979). *J. Clin. Microbiol.* **9,** 575–578.

Blackmon, J. A., Chandler, F. W., Cherry, W. B., England, A. C., III, Feeley, J. C., Hicklin, M. D., McKinney, R. M., and Wilkinson, H. W. (1981). *Am. J. Pathol.* **103,** 429–465.

Broome, C. V., Cherry, W. B., Winn, W. C., Jr., and MacPherson, B. R. (1979a). *Ann. Intern. Med.* **90,** 1–4.

Broome, C. V., Goings, S. A. J., Thacker, S. B., Vogt, R. L., Beaty, H. N., Fraser, D. W., and the Field Investigation Team (1979b). *Ann. Intern. Med.* **90,** 573–577.

Cherry, W. B., Pittman, B., Harris, P. P., Hebert, G. A., Thomason, B. M., Thacker, L., and Weaver, R. E. (1978). *J. Clin. Microbiol.* **8,** 329–338.

Cohen, M. L, Broome, C. V., Paris, A. L., Martin, W. T., and Allen, J. R. (1979). *Ann. Intern. Med.* **90,** 611–613.

Dochez, A. R., and Avery, O. T. (1917). *J. Exp. Med.* **26,** 447–493.

Edelstein, P. H. (1981). *J. Clin. Microbiol.* **14,** 298–303.

Edelstein, P. H., McKinney, R. M., Meyer, R. D., Edelstein, M. A. C., Krause, C. J., and Finegold, S. M. (1980a). *J. Infect. Dis.* **141,** 652–655.

Edelstein, P. H., Meyer, R. D., and Finegold, S. M. (1980b). *Am. Rev. Respir. Dis.* **121,** 317–327.

England, A. C., and Fraser, D. W. (1981). *Am. J. Med.* **70,** 707–710.

Farshy, C. E., Klein, G. C., and Feeley, J. C. (1978). *J. Clin. Microbiol.* **7,** 327-331.

Feigin, R. D., Wong, M., Shackelford, P. G., Stechenberg, B. W., Dunkle, L. M., and Kaplan, S. (1976). *J. Pediatr. (St. Louis)* **89,** 773–677.

Flesher, A. R., Kasper, D. L., Modern, P. A., and Mason, E. O., Jr. (1980). *J. Infect. Dis.* **142,** 313–318.

Ford, D. J., Radin, R., and Pesce, A. J. (1978). *Immunochemistry* **15,** 237–243.

Foy, H. M., Hayes, P. S., Cooney, M. K., Broome, C. V., Allan, I., and Tobe, R. (1979). *Lancet* **I,** 767–770.

Grist, N. R., Reid, D., and Najera, R. (1979). *Ann. Intern. Med.* **90,** 563–564.

Halbert, S. P., and Auken, M. (1977). *J. Infect. Dis.* **136,** S318–S323.

Haley, C. E., Cohen, M. L., Halter, J., and Meyer, R. D. (1979). *Ann. Intern. Med.* **90,** 583–586.

Kacaki, J., Wolters, G., Kuijpers, L., and Stulemeyer, S. (1978). *Vox Sang.* **35,** 65–74.

Kirby, B. D., Snyder, K. M., Meyer, R. D., and Finegold, S. M. (1980). *Medicine (Baltimore)* **59,** 188–204.

Kohler, R. B., Wheat, L. J., and White, A. (1979). *J. Clin. Microbiol.* **9,** 253–258.

Kohler, R. B., Wheat, L. J., and White, A. (1980). *J. Clin. Microbiol.* **12,** 39–43.

Kohler, R. B., Zimmerman, S. E., Wilson, E., Allen, S. D., Edelstein, P. H., Wheat, L. J., and White, A. (1981). *Ann. Intern. Med.* **94,** 601–605.

Kohler, R. B., Winn, W. C., Jr., Girod, J. C., and Wheat, L. J. (1982). *J. Infect. Dis.* **146,** 144.

Kozaki, S., Dufrenne, J., Hagenhaars, A. M., and Notermans, S. (1979). *Jpn. J. Med. Sci. Biol.* **32,** 199–205.

McConahey, P. J., and Dixon, F. J. (1966). *Intern. Arch. Allergy Appl. Immunol.* **29,** 185–189.

Mangiafico, J. A., Hedlund, K. W., and Knott, A. R. (1981). *J. Clin. Microbiol.* **13,** 843–845.

Meyer, R. D., and Finegold, S. M. (1980). *Annu. Rev. Med.* **31,** 219–232.

Neill, J. M., Sugg, J. Y., and McCauley, D. W. (1951). *Proc. Soc. Exp. Biol. Med.* **77,** 775–783.

Politi, B. D., Fraser, D. W., Mallison, G. F., Mohatt, J. V., Morris, G.K., Patton, C. M., Feeley, J. C., Telle, R. D., and Bennett, J. V. (1979). *Ann. Intern. Med.* **90,** 587–591.

Sathapatayavongs, B., Kohler, R. B., Wheat, L. J., White, A., Winn, W. C., Jr., Girod, J. C., and Edelstein, P. H. (1982). *Am. J. Med.* **72,** 576–582.

Sathapatayavongs, B., Kohler, R. B., Wheat, L. J., and Winn, W. C., Jr. (1983). *Am. Rev. Resp. Dis.* (in press).

Tabbarah, Z. A., Kohler, R. B., Wheat, L. J., Griep, J. A., and White A. (1979). *J. Infect. Dis.* **140,** 822–825.

Tilton, R. C. (1979). *Ann. Intern. Med.* **90,** 697–698.

Voller, A., Bidwell, D. E., and Bartlett, A. (1976). *Bull. WHO* **53,** 55–65.

Wheat, L. J., Kohler, R. B., and White, A. (1978). *J. Infect. Dis.* **138,** 174–180.

Wilkinson, H. W., Cruce, D. D., Fikes, B. J., Yealy, L. P., and Farshy, C. E. (1979). *In* "Legionnaires'" (G. L. Jones and G. A. Hebert, eds.), pp. 111–116, HEW Publ. No. (CDC) 79-8375. U. S. Dept of Health, Education and Welfare, Washington, D.C.

Wolters, G., Kuijpers, L., Kacaki, J., and Schuurs, A. (1976). *J. Clin. Pathol.* **29,** 873–879.

Wolters, G., Kuijpers, L., Kaçaki, J., and Schuurs, A. (1977). *J. Infect. Dis.* **136,** S311–S317.

Yolken, R. H., and Stopa, P. J. (1980). *J. Clin. Microbiol.* **11,** 546–551.

17

Detection of Fungal Antigens in Clinical Samples

MARC H. WEINER

Audie L. Murphy Memorial Veterans' Administration Hospital
and
Department of Medicine
University of Texas Health Sciences Center
San Antonio, Texas

I. Introduction ... 207
II. Candidiasis ... 208
A. Detection of Antigens ... 208
B. *Candida* Radioimmunoassay Protocol ... 213
C. Enzyme-Linked Immunosorbent Assay Protocol ... 214
III. *Aspergillus* ... 215
A. Detection of Antigens ... 215
B. Detailed Protocol for the Detection of *Aspergillus* Antigen by Radioimmunoassay ... 219
References ... 221

I. INTRODUCTION

Invasive opportunistic mycoses are common infections in patients with myeloproliferative diseases, renal transplant recipients, and postsurgical patients. Although the fatality rate is high, early diagnosis and therapy can improve the chances of survival. However, difficulties are encountered in the traditional methods of clinical, microbiological, and serologic diagnosis of aspergillosis and candidiasis.

An alternative approach to diagnosis is the detection of a fungal antigen or metabolic products in serum and other body fluids. This approach has two possible advantages: a direct correlation can be established between antigenemia and disease activity, and the assay is not dependent on a patient's humoral

The Direct Detection of Microorganisms in Clinical Samples

ISBN 0-12-187780-9

response, which can be attenuated or delayed due to underlying disease or chemotherapy. The demonstration in body fluids of capsular pneumococcal (Coonrod and Rytel, 1973), meningococcal (Edwards, 1971), and cryptococcal (Goodman *et al.*, 1971) polysaccharides proved useful for the serodiagnosis of these infections and has since been incorporated into the routine laboratory data base. Subsequently, sensitive immunoassay and biochemical techniques have been applied to the detection of *Candida* and *Aspergillus* infections. With refinement of assay procedures and the commercial availability of immunologic reagents, widespread use of the methodology is anticipated.

II. CANDIDIASIS

A. Detection of Antigens

Although recognized as a significant infectious complication in immunocompromised patients, systemic candidiasis is often difficult to diagnose antemortem. The routine microbiological data are often misleading. Invasive candidal infection without documented candidemia occurs 28–64% of the time in reported series (Gaines and Remington, 1973; Rosner *et al.*, 1971); conversely, transient candidemia without later evidence of systemic disease also occurs. The variable utility of diagnostic cell-mediated and humoral immune assays has likewise been emphasized. Skin testing with candidal antigens has not been useful for the diagnosis of acute infections because of the high incidence of positive tests in healthy populations (Shannon *et al.*, 1966) and because of possible anergy in those with overwhelming infection. Measurement of serum agglutinins to nonviable *Candida* has, in general, been unrewarding (Jones *et al.*, 1973). The serologic demonstration of precipitins to candidal antigens has been widely applied. Difficulties with the immunoprecipitin assays include false negative tests in up to 50% of cases with disseminated candidiasis (Filice *et al.*, 1977) and false positive reactions of 10 to 44% in selected patient groups (Kozinn *et al.*, 1976; Murray *et al.*, 1969). In any case, mortality remains high. Gaines and Remington (1973), for example, reported a 92% antemortem diagnosis using precipitin techniques but 42% mortality secondary to disseminated fungal infection; 9 of 11 patients died before an adequate course of antifungal therapy was completed. Thus, even under the best reported circumstances, current methods of diagnosis allow insufficient time for treatment, and significant mortality is encountered.

To develop an antigen immunoassay for *Candida*, a fungal pathogen without a known capsule, we (Weiner and Yount, 1976) studied a candidal carbohydrate that frequently elicited humoral responses after infection. We found that in 9 of 14 cases of systemic candidiasis, anticandidal antibody directed to a cathodally migrating antigen developed (Fig. 1). The antigen was identified as a cell wall polysaccharide, mannan. Mannan was isolated as a carbohydrate–protein conjugate. When the protein component was enzymatically digested, biochemically extracted, or heat-denatured, the mannan maintained immunoreactivity with specific antisera. To cell wall mannan obtained from *C. albicans* serogroup A we developed a hemagglutination inhibition assay using conconavalin A to bind the carbohydrate to sheep erythrocytes. In the clinical evaluation of the efficacy of the hemagglutination inhibition assay, candidal antigenemia was detected in 4 of 14 patients with systemic candidiasis and 2 of 5 patients with invasive gastrointestinal candidiasis. In contrast, mannan was not detected in 48 patients with noninvasive candidal or with other mycotic infections or in 99% of 234 patients in other control groups. Antimannan antibodies were detected in nearly all patients and normal controls. In patients with invasive candidiasis an early antigenemic period was followed by a rapid rise in antimannan antibodies. These findings suggested that candidal serodiagnosis could be improved by the detection of mannan antigenemia and that a positive test was an early and specific signal of invasive disease. Using the hemagglutination inhibition assay to group A *C. albicans* mannan, Meunier-Carpentier and Armstrong (1981) reported similar findings. Antigenemia was detected in 51% of 57 patients with candidemia. In patients with disseminated candidiasis, antigenemia was detected a mean of 4.4 days before positive blood cultures.

In all, candidal antigenemia has been documented in 13 of the 14 studies of invasive candidiasis published to date (Table I). Also, mannan or mannan–protein conjugates were described as the predominant antigen detected in 10 of 13 studies. Nearly all published work has been with *C. albicans* serogroup A. We have found, however, that diagnostic sensitivity is improved by testing for *C. albicans* serogroups A and B, as well as non*Candida albicans* species. To

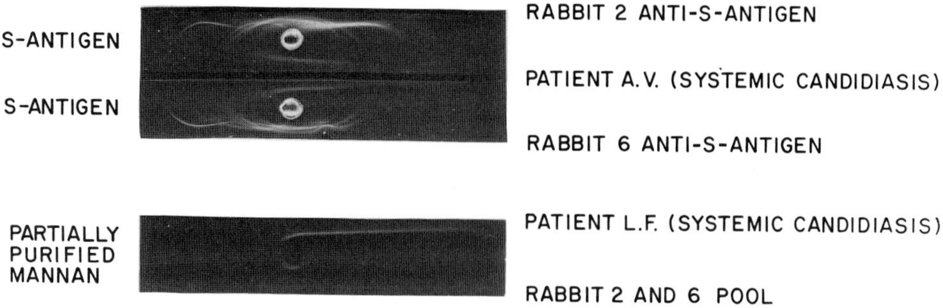

Fig. 1. Immunoelectrophoresis of sonicate of whole *Candida albicans* (S-antigen) in upper and middle wells. The anode is on the left. Rabbits immunized with S-antigen exhibit specificity for 15 *Candida* antigens. Sera from patients A. V. and L. F. with systemic candidiasis recognize a cathodally migrating antigen, which was identified as cell wall mannan (Weiner and Yount, 1976).

TABLE I

Antigen Serodiagnosis of Invasive Candidiasis

Immunoassay protocol									
Technique[a]	Preliminary extraction	Competitive inhibition	Antigen	Type of study	Invasive candidiasis (*n*)	Sensitivity (%)	Controls (*n*)	Specificity (%)	Reference
HI	56°C × 1 hr	+	Mannan, *C. albicans* group A, strain B311	Clinical	17	35	282	99	Weiner and Yount (1976)
ELISA	None	−	*C. albicans* strain 6224	Experimental	2	100	2	100	Warren *et al.* (1977)
				Clinical	3	100	[b]	—	
RIA	None	−	Phenol extract, *C. albicans* strain 9938	Experimental	27	70	14	100	Poor and Cutler (1979)
CIE	None	−	Formamide extract, *C. albicans* group A	Clinical	8	100	43	100[c]	Kerkering *et al.* (1979)
ELISA–I	56°C ×2 hr in 0.5 *N* NaOH	+	Mannan, *C. albicans* group A, strain B311	Clinical	7	100[d]	20	100	Segal *et al.* (1979)
RIA	96°C × 20 min in 0.04 *M* citrate, pH 2.7	+	Mannan, *C. albicans* group A	Experimental	29	52	91	100	Weiner and Coats-Stephen (1979b)
			Mannan, *C. tropicalis*	Clinical	15	47	51	100	
ELISA, CIE	None	−	Mannan, *C. albicans* group A, strain 20	Experimental	—[e]	0	—	—	Lehmann and Reiss (1980)
				Clinical	6	0	—	—	

ELISA	56°C × 0.5 hr	–	Mannan, *C. albicans* group A	Experimental	12	83	27	100	Harding *et al.* (1980)
RIA	None	–	Cytoplasmic antigens, *C. albicans*	Clinical	22	55	4[f]	100	Stevens *et al.* (1980)
ELISA	56°C × 0.5 hr	–	Mannan, *C. albicans*, group A	Experimental	16	75	34	100	Scheld *et al.* (1980)
HI	56°C × 0.5 hr	+	Mannan, *C. albicans*, group A	Clinical	32	50	71	100	Meunier-Carpentier and Armstrong (1981)
ELISA–I	56°C × 2 hr in 0.5 *N* NaOH	+	Mannan, *C. albicans* group A strain 20A	Clinical	7	100	82	91	Meckstroth *et al.* (1981)
ELISA	121°C × 5 min in PBS, 0.05% Tween 20, and 0.02 *M* Na_2-EDTA	–	Mannan, *C. albicans* group A, strain B311	Experimental Clinical	30 15	100 53	30 449	100 93[g]	Lew *et al.* (1982)

[a] HI, Hemagglutination inhibition; ELISA, enzyme-linked immunosorbent assay (double antibody); RIA, radioimmunoassay; CIE, counterimmunoelectrophoresis; ELISA–I, enzyme-linked immunosorbent assay–inhibition.

[b] Mean value of nine normal sera reported as negative control.

[c] Of 13 antigen-positive patients, 8 were proven to have invasive disease. Histologic data were not obtained in the remaining 5 patients.

[d] Author's criteria for positive and negative antigenic activity not stated, although discrimination between infected cases and controls was achieved.

[e] Number of experimental rabbits not reported.

[f] Unspecified number of patients with gram-positive and gram-negative bacteremias were negative with the RIA.

[g] False positive controls were negative (i.e., true negative) on retest or failed to show neutralization with anticandidal antibody.

increase immunoassay sensitivity, we have employed competitive inhibition methods in developing the mannan hemagglutination inhibition assay (Weiner and Yount, 1976) and radioimmunoassay (RIA) (Weiner and Coats-Stephen, 1979b), and Segal *et al.* (1979) have adapted enzyme-linked immunosorbent assay–inhibition (ELISA–I) to mannan detection. Technically simpler, noncompetitive double-antibody RIA (Poor and Cutler, 1979) and ELISA (Harding *et al.*, 1980) have also been used successfully for candidal antigen serodiagnosis. However, a comparison of diagnostic sensitivity between competitive and noncompetitive techniques using the same immunologic reagents has not been reported.

In developing a competitive inhibition RIA to mannan antigen, we labelled the nonantigenic protein component of the peptidomannan conjugate with ^{125}I. With an RIA to a cross-reactive candidal mannan, we found that the diagnostic sensitivity was 89% in patients with disseminated candidiasis but 75% in all patients with candidal infections or candidemia. False negative cases were related in part to documented serotypic differences in the infecting candidal species and strains. Also, serum concentrations of mannan were lower in patients with localized infections than in patients with disseminated disease. The mean mannan concentration was 78 ng/ml in disseminated candidiasis but 29 ng/ml in localized infections of the kidney, peritoneum, or gastrointestinal tract. Thus, in comparing the diagnostic efficacy of different assays, the study patients and their extent of infection must be taken into account.

Serum binding of fungal antigens in immune and nonimmune complexes should be considered in the development and application of fungal antigen assays. We have been unable to detect unbound, free mannan in serum and have had to extract specimens before assay in order to dissociate candidal antigen from serum complexes. Various extraction procedures involve heating from 56 to 100°C in neutral, acidic, or basic buffers (Table I) and/or proteolytic digestion (Weiner, 1980; Lew *et al.*, 1982).

In the only published study (Lehmann and Reiss, 1980) in which candidal antigenemia could not be detected, sera were not extracted before immunoassay. Subsequently, when the same investigators (Meckstroth *et al.*, 1981) modified their ELISA by employing an extraction protocol (Segal *et al.*, 1979) and a competitive inhibition technique, mannan antigenemia was detected. In this study 10 of 92 patients with leukemia developed candidiasis. Mannan antigenemia was detected in 7 patients with disseminated infections. The specificity was 91% in 82 patients with noninvasive candidal colonization. In another study by Lew *et al.* (1982) a preliminary extraction and double-antibody ELISA was employed. In 30 rats lethally infected with candidiasis, antigenemia was detected within 24 hr of infection and persisted until death. In clinical studies the diagnostic sensitivity of the ELISA was 53% and the specificity was 100% when specimens were retested with and without a neutralizing antibody control.

In another approach to diagnosis, the biochemical detection in body fluids of candidal sugars and fatty acids has been studied. Miller *et al.* (1974) were the first to employ gas–liquid chromatography successfully for fungal diagnosis. By the use of trimethylsilyl ether derivatives of sera and flame ionization detectors, abnormal chromatograms were detected in serum from 6 patients with candidemia and from 2 of 4 patients with invasive candidal infections. Two of the four abnormal peaks cochromatographed with mannose. The two independent findings of candidal mannan antigenemia by hemagglutination inhibition and circulating mannose by gas–liquid chromatography were in accordance because candidal mannan is a homopolysaccharide of mannose. Using aldonitrile acetate derivitization, Monson and Wilkerson (1981) confirmed the finding of elevated mannose levels in candidiasis. In a study of 110 hospital patients, free mannose levels were elevated in 6 patients with invasive candidiasis. The test appeared to be specific with only one exception; when the blood glucose concentration was greater than 300 mg/dl in poorly controlled diabetic patients, free mannose levels appeared to be elevated and mannose concentrations overlapped with patients with invasive candidiasis.

The candidal metabolite arabinitol has also been reported to be a biochemical marker of invasive candidal infection. In two different studies Kiehn *et al.* (1979) employed gas–liquid chromatography and trimethylsilyl ether derivatives, whereas Roboz *et al.* (1980) used combined gas chromatography–mass spectroscopy with selected ion monitoring. Because of the accumulation of arabinitol in renal failure, the use of an arabinitol/creatinine ratio to increase specificity was proposed (Gold *et al.*, 1981). The diagnostic sensitivity of the ratio was 66% in patients with normal renal function and 54% in patients with elevated serum creatinine levels.

B. *Candida* Radioimmunoassay Protocol

1. Protocol for RIA Antigens

A complex mixture of somatic and cell wall antigens was prepared by homogenization of 2-day cultures of *C. albicans* group A (strain B311) or *C. tropicalis* in a cell homogenizer (MSK, B. Braun Instruments, San Francisco) at 4°C. The homogenate was centrifuged at 12,000 *g* for 20 mins at 4°C. The supernatants were pooled, dialyzed, and stored at −80°C until used.

Candidal peptidomannans used for the RIA were purified by ultrafiltration and bioaffinity chromatography. The pooled supernatant was fractionated by ultrafiltration with the use of an XM100A Diaflo membrane (Amicon, Lexington, Massachusetts). The permeable solute was collected and concentrated with an XM50 Diaflo membrane. The retenate was then fractionated with an immunosorbent antibody to mannan. *Candida tropicalis* mannan used for a second RIA was prepared by the method of Peat *et al.* (1961). Purified *C. albicans* mannan (25 μg) was radiolabeled with ^{125}I by the use of 50 μg of chloramine-T. The labeled mannans were fractionated by gel filtration chromatography (Sephadex G-10) and by bioaffinity chromatography with a mannan immunosorbent.

2. *RIA Protocol*

Radioimmunoassays for each sample were performed in triplicate. Total serum mannan was determined as follows: 250 μl sera diluted 1 : 5 in 0.9% NaCl was acidified with 250 μl 0.04 *M* citrate buffer, pH 2.7, heated at 96°C for 20 min, and clarified by centrifugation at 3900 *g* for 12 mins. In triplicate, 50 μl supernatant was neutralized with 50 μl 0.35 *M* phosphate buffer. Neutralized sera were incubated at 20°C for 2 hr with 100 μl antisera to *C. albicans* serotype A (strain B311) diluted in 0.1 *M* Trizma buffer (pH 7.3; Sigma Chemical Co., St. Louis, Missouri), 0.9% NaCl with 2% bovine serum albumin, and 0.1% NaN_3 (RIB) to bind 25–35% of counts in the absence of inhibiting carbohydrate, and then incubated at 4°C for 14 hr with ^{125}I-labeled antigen at 10,000 cpm diluted in 100 μl RIB. Radiolabeled *C. albicans* and *C. tropicalis* mannans were used in the RIA analysis of human sera. Incubation solutions were precipitated with 280 μl of 90% saturated ammonium sulfate, to which was added 100 μl of newborn calf serum absorbed on a mannan immunosorbent and diluted 1 : 5 in RIB with 6% dextran [molecular weight 150,000 (Sigma)]. After incubation for 20 min at 4°C, tubes were centrifuged at 5800 *g* for 12 min. The supernatants were aspirated, and the precipitates were counted in an Auto-Gamma scintillation spectrometer (Packard Instruments, Downers Grove, Illinois). Alternatively, antibody-bound radiolabeled mannan can be separated from unbound label by solid-phase absorption with staphylococcal protein A. Purified mannan added to normal human serum absorbed over a Sepharose (Pharmacia Fine Chemicals, Uppsala, Sweden) mannan immunosorbent was employed for standards. Standard curves were derived from linear transformation of raw counts in known standards as described by Groner *et al.* (1977). Assays were rejected if the correlation coefficient of the standard curve was <0.975. Concentrations of candidal mannan in unknowns were calculated from the standard curves when the bound/bound at zero concentration ratio (B/B_0) was <0.91.

C. Enzyme-Linked Immunosorbent Assay Protocol

Enzyme-linked immunosorbent assays using noncompetitive double-antibody "sandwich" protocols or competitive inhibition techniques have been applied to candidal antigen serodiagnosis (Table I). To improve assay sensitivity to heat-stable mannans, clinical specimens have been heat-extracted to denature serum proteins before immunoassay. In addition to preliminary extraction by heating in citric acid (Weiner, 1979), as described in the RIA protocol, and in sodium hydroxide (Segal *et al.*, 1979), proteolytic digestions (Weiner, 1979; Lew *et al.*, 1982) and EDTA–heat extraction (Lew *et al.*, 1982) have been employed. For ELISA, conjugates of alkaline phosphatase (Type VII, Sigma) and antiserum were prepared with 0.2% glutaraldehyde (Voller *et al.*, 1976). The ELISA double-antibody assay described below is less sensitive than the comparable competitive inhibition RIA but is simpler and more easily performed.

Antimannan antibody in 0.02 *M* bicarbonate buffer, pH 9.6, is adsorbed overnight at 4°C to polystyrene microtitration substrate plates (Dynatech Laboratories, Alexandria, Virginia). The optimal dilution of the first "coat" antisera is established by cross-titration with positive standards.

Test wells are washed three times with phosphate-buffered saline (PBS)–Tween buffer made from 0.01 *M* phosphate buffer, pH 7.4, 0.14 *M* NaCl, and 0.05% Tween 20. After preliminary extraction, duplicate 200-μl samples of positive and negative reference sera and unknown test samples are incubated at 37°C for 3 hr in antibody-coated wells and control uncoated wells. After incubation the plate is washed as before with PBS–Tween buffer. Then 200 μl of an alkaline phosphatase–antimannan globulin conjugate diluted in PBS–Tween buffer is added and incubated for 1 hr at 37°C. The optimal dilution of conjugate is determined by cross-titration with reference positive and negative sera. After the plate is washed, 200 μl of the substrate *p*-nitrophenyl phosphate (Sigma 104 phosphate substrate) is added and incubated for 30 min at room temperature. Fifty microliters of 3 *M* NaOH is used to stop the reaction, and the color change is read at 400 nm. Baseline levels of activity recorded in sera from uninfected controls are recorded. Significant levels of antigenic activity are found 2 SDs above baseline control activity levels. Concentrations in test samples are estimated from standard curves.

III. *ASPERGILLUS*

A. Detection of Antigens

Invasive *Aspergillus* infections are frequently difficult to diagnose antemortem. The presentation of a febrile patient with pulmonary aspergillosis is not specific, and the microbiology is often not helpful. Blood cultures are rarely positive, and *Aspergillus* is recovered from other sites only 12–34% of the time, despite vigorous attempts at fungal culture (Young *et al.*, 1970; Meyer *et al.*, 1972). Furthermore, if *Aspergillus* is isolated, it can represent colonization without infection or specimen contamination. Although the classical assays for *Aspergillus* antibody are of proven diagnostic value for noninvasive forms of aspergillosis, that is, pulmonary allergic aspergillosis and aspergilloma, the serologic data for invasive aspergillosis are limited and conflicting. Young and

Bennett (1971) were unable to detect antibodies to *Aspergillus fumigatus* in sera from 15 patients within 3 weeks of death from systemic aspergillosis. Double immunodiffusion, complement fixation, immunoelectrophoresis, and indirect fluorescent antibody assays were used. In contrast to the negative results in patients with widespread invasive infection, anti-*Aspergillus* antibody was detected in a patient with noninvasive bronchopulmonary aspergillosis. In one study (Weiner *et al.*, 1982), we found that antibody serology detected by ELISA to a polyvalent mixture of *Aspergillus* antigens had relatively low diagnostic efficacy. In sera from 14 patients with acute leukemia and 24 normal controls, the diagnostic sensitivity of *Aspergillus* antibody detection for the diagnosis of invasive aspergillosis was 33%, the specificity was 75%, the predictive value of a positive result was 20%, and the predictive value of a negative result was 86%. In contrast, Schaefer *et al.* (1976) detected anti-*Aspergillus* antibodies by prospectively following sera obtained biweekly from 80 patients with leukemia. Of 10 documented cases of systemic aspergillosis, 7 developed a positive double immunodiffusion antibody test. Only 4 were diagnosed early and treated successfully. An additional 6 patients seroconverted, but these were not included because of inadequate documentation of aspergillosis. Thus, even with the best reported experience with systemic aspergillosis, current serologic methods of diagnosis may not detect the disease or may allow insufficient time for treatment.

In our initial studies (Weiner and Coats-Stephens, 1979a) we found that the predominant humoral response of infected animals was directed against four *Aspergillus* antigens identified by crossed immunoelectrophoresis. Two of the mycelial antigens were contained in a high molecular weight preparation. When an RIA was developed with this crude preparation, antigenic activity was detected in sera from heavily infected animals before death from disseminated aspergillosis. In light of this promising initial result, the preparation was further fractionated by gel filtration and bioaffinity chromatography, and one component was used to develop a sensitive RIA.

When the RIA was evaluated in an animal model, antigenemia was detected in 86% of 42 rabbits with lethal infection. The RIA appeared to have a high specificity: Negative controls included sera from 76 normal rabbits and from 25 rabbits with systemic candidiasis. The candidal control group was included because 48% of these rabbits had candidal antigenemia detected by a mannan RIA. These studies demonstrated that *Aspergillus* antigenemia occurred antemortem during the course of experimental aspergillosis and illustrated the potential of antigen serodiagnosis for systemic aspergillosis.

Initial clinical studies were encouraging as well (Weiner, 1979, 1980). Antigen was detected in 4 of 7 patients who died of systemic aspergillosis documented at autopsy. Antigen was also detected in a pleural fluid from an *Aspergillus* empyema. In the analysis of control groups, *Aspergillus* antigenemia was not detected. Negative controls included 6 patients colonized with *As-*

pergillus or convalescent from systemic aspergillosis, 4 patients with chronic mucocutaneous candidiasis, 10 patients with systemic candidiasis, and 3 patients with histoplasmosis, 14 patients with bacterial sepsis, and 27 normal donors. In a patient whose serum had both candidal mannan and *Aspergillus* antigenic activity, concurrent systemic candidiasis and aspergillosis were documented.

In other studies Shaffer *et al.* (1979b) described an RIA to an alkali extract of *A. fumigatus*. Antigenic activity was detected in six sera obtained from four of six rabbits with disseminated aspergillosis. A control group with a second infection was not studied. In a follow-up clinical study by the same investigators (Shaffer *et al.*, 1979a) *Aspergillus* antigenemia was detected in two patients with disseminated aspergillosis and in a patient with possible fungal pneumonia. Negative control sera were obtained from 8 uninfected subjects and 12 patients with other mycotic diseases. Reiss and Lehmann (1979) also detected antigenemia to an *Aspergillus* galactomannan in sera from eight rabbits and from three confirmed and three possible cases of aspergillosis out of 40 to 60 individuals suspected of having this disease.

In blind, controlled trials the utility of fungal antigenemia for invasive aspergillosis was evaluated (Weiner *et al.*, 1982) and correlated with the detection of *Aspergillus* antibody by ELISA and by routine, clinical diagnostic procedures. In one blind collaborative trial, sera were collected prospectively from patients with acute leukemia hospitalized for induction chemotherapy. Coded specimens from 6 patients (53 sera) with histopathologic evidence of invasive pulmonary aspergillosis, from 8 patients (29 sera) without pulmonary aspergillosis, and from 24 normal controls were tested. Antigenemia was detected in 17 sera from 4 patients with invasive pulmonary aspergillosis due to *A. flavus*. Antigen appeared early in the course of infection of 3 patients, concurrent with the onset and evolution of lung infiltrates; it persisted from 8 to 75 days and remitted with antifungal chemotherapy. The specificity of the RIA was 100%; there were no false positive reactions. Antigenemia was not detected in 3 patients before the onset of invasive pulmonary aspergillosis, in 8 leukemic control subjects, or in the 24 normal subjects. In comparison, *A. fumigatus* antibody was detected in 2 patients with invasive pulmonary aspergillosis but also in 2 leukemic control patients and in 6 normal subjects. *Aspergillus* was isolated from nose cultures in 4 of 6 patients with invasive pulmonary aspergillosis. Seven bronchoscopies were performed in 5 patients with *Aspergillus* pneumonia; from 3, fungi were not identified; and in 2, false positive AFB smears were obtained. In the second blind study, antigenemia was detected from a case of *Aspergillus* sinusitis and in a case of invasive pulmonary aspergillosis documented by culture of a percutaneous lung aspirate but was not detected in 8 controls. These trials have shown that the RIA for *Aspergillus* antigen is highly specific and of moderate sensitivity for the diagnosis of invasive aspergillosis.

In all, we detected antigenemia in 84% of the documented cases of invasive aspergillosis that were tested with the RIA for *A. fumigatus*. Antigenemia was found in 77% of 21 cases of invasive pulmonary aspergillosis, in 4 cases of widely disseminated aspergillosis, in 2 cases of cutaneous dissemination, in 3 cases of endocarditis, in 2 cases of osteomyelitis, and in 2 of 3 episodes of invasive sinusitis, in a case of peritonitis, and in a case of meningitis. Some cases in which antigenemia could not be detected were caused by non *A. fumigatus* species or were not speciated. The RIA for *A. fumigatus* antigen has been found to be cross-reactive in infections from patients with *A. flavus* and *A. terreus*.

Of clinical importance is the temporal characterization of antigenemia in clinical material. Of nine cases in which available sequential specimens adequately covered the clinical course of disease, antigenemia was detected early in the course of infection in six patients. Thus, finding *Aspergillus* antigenemia would result in earlier antifungal chemotherapy, which is critical for successful treatment.

Because invasive aspergillosis characteristically presents as a pneumonia in an immunocompromised patient, it seemed possible that the diagnostic sensitivity of antigen immunodiagnosis could be improved by direct analysis of pulmonary secretions. Bronchoalveolar lavage fluid was tested because it can be obtained clinically by a relatively noninvasive procedure and because it contains constituents representative of the tracheobronchial tree and alveolar space. In an experimental model of disseminated aspergillosis (Andrews and Weiner, 1981), 10 of 11 rabbits with major histologic pulmonary involvement with aspergillosis had *Aspergillus* antigen detected in unconcentrated bronchoalveolar lavage fluid. In an additional 10 rabbits with only minor pulmonary involvement, antigen was detected in 40% of concentrated lavage samples. In contrast, in 35 control animals, which were normal or had candidal or staphylococcal pneumonia, antigen was not detected in bronchoalveolar lavage fluid. One false positive reaction with low-level antigenic activity was detected in one rabbit with candidal pneumonia. Although antigen was present in the serum of 76% of all animals infected with *Aspergillus*, 27% of those with major pulmonary involvement had antigen detected only in bronchoalveolar lavage fluid. In this experimental study an extracellular microbial antigen was detected in bronchoalveolar lavage fluid. Furthermore, pulmonary fluid analysis appeared to be relatively sensitive to disease and augmented the diagnostic information from serum analysis alone. These data were confirmed in clinical studies (Andrews and Weiner, 1982). Antigen was detected in bronchoalveolar lavage fluids from two patients with invasive *Aspergillus* pneumonia and from two patients with aspergillomas. Antigen was not detected in 35 control bronchoalveolar lavage fluids obtained from patients with adult respiratory distress syndrome, immunocompromised patients with pneumonia, and other patients with a variety of pulmonary disorders. In the cases of invasive aspergillosis, antigen was detected in bronchoalveolar lavage fluid 3 days before serum antigenemia was detected in one patient. In the second patient, bronchoalveolar lavage fluid was positive, although antigen was not detected in serum. In a similar manner, it may be that fungal antigen analysis of cerebrospinal fluid could be of immediate diagnostic utility, as was capsular antigen analysis for cryptococcal meningitis.

Finally, antigen immunodiagnosis may be applicable to the diagnosis of other systemic mycoses. With a protocol similar to those described for *Candida* and *Aspergillus,* an RIA was developed for an antigen of *Coccidoides immitis* (Weiner, 1983). In a clinical trial antigenemia was detected in 5 of 9 patients with active coccidioidomycosis but not in 33 patients with other systemic mycoses, in 19 patients with bacterial, mycobacterial, or nocardial infections, or in 50 normal donors.

B. Protocol for the Detection of *Aspergillus* Antigen by Radioimmunoassay

1. Preparation of RIA Antigen

Aspergillus fumigatus mycelia for antigen lots were grown in 2-liter flasks in 1% neopeptone–2% dextrose broth at 35°C for 72 hr. Hyphae were collected by filtration, washed twice with normal saline, and stored at −80°C until used.

A complex mixture of somatic and cell wall antigens was prepared by homogenizing *A. fumigatus* hyphae in a Braun cell homogenizer MSK with 0.5-mm glass beads for 120 sec at 4°C. The homogenate was centrifuged at 12,000 *g* for 20 min at 4°C. The supernatants were pooled, dialyzed, and stored at −80°C until used.

An *A. fumigatus* cell wall antigen used for the RIA was purified by gel filtration chromatography (Weiner and Coats-Stephen, 1979a). The complex *A. fumigatus* mixture was fractionated on Sephacryl S-200 Superfine gel (Pharmacia) under the following conditions: bed dimensions, 2.6 × 100 cm; flow rate, 36 ml/hr; eluant, 0.10 *M* phosphate buffer, pH 7.4, 0.9% NaCl (Fig. 2). The protein content was detected by light absorption at 278 nm. Antigenic analysis was performed with double immunodiffusion (Fig. 3) and rocket electrophoresis. An immunoreactive pool with molecular weights estimated to be between 150,000 and 180,000 was dialyzed, concentrated by ultrafiltration, and conjugated to tyramine with cyanogen bromide used as an activating agent (Porath *et al.*, 1967). A 50-μg amount of the conjugated carbohydrate antigen was radiolabeled with ^{125}I, 50 μg of chloramine-T being used as an oxidizing agent (Hunter, 1973). The labeled antigen was fractionated by gel filtration and bioaffinity chromatography with a Sepharose–*Aspergillus* antibody immunosor-

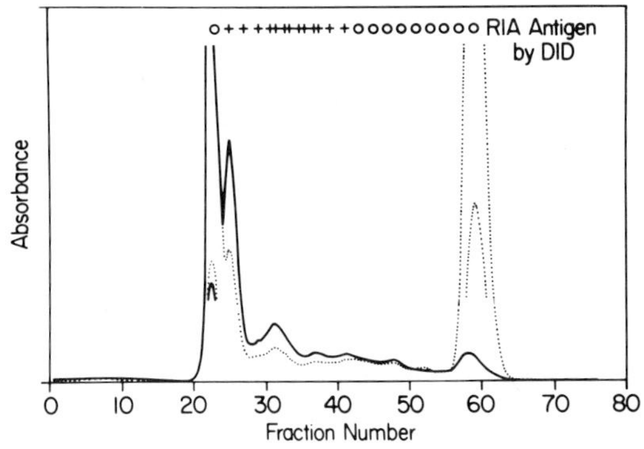

Fig. 2. Chromatography of the complex mixture of *Aspergillus* antigens over Sephacryl S-200 Superfine gel. Antigenic analysis was performed by double immunodiffusion (DID) for the cell wall antigen used to develop the RIA. Fractions positive (+) and negative (○) for the antigen are shown. Void and bed volumes are indicated by blue dextran 2000 (Pharmacia) and barbital absorbance at 278 nm (——) and 254 nm (···) (Weiner and Coats-Stephen, 1979a).

bent. The labeled antigen was 86% precipitable with excess antisera and could be used for 2 to 3 months.

2. *RIA Procedure*

Radioimmunoassays for each sample were performed in triplicate as follows: 500 μl serum diluted 1 : 5 in 0.9% saline was acidified with 500 μl of 0.04 *M* citrate buffer, pH 2.7, heated at 96°C for 15 min, and clarified by centrifugation; 50 μl of supernatant was neutralized in 50 μl of 0.35 *M* sodium phosphate dibasic buffer. Neutralized sera were incubated at 20°C for 2 hr with *Aspergillus* antisera diluted in 100 μl RIB to bind 25–35% of counts in the absence of inhibiting carbohydrate and then at 4°C for 14 hr with ^{125}-I-labeled antigen at 10,000 cpm per 100 μl diluted in 0.1 *M* Trizma buffer (Sigma), pH 7.3, 0.9% NaCl with 2% bovine serum albumin and 0.1% NaN_3 (RIB). Incubation solutions were precipitated with 360 μl of 90% saturated ammonium sulfate, to which 100 μl absorbed newborn calf serum diluted 1 : 5 in RIB was added. After incubation for 20 min at 4°C, tubes were centifuged at 5800 *g* for 12 min. The supernatants were aspirated, and the precipitates were counted in an Auto-Gamma spectrometer. Normal rabbit serum, absorbed over a Sepharose–*Aspergillus* antigen immunosorbent, was employed for standards. Standard curves were derived from linear transformation of raw counts in standards as described by Groner *et al.* (1977). Assays were rejected if the correlation coefficient of the standard curve was <0.975. Concentrations of *A. fumigatus* antigen in unknowns were calculated from the standard curve when bound/bound at zero

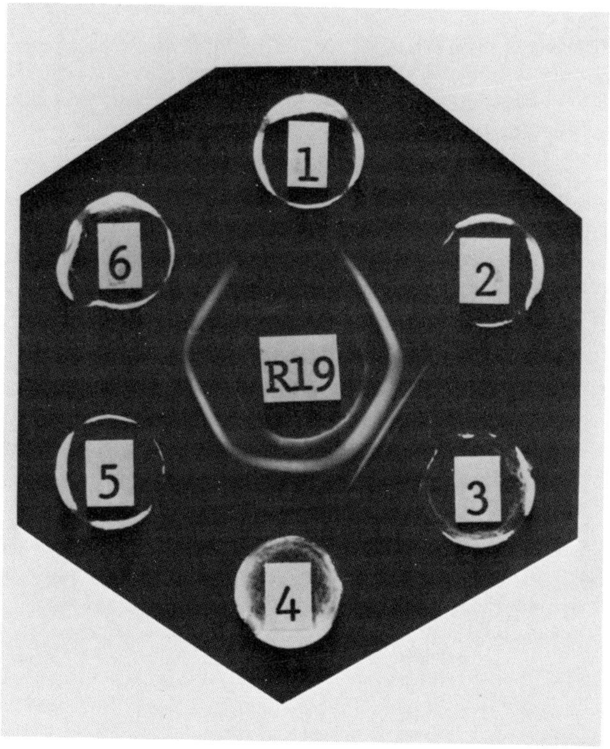

Fig. 3. Double immunodiffusion plate with a polyvalent rabbit antiserum placed in the center well. A line of identity is present between a component of the complex antigen mixture of the *Aspergillus* homogenate (well 3), a carbohydrate extract of *Aspergillus* cell wall (well 4), its 1-*O*-methyl α-D-glucopyranoside eluate from a concanavalin column (well 5), and the fraction obtained by Sephacryl S-200 chromatography that was conjugated to tyramine (well 6). The antigen in the Sephacryl S-200 fraction was removed by periodate oxidation (well 1) (Weiner and Coats-Stephen, 1979a).

concentration (B/B_0) was <0.90. Equivocal activity was found in the B/B_0 range from 0.90 to 0.94. No activity was detected if B/B_0 was greater than 0.94.

REFERENCES

Andrews, C. P., and Weiner, M. H. (1981). *Am. Rev. Respir. Dis.* **124,** 60–64.
Andrews, C. P., and Weiner, M. H. (1982). *Am. J. Med.* **73,** 372–380.
Coonrod, J. D., and Rytel, M. W. (1973). *J. Lab. Clin. Med.* **81,** 770–777.
Edwards, E. A. (1971). *J. Immunol.* **196,** 314–317.
Filice, G., Yu, B., and Armstrong, D. (1977). *J. Infect. Dis.* **135,** 349–357.

Gaines, J. D., and Remington, J. S. (1973). *Arch. Intern. Med.* **132,** 699–702.

Gold, J. W. M., Wong, B., Bernard, E. M., McKean, S. W., Kiehn, T., and Armstrong D. (1981). *Abstr. 21st Interscience Conf. Antimicrob. Agents Chemother.* No. 688.

Goodman, J. S., Kaufman, L., and Koenig, M. G. (1971). *N. Engl. J. Med.* **285,** 434–436.

Groner, G. F., Palley, N. A., Hopwood, M. D., Sibley, W. L., and Fishman, B. (1977). "Clinfo User's Guide," pp. 130–141. Rand Corp., Santa Monica, California.

Harding, S. A., Brody, J. P., and Normansell, D. E. (1980). *J. Lab. Clin. Med.* **95,** 959–966.

Hunter, W. M. (1973). *In* "Handbook of Experimental Immunology" (D. M. Weir, ed.), pp. 17.1–17.36. Blackwell, Oxford.

Jones, S. A., Brennan, M., and Kundsin, R. B. (1973). *J. Surg. Res.* **14,** 235–237.

Kerkering, T. M., Espinel-Ingroff, A., and Shadomy, S. (1979). *J. Infect. Dis.* **140,** 659–664.

Kiehn, T. E., Bernard, E. M., Gold, J. W. M., and Armstrong, D. (1979). *Science* **206,** 577–580.

Kozinn, J. P., Galen, R. S., Taschdjian, C. L., Goldberg, P. K., Protzman, W., and Kozinn, M. A. (1976). *J. Am. Med. Assoc.* **235,** 628–629.

Lehmann, P. F., and Reiss, E. (1980). *Mycopathologia* **70,** 83–88.

Lew, M. A., Siber, G. R., Donahue, D. M., and Maiorca, F. (1982). *J. Infect. Dis.* **145,** 45–56.

Meckstroth, K. L., Reiss, E., Keller, J. W., and Kaufman, L. (1981). *J. Infect. Dis.* **144,** 24–32.

Meunier-Carpentier, F., and Armstrong, D. (1981). *J. Clin. Microbiol.* **13,** 10–14.

Meunier-Carpentier, F., Kiehn, T. E., and Armstrong, D. (1981). *Am. J. Med.* **71,** 363–370.

Meyer, R. D., Young, L. S., Armstrong, D., and Yu, B. (1972). *Am. J. Med.* **54,** 6–15.

Miller, G. G., Witwer, M. W., Braude, A. I., and Davis, C. E. (1974). *J. Clin. Invest.* **54,** 1235–1240.

Monson, T. P., and Wilkerson, K. P. (1981). *J. Clin. Microbiol.* **14,** 557–562.

Murray, I. G., Buckley, H. R., and Turner, G. C. (1969). *J. Med. Microbiol.* **2,** 463–469.

Peat, S., Whelan, W. J., and Edwards, T. E. (1961). *J. Chem. Soc. B.* No. 1, pp. 29–34.

Poor, A. H., and Cutler, J. E. (1979). *J. Clin. Microbiol.* **9,** 362–368.

Porath, J., Axen, R., and Ernbach, S. (1967). *Nature (London)* **215,** 1491–1492.

Reiss, E., and Lehmann, P. F. (1979). *Infect. Immun.* **25,** 357–365.

Roboz, J., Suzuki, R., and Holland, J. F. (1980). *J. Clin. Microbiol.* **12,** 594–602.

Rosner, F., Gabriel, F. D., Taschdjian, C. L., Caresta, M. A., and Kozinn, J. P. (1971). *Am. J. Med.* **51,** 54–62.

Schaefer, J. C., Yu, B., and Armstrong, D. (1976). *Am. Rev. Respir. Dis.* **113,** 325–329.

Scheld, W. M., Brown, R. S., Jr., Harding, S. A., and Sande, M. A. (1980). *J. Clin. Microbiol.* **12,** 679–683.

Segal, E., Berg, R. A., Pizzo, P. A., and Bennett, J. E. (1979). *J. Clin. Microbiol.* **10,** 116–118.

Shaffer, P. J., Kobayashi, G. S., and Medoff, G. (1979a). *Am. J. Med.* **67,** 627–630.

Shaffer, P. J., and Medoff, G., and Kobayashi, G. S. (1979b). *J. Infect. Dis.* **239,** 313–319.

Shannon, D. C., Johnson, G., Rosen, F. S., and Austen, K. F. (1966). *N. Engl. J. Med.* **275,** 690–693.

Stevens, P., Huang, S., Young, L. S., and Berdischewsky, M. (1980). *Infection* **8** Suppl. 3, 5334–5338.

Voller, A., Bidwell, D., and Bartlett, A. (1976). *In* "Manual of Clinical Immunology" (N. R. Rose and H. Friedman, eds.), pp. 506–512. Am. Soc. Microbiol., Washington, D.C.

Warren, C., Bartlett, A., Bidwell, D. E., Richardson, M. D., Voller, A., and White, L. O. (1977). *Br. Med. J.* **1,** 1183–1185.

Weiner, M. H. (1979). *In* "Proceedings of the Congress of the International Society for Human and Animal Mycology, 7th" (E. S. Kettin and G. L. Baum, eds.), pp. 165–168. Excerpta Medica, Amsterdam.

Weiner, M. H. (1980). *Ann. Intern. Med.* **92,** 793–796.

Weiner, M. H. (1983). *J. Clin. Microbiol.* **18,** 136–142.

Weiner, M. H., and Coats-Stephen, M. (1979a). *J. Lab. Clin. Med.* **93,** 111–119.
Weiner, M. H., and Coats-Stephen, M. (1979b). *J. Infect. Dis.* **140,** 989–993.
Weiner, M. H., and Yount, W. J. (1976). *J. Clin. Invest.* **58,** 1045–1053.
Weiner, M. H., Talbot, G. H., Gerson, S. L., Filice, G., and Cassileth, P. A. (1982). *Clin. Res.* **30,** 381A.
Young, R. C., and Bennett, J. E. (1971). *Am. Rev. Respir. Dis.* **104,** 710–716.
Young, R. C., Bennett, J. E., Vogel, C. L., Carbone, P. P., and DeVita, V. T. (1970). *Medicine (Baltimore)* **49,** 147–173.

18

Prospects for Solid-Phase Immunoassays in the Diagnosis of Respiratory Infections

ROBERT H. YOLKEN

Department of Pediatrics
Johns Hopkins University School of Medicine
Baltimore, Maryland

I. Introduction 225
II. Antibody Labels 226
III. Assay Formats 229
IV. Support Systems 233
V. Reagents 235
VI. Collection of Specimens 236
VII. Sensitivity of Assay Systems 237
Appendix 238
References 239

I. INTRODUCTION

The rapid, accurate diagnosis of viral respiratory infection is desirable for optimal patient management and for the prevention of the secondary spread of infection in closed populations. Accurate diagnosis is hampered by the time required for the cultivation and identification of infecting viruses. In addition, the fastidious nature of many respiratory viruses can hinder their recovery, especially if a specimen must be transported to a diagnostic facility (Gardner, 1977; Rytel, 1979; Hawkes, 1979). There has been, therefore, a great deal of interest in the development of practical techniques for the rapid diagnosis of respiratory infections. Most of these techniques are immunoassays that rely on the specific identification of infecting agents by labeled antibodies. The utility of immunoassays is based on the fact that the number of different antigenic species

The Direct Detection of Microorganisms in Clinical Samples

ISBN 0-12-187780-9

capable of causing severe respiratory disease is limited. Thus, labeled antibodies to influenza viruses, parainfluenza viruses, respiratory syncytial virus, and adenoviruses react with a large majority of the pathogens causing serious respiratory infections, especially in the pediatric population (Chanock and Parrott, 1965; Monto, 1978; Glezen and Denny, 1973; Fox and Kolbourne, 1973).

One immunoassay that can provide for efficient diagnosis of respiratory infections is immunofluorescence microscopy. This technique, described in Chapter 4 of this volume, involves the incubation of nasal secretions with fluorescein-labeled antibodies and microscopic examination of the stained specimen to visualize infected cells (Gardner, 1977; Emmons and Riggs, 1977). Although this technique can provide for the accurate diagnosis of many respiratory infections, it suffers from the need for highly trained personnel to distinguish infected cells from artifacts and requires specimens with sufficient cellular material for an adequate examination (Gardner and McQuillian, 1968). This technique is also difficult to quantify or to automate. For this reason a number of quantitative immunoassays have been developed. Most of these techniques involve the immobilization of antigens onto a solid matrix and reaction of the solid-phase antigen with specific labeled antibodies. The measurement of the labeled antibody bound to a solid phase allows for the objective and quantitative determination of the antigen in clinical specimens. Quantification facilitates the delineation of specific and nonspecific activity. In addition, it allows for objective screening of immunoreagents (Yolken, 1980; Engvall and Perlmann, 1972).

Solid-phase assays also allow the measurement of extracellular viral antigens and thereby facilitate the detection of viral antigens in body fluids containing only a small amount of cellular material. Quantification of antigen during the course of an illness (Pepple *et al.*, 1980) might be utilized to monitor clinical improvement or to gauge the response to antiviral chemotherapy.

II. ANTIBODY LABELS

There are numerous markers that can be used to label antibodies for immunoassays. Radioiodine is widely used because it can be linked with antibody molecules by numerous techniques and because it can be measured in small quantities by widely available gamma counters (Ferber, 1978). Efficient radioimmunoassays have been formulated for the detection of many respiratory antigens (Sarkkinen *et al.*, 1981a) but suffer from the short half-life of radioiodine. In addition, radioimmunoassays pose some biohazard (Lancet, 1976) and are not applicable outside of central laboratories. For these reasons numerous nonisotopc labels have been devised. One group of nonisotopic labels that are widely used for immunoassays are enzymes. The use of enzymes as antibody markers has a number of advantages including good sensitivity (due to the high turnover rate possible in enzyme–substrate reactions) and reagent stability with a long shelflife. In addition, enzyme–substrate reactions can yield visible color and thus be monitored by the naked eye or measured by simple instrumentation (Yolken, 1980; Engvall and Perlmann, 1972; Wisdom, 1976; Voller *et al.*, 1980; Clem and Yolken, 1978). The use of enzymes, however, has some inherent disadvantages. Enzyme molecules are large in relation to other markers and to antibody molecules. Many of the available methods of directly linking immunoglobulin molecules with enzymes thus result in the generation of high molecular weight polymers (Avrameas and Ternynck, 1971; Pesce *et al.*, 1976). These conjugates often have less enzymatic and specific antibody activity than the native enzyme molecules. In addition, high molecular weight conjugates often exhibit more nonspecific activity than do unlabeled antibodies due to a higher degree of nonspecific binding to the solid phase and reactivity with antiglobulins such as rheumatoid factors (Cremer *et al.*, 1978; Salonen *et al.*, 1980). It is therefore necessary that the antibody and enzyme preparations used to form conjugates have a high specific activity to compensate for the loss of activity on conjugation.

This problem can be overcome by labeling antibodies with low molecular weight cofactors instead of enzymes. Binding of labeled antibody to solid-phase antigen can be efficiently quantified by reaction with enzyme that has been modified to react with the cofactor. One cofactor system that has been successfully adapted to enzyme immunoassays is the biotin–avidin system. In this system, biotin-labeled antibody is reacted with solid-phase antigen and, following the removal of unreacted biotinylated antibody, enzyme-labeled avidin–biotin complex is added (Yolken and Leister, 1983). Because biotinylated antibody is monomeric and has favorable diffusion characteristics, these systems can offer greater sensitivity than ones that utilize polymeric enzyme–antibody conjugates (Bayer and Wilcheck, 1977).

Other markers can be used in immunoassay systems. For example, fluorescent molecules can be efficiently linked to immunoglobulins by a number of techniques. Although, in the past, fluorescent antibodies have been used largely in microscopy, the development of accurate instrumentation for the measurement of fluorescence in small volumes offers the possibility of performing quantitative solid-phase fluorescent immunoassays (Soini and Hemmila, 1979; Rietz and Guilbault, 1975). Fluorescence can also be generated by the use of enzyme-labeled antibodies that yield fluorescent products (Yolken and Stopa, 1979). These systems offer the potential for high sensitivity because they combine the low detection limit of fluorescence measurements with the magnification inherent in enzyme–substrate reactions. The techniques of time-resolved fluorescence are particularly applicable to the detection of antigens in clinical specimens utilizing rare earth metals as labels, because these techniques distinguish fluorescent-labeled antibodies from fluorescence due to protein (Pesce *et al.*, 1976;

Soini and Hemmila, 1979). Fluorescence measurements also have the potential of allowing for the quantification of antigen–antibody reactions without the need to remove unreacted labeled antibodies. The availability of such "homogeneous" systems would markedly improve immunoassay technology by eliminating the need for washing and other problems inherent in solid-phase systems.

Another set of labels that can be used in immunoassays are those that generate measurable pulses of light on reaction with an appropriate indicator (Whitehead *et al.*, 1979). Such chemiluminescent markers have the advantages of sensitivity and rapidity. Although instrumentation for the measurement of light is not widely available, scintillation counters can often be modified to measure chemiluminescence. Chemiluminescence utilizing peroxidase and luminol has been used for the measurement of a number of antigens in clinical specimens (Konishi *et al.*, 1980; Pronovost, 1980). The disadvantage of chemiluminescent reactions is that the duration of light emission is very short. Thus, the measurement of luminescence in specimens must be made in a sequential fashion. However, this is not a serious problem when a small number of specimens are being tested.

Another form of immunoassay involves the use of particles coated with antibody. The particles can be made of a number of materials including latex, red blood cells, Sepharose, bacteria (such as protein A-bearing staphylococci), or metal chelates. The sensitivity of these systems is based on the fact that the surface of each particle is coated with many molecules of antibody, thus allowing for favorable reaction kinetics. Agglutination reactions are quite rapid, often reaching completion in less than 10 mins. In addition, the reaction of a single molecule of antigen with two molecules of antibody results in a cross-linking of a relatively large particle, leading to a magnification of the antigen–antibody reaction. Although the cross-linking of particles can be monitored with the naked eye or microscopically, the development of particle-sizing equipment has markedly expanded the scope of this technology by providing for accurate quantification (Grange *et al.*, 1977; Collette-Cassart, 1981). Quantification is necessary to distinguish low-level specific reactions from nonspecific ones. However, the expense and complexity of particle-sizing equipment limit its utility at present. In contrast, we have found that microplate colorimeters developed for the quantification of enzymatic reactions (Clem and Yolken, 1978) can also be used to measure agglutination reactions. This is because agglutination results in a decrease in light scattering and optical density over a wide range of wavelengths. The availability of practical quantitative measures of agglutination may markedly expand the scope of rapid diagnostic techniques for respiratory viruses.

The choice of label depends to a great extent on the availability of instrumentation and the particular needs of a given diagnostic laboratory. In addition, the techniques using higher-energy nonisotopic labels such as fluorescent and chemiluminescent labels offer the possibility of sensitivity beyond that of radioactivity or standard enzyme immunoassays if antibodies of sufficient sensitivity and specificity are available. The availability of these assays might thus markedly expand the sensitivity of the solid-phase assay systems for the detection of viral antigens in respiratory secretions.

III. ASSAY FORMATS

There are numerous ways to use labeled antibodies in solid-phase systems. One of the most common formats is the direct assay, as depicted in Fig. 1 (Yolken *et al.*, 1980). In this format antibodies are bound to the solid phase by means of hydrophobic absorption or covalent linkage (see following discussion). After the unreacted antibody is removed, the specimen is diluted in a buffer such as phosphate-buffered saline containing 0.05% Tween 20 and 1% gelatin to maximize antigen–antibody interactions and minimize nonspecific absorption. If antigen is present in the specimen, it will react with antibody. After washing to remove unreacted material, labeled antibody is added. It is preferable that the solid-phase and liquid-phase antibodies be obtained from different sources to

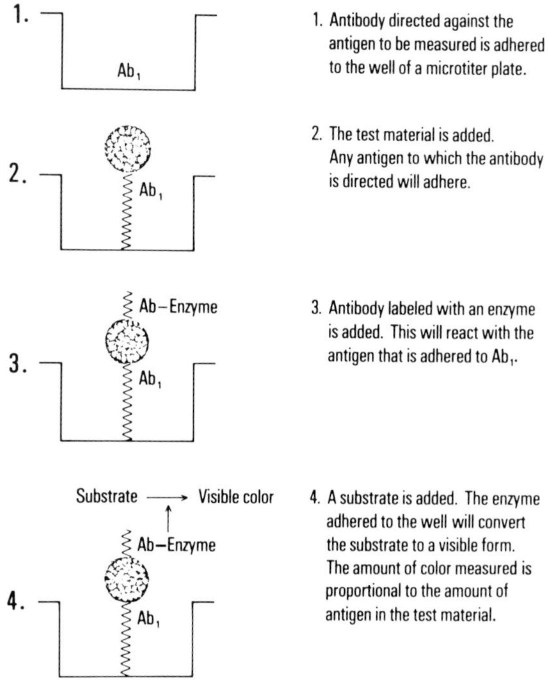

Fig. 1. Format of the direct ELISA for detecting antigens.

minimize nonspecific binding of nonviral antigens to the solid phase and to labeled antibody. The labeled antibody will react with antigen bound to the solid phase in previous steps. After unreacted antibody is removed by washing, the amount of label bound to the solid phase is quantified by means of appropriate instrumentation. In the case of an enzyme immunoassay, substrate is added, and the amount of color generated is measured. This assay format is simple and can be fairly efficient for antigen detection if the antigen has at least two binding sites to allow for binding to the solid phase and to the labeled antibody. One disadvantage of this assay is that it requires a separate labeled antibody for each antigen. Because the clinical laboratory is often called on to examine a specimen for several different antigens, it is desirable to use an indirect assay system.

In the indirect assay format, unlabeled second antibody is substituted for the labeled antibody in the direct system. After reaction of the antibody to the solid phase and removal of unreacted antibody, an additional immunoreactant is added to combine with the second antibody. This immunoreactant is usually an antibody. However, other materials, such as labeled staphylococcal protein A (Yolken and Leister, 1981), can be used. The reaction is quantified as in the direct assay. The widespread availability of the labeled antiglobulins eliminates the need for preparing labeled reagents directed at a specific virus. In addition, indirect assays are more sensitive than direct assays, because a single molecule of antibody can react with a number of molecules of labeled antiglobulin (Yolken and Stopa, 1980). In most cases antibody from two species must be used as the solid-phase and liquid-phase antibodies to prevent nonspecific interaction, and this can pose a problem. However, antibodies from the same species can be used if conditions are adjusted to minimize nonspecific interaction. For example, the $F(ab_2)$ fraction can be used to coat the solid phase, and intact immunoglobulin can be used as the liquid-phase antibody. The interaction of antibody with the solid phase can be measured by the addition of a labeled immunoreactant capable of reacting only with the Fc portion of the antibody molecule. These immunoreactions will thus display little if any reactivity with the solid-phase antibody, which lacks Fc determinants. Such labeled reactants can be formulated from affinity-purified or monoclonal antibodies or from staphylococcal protein A.

Alternatively, one class of antibody (such as IgM) can be used to coat the solid phase and another class of antibody (such as IgG) used as the liquid phase reactant. Specific binding is then measured by the addition of a labeled antiglobulin, which reacts only with the immunoglobulin class of the liquid-phase antibody. This method is particularly applicable to monoclonal antibodies because clones of different immunoglobulin classes can often be isolated (Lazekas *et al.*, 1980). Alternatively, IgG and IgM antibodies can be obtained from a single immunized animal at appropriate times after immunization and purified by standard methodology.

One problem with these immunoassays is that they require several lengthy incubations and washing steps, making them long and tedious. A modification

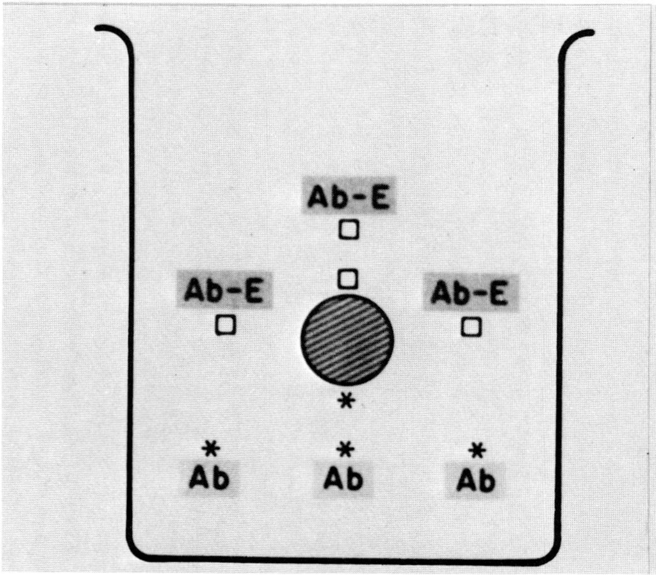

1. Antigen is added to a solid phase coated with antibody directed against one antigenic site (* Ab). Enzyme-labeled antibody directed against a different site on the antigen (□ Ab–E) is then added. This will react with unbound sites on the antigen.

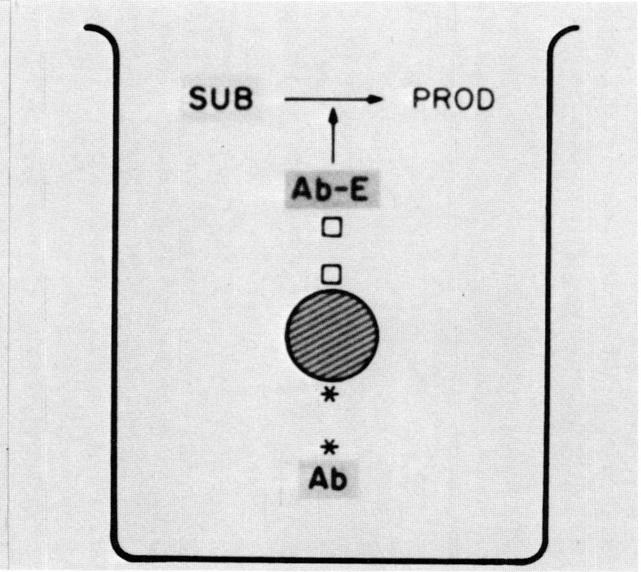

2. Following a washing step to remove unreacted □ Ab–E, substrate is added. This will be converted by bound □ Ab–E to a measureable product. The amount of product formed will be proportional to the concentration of antigen in the specimen.

that appears to be simpler is the double determinant assay (Fig. 2) (Yolken and Leister, 1982). In this assay, antibody directed at one viral determinant is used to coat the solid phase, and labeled antibody directed at another determinant is added to the liquid phase. The specimen and label can be added simultaneously to the solid phase because competition between solid-phase and liquid-phase antibody will not occur. Following incubation the solid phase is simply washed and the substrate is added. These assays are both simpler and more rapid than standard solid-phase immunoassays. Also, because they require only a single washing step, they are more easily automated. Double determinant assays are particularly suitable for use with monoclonal antibodies because different clones are often directed at distinct viral determinants (Gerhard *et al.*, 1978).

Solid-phase immunoassays can also be performed in a competitive manner. Because sufficient quantities of labeled antigens are generally not available, the assays are usually performed with labeled antibodies or antiglobulins. One format is depicted in Fig. 3. In this assay, labeled antibody is reacted with the specimen. The resulting antigen–antibody mixture is added to the solid phase coated with antigen. Antigen present in the clinical specimen will compete with the labeled antibody for binding sites on the solid-phase antigen. The presence of antigen in the specimen will be manifested by a decrease in labeled antibody bound to the solid phase. Alternatively, unlabeled antibody can be used and the reaction quantified by the addition of labeled antiglobulin.

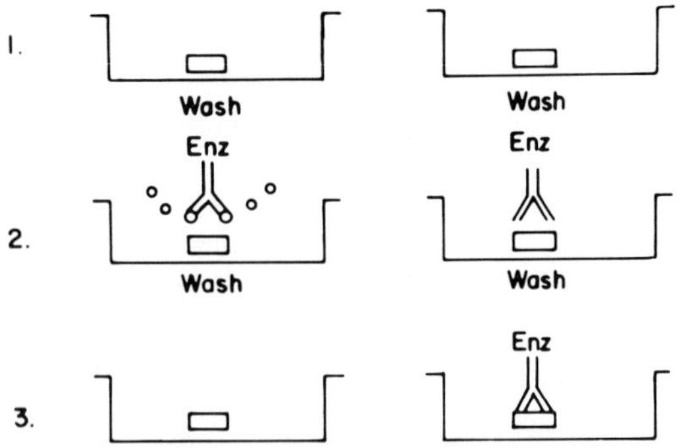

Fig. 3. Format of a competitive solid-phase immunoassay for antigens. (1) Antigen (▭) is bound to the solid phase. Unbound antigen is removed in a washing step. (2) The specimen is reacted with enzyme-labeled antibody (Enz-labeled antibody symbol), and the mixture is added to the solid phase. If the specimen contains antigen (○), it will react with the enzyme-labeled antibody, thus preventing the enzyme-labeled antibody from reacting with the solid phase. (3) Unbound enzyme-labeled antibody is removed in a washing step and the bound enzyme is quantitated by the addition of substrate. The amount of substrate product is inversely proportional to the amount of antigen in the specimen.

Unlabeled antibody can also be used in place of the enzyme-labeled antibody. In this case, enzyme-labeled antiglobulin or staphylococcal protein A is used to quantify the antibody bound to the solid phase.

Competitive assays have a number of advantages, including the fact that the initial antigen–antibody reaction is performed in a liquid phase. The assays thus take advantage of the more favorable kinetics offered by liquid phase reactions (Rubin *et al.*, 1980). In addition, competitive assays generally require fewer incubations than the noncompetitive ones and are simpler to perform.

One problem with competitive assays is that they require a fairly large degree of saturation of labeled antibody with antigen before the antigen–antibody reaction can be accurately measured. This decreases the sensitivity in comparison with noncompetitive immunoassays (Yolken *et al.*, 1980). Competitive assays also suffer from the fact that endogenous antibody in the specimen can compete with the labeled antibody for antigenic sites, which can cause a false positive reaction (Segal *et al.*, 1979). It is therefore necessary to perform these assays with a large excess of antigen on the solid phase or treat the specimens to remove endogenous antibody.

IV. SUPPORT SYSTEMS

An important determinant of the sensitivity of the solid-phase immunoassays is the nature of the solid-phase system. The solid phase should have a high capacity to immobilize antibody and a minimum nonspecific interaction with nonviral materials and labeled immunoreactants. In addition, the solid phase should be convenient to use.

The importance of the antibody-binding capacity of the solid phase is evident from the law of mass action, which indicates that an increase in antibody on the solid phase will lead to an increase in the amount of antigen bound. Most immunoassays have relied on a plastic support to immobilize antibody. Beads, microtitration plates, cuvettes, and rods have been used (Lehtonen and Viljanen, 1980; Pesce *et al.*, 1977). Plastic supports have the advantage of being inexpensive; they have a relatively low amount of nonspecific binding in the presence of nonionic detergents and can bind antibodies by simple absorption. Microtiter plates are particularly convenient for the development of immunoassay systems because they permit the evaluation of a large number of specimens with a minimum of reagents and manipulation (Purcell *et al.*, 1976). However, the standard microtiter plates may be inconvenient for general clinical use because the number of specimens tested varies. For this reason individual test units, such as strips of wells that can fit into a microtiter format or be utilized as individual test units, are often used. These allow for the direct transfer of contents to measurement instruments such as gamma counters, fluorescent readers, or spectrophotometers.

Although plastic supports have been widely used, they have a number of disadvantages. The most important is that the antibody-binding capacity of commonly used supports is limited to approximately 0.5 $\mu g/cm^2$ (Rubin *et al.*, 1980;

Lehtonen and Viljanen, 1980; Pesce *et al.*, 1977). Attempts to put larger amounts of antibody on untreated plastic surfaces generally result in a decrease in sensitivity due to the elution of antibody from the solid phase and/or the layering of the solid-phase antibody. This low binding capacity limits the sensitivity of the assay. Another disadvantage is that the diffusion of antibody is limited, leading to a slower reaction compared with liquid-phase interactions. Another problem with plastics is the elution of solid-phase antibody into the liquid phase. This results in a competition between solid-phase and liquid-phase antibody, with a resulting decrease in sensitivity. If the amount eluted is greater than 10% of that which is bound, competition becomes significant because the kinetics of the antigen–antibody reaction in the liquid phase are more favorable than the kinetics in the solid phase.

There are several ways to overcome these problems. The limited antibody-binding capacity of the solid phase and the problem of antibody elution can be overcome by the use of an antibody covalently linked to the solid phase. An effective means of accomplishing this is to add amino groups to the solid phase by treating polystyrene with sulfuric and nitric acids (Nevrath and Strick, 1981). The antibody can then be covalently linked with the aminated solid phase with glutaraldehyde. Alternatively, the solid phase can be carboxylated and the antibody linked to the carboxyl groups with carbodiimide. Studies in our laboratory indicate that the use of covalently linked antibodies can markedly improve the sensitivity of solid-phase immunoassays. An additional modification that can improve the sensitivity of enzyme immunoassays is a mobile solid phase in the form of a resin or latex particles rather than a stationary solid phase. Because both the antibody-binding capacity and the diffusion characteristics of a mobile phase are more favorable for reaction than a stationary solid phase, the result can be a more rapid and more sensitive reaction. One problem with particulate solid phases is that they are more difficult to manipulate and wash. The availability of simple instrumentation to perform immunoassays that make use of particulate solid phases might markedly expand the utility of solid-phase immunoassay systems.

Solid-phase immunoassays can also be performed using nonplastic supports. Nitrocellulose and other papers of high binding capacity have potential as solid-phase matrices (Watanabe *et al.*, 1977). These materials have binding capacities of up to 500 $\mu g/cm^2$. Because higher concentrations of immunoglobulin can be bound, paper matrices have the potential for greater sensitivity than do plastic ones. The availability of filtration manifolds in a microtiter format allows for the manipulation of paper matrices with microtiter equipment and for the performance of multiple assays in a single run. The main problem with these solid phases stems from their high binding capacity; namely, a high degree of non-specific binding. When these solid phases are used, extensive washing is needed and, after the solid-phase antibody is bound, the matrices must be reacted with protein solutions to prevent nonspecific absorption of labeled reagents. Nonetheless, this system could lead to immunoassays of far greater sensitivity than those currently in use.

V. REAGENTS

The most important determinants of sensitivity and specificity in an enzyme immunoassay are the immunoreagents. The sensitivity of immunoassays is dependent on antibody concentration and affinity. Because antibody concentration is relatively constant for many polyclonal antisera, the main determinant of assay sensitivity will be antibody avidity. Because solid-phase assays involve multiple incubation and washing steps, they are especially dependent on antibody avidity. In our experience only antibodies of high avidity can be used for sensitive immunoassay systems. In addition, the reagents must have a high degree of specificity. In the case of reagents directed at respiratory viruses, one cause of nonspecific activity is the presence of antibody to host components derived from the system used to propagate the virus. These antibodies are capable of reacting with nonviral antigens in human body fluids, leading to false positive results. One practical approach to limiting nonspecific reactions is to use antigens prepared in different systems for immunoization (Decker and Overby, 1977). Cross-reactive antigens must react both with the solid-phase antibody and with the second antibody to cause problems. One approach with many myxoviruses and paramyxoviruses is to use one antibody prepared from virus grown in tissue culture and another from virus grown in eggs. With tissue culture antigens, it is advisable to use serum-free media, if possible, because fetal calf serum or other animal sera may contaminate viruses purified from cultures. A more general approach to this problem consists of absorbing antisera with uninfected tissue culture cells or egg to remove cross-reacting antibodies. In our experience absorption can improve the specificity of reagents, but often it does not eliminate nonspecific reactions. A more effective approach is to use affinity chromatography (Robbins *et al.*, 1967). Although affinity chromatography specificity yields can be low, Hornsleth and co-workers described a technique similar to affinity chromatography that is potentially more practical (Hornsleth *et al.*, 1981). Antigen–antibody precipitates formed in agarose gel are eluted and isolated and tested for antibody activity. The authors used this method to prepare highly specific antibody to respiratory syncytial virus. In addition, these purified antibodies can increase assay sensitivity due to an increase in the proportion of antibody directed at a particular antigen. It appears that this method can markedly improve the reaction characteristics of available antisera to respiratory viruses.

Another method for the assurance of high specificity is the use of monoclonal antibodies (Kohler and Milstein, 1975; Zurawski *et al.*, 1980). Monoclonal antibody technology involves the isolation of antibody directed at a single determinant, decreasing the chance of producing antibodies that will react with nonviral antigens. In addition, because each molecule of antibody is directed at the antigen to be measured, monoclonal antibodies can potentially increase the sensitivity of immunoassay systems. In addition, antibody-producing cells are often capable of unlimited replication, ensuring a large supply of a consistent reagent. This should provide immunoassays of uniform reaction characteristics. There are some potential problems in monoclonal antibody technology, however. One is that the specificity of monoclonal antibodies might preclude their use as general diagnostic reagents. That is, monoclonal antibodies might fail to recognize common determinants of viral antigens in clinical specimens and thus not detect the presence of virus in some infected patients (Gerhard *et al.*, 1978). In addition, the affinity characteristics of some monoclonal antibodies might not be as favorable as that of polyclonal antibodies (Frankel and Gerhard, 1979). Lower affinity might counterbalance some of the advantage of monoclonal antibodies.

Monoclonal antibodies have been produced to adenoviruses, respiratory syncytial viruses, and influenza viruses. Studies in our laboratory indicate that in all cases it is possible to select clones that are directed at common antigens and that can react with antigens present on most strains of the viruses. In addition, assays can be developed that appear to have sensitivity equal to or greater than those assays utilizing polyclonal antibodies. The availability of monoclonal antibodies to respiratory viruses should markedly improve the sensitivity and specificity of immunoassays.

VI. COLLECTION OF SPECIMENS

Before an antigen can be detected by solid-phase immunoassay, it must be obtained in a reactive form in a clinical sample. The most efficient way of obtaining respiratory secretions is by the instillation of saline in the nares and extraction with a vacuum extractor (Gardner, 1977; Gardner and McQuillian, 1974). The mucus is then liquefied by sonication or by treatment with mucolytic agents, such as *N*-acetylcysteine. These procedures can be very difficult to perform, especially in young children. When nasal washes cannot be obtained, we have found that nasal or nasopharyngeal swabs can be used, although the sensitivity might be somewhat less than with aspirates. The swabs are moistened with saline before insertion, and after an adequate amount of material is obtained the specimen is immediately put into a holding medium consisting of buffered saline and a protein carrier such as 0.5% gelatin. It is advisable to avoid the use of balanced salt solutions because they have been reported to decrease the antigenic reactivity of some viruses. The development of immunoassay systems for detecting viral antigens in easily obtainable specimens would markedly improve the practicality of the immunodiagnosis of viral infection.

VII. SENSITIVITY OF ASSAY SYSTEMS

Although a number of solid-phase immunoassay systems have been proposed for the detection of respiratory antigens in clinical specimens, published comparative data are available largely for solid-phase radioimmunoassay and enzyme immunoassay. Such systems have been developed for influenza viruses a and b, parainfluenza virus, adenoviruses, and respiratory syncytial virus (RSV). The largest body of information is that pertaining to RSV infection. Several investigators have established solid-phase immunoassay systems for this virus that are as sensitive or slightly more sensitive than immunofluorescence assays. Several of the reported assays use reagents that are commercially available, including horse and guinea pig antisera from Flow Laboratories (Alexandria, Virginia) and bovine anti-RSV (Wellcome Laboratories, Research Triangle, North Carolina). McIntosh *et al.* (1982) and Hendry and McIntosh (1982) used available reagents to devise an enzyme immunoassay for the detection of RSV antigen in nasopharyngeal specimens in children. They found significant enzyme-linked immunosorbent assay (ELISA) activity in 89 of 106 culture-positive specimens. They also found antigenic activity in 5 of 98 specimens that were culture-negative. They found that specimens obtained from nasopharyngeal swabs gave less activity than nasal wash specimens, highlighting the importance of specimen collection. The sensitivity and specificity of ELISA were slightly less than those of fluorescence microscopy. Of particular interest, specimens known to contain virus remained positive by ELISA following storage at room temperature while the viability of the virus declined. This suggests that the enzyme immunoassay might be particularly useful in situations in which specimens cannot be cultured quickly.

Sarkkinen *et al.* (1981b,c) found positive enzyme immunoassay results in 28 of 29 specimens that were positive by indirect immunofluorescence, and they obtained negative results in 81 of 82 specimens that were negative by indirect immunofluorescence. They also developed an enzyme immunoassay to detect parainfluenza viruses in nasopharyngeal secretions. They detected antigen by enzyme immunoassay in 26 of 28 specimens positive for parainfluenza virus by immunofluorescence. They detected antigen in 7 of 146 specimens negative by immunofluorescence. These results demonstrate that enzyme immunoassay is a reasonable alternative to immunofluorescence for detecting certain respiratory viruses in clinical specimens.

Harmon and Pawlick developed an indirect enzyme immunoassay for the detection of influenza a virus in clinical specimens. This assay was capable of detecting approximately $10^{3.5}$ tissue culture infectious doses of purified virus and detected virus in 21 of 40 nasal washes specimens obtained from infected individuals (Harmon and Pawlick, 1982).*

Berg *et al.* (1980) found that the sensitivity of enzyme immunoassay systems for the detection of influenza a could be increased by the use of high-energy fluorescent or radioactive substrates, but the improved sensitivity of these systems is somewhat offset by an increased tediousness.

The success of solid-phase immunoassay systems for the detection of respiratory viruses is currently somewhat less than that of tissue culture. However, the lower sensitivity of immunoassays is counterbalanced by the greater rapidity of these assays. In addition, quantitative solid-phase assay systems can be performed in settings in which viral isolation is not available. Solid-phase immunoassays utilizing current technology could have an important role in monitoring the course of infection due to respiratory viruses. The improvement of the sensitivity of the assay systems by the methods described in this chapter might markedly increase their utility in the future.

APPENDIX

Protocol for Enzyme Immunoassay for Viruses: Indirect Assay for Antigen

1. Coat alternate rows of wells of the microtiter plate with a dilution of antivirus serum (or IgG) and an equal dilution of serum (or IgG) that does not contain measurable antibody to virus.
2. Incubate the plate at least overnight at 4°C. If the plate is not used the next day, it should be covered with Parafilm and stored at 4°C until used.
3. Wash the plate five times with phosphate-buffered saline (PBS)–Tween.
4. Add 50 μl of *N*-acetylcysteine (adjusted to pH 7) to each of the wells. Add an equal amount of specimen to two wells coated with antivirus serum and two wells coated with nonimmune serum. Include a weekly positive control and four negative controls in each test.
5. Incubate the plate for 2 hr at 37°C or overnight at 4°C.
6. Wash the plate five times with PBS–Tween.
7. Add unlabeled antivirus serum (or IgG) diluted in PBS–Tween–fetal calf serum.
8. Incubate the plate for 1 hr at 37°C.
9. Wash the plate five times with PBS–Tween.

*These and other investigators have been able to improve the sensitivity of their assay substantially by the use of a fluorescent substrate in place of the color-generating ones usually utilized in ETA systems (Harmon *et al.*, 1983; Berg *et al.*, 1980).

10. Add enzyme-labeled antiglobulin directed at the species of IgG added in step 7.
11. Incubate the plate for 1 hr at 37°C.
12. Wash the plate five times with PBS–Tween.
13. Add appropriate substrate. Calculate a specific activity by subtracting the mean activity of the specimen in wells coated with the virus-negative serum from the mean activity of the wells coated with the antivirus serum. This procedure distinguishes true positive reactions due to the presence of virus from false positive reactions due to antiglobulins. To ensure accurate quantification, specimens giving readings of greater than 1.2 optical density units should be diluted 1 : 4 and retested. A specimen is considered positive if its mean activity is greater than 2 SD units above the mean of the negative controls. Alternatively, a specimen can be considered positive if its activity is greater than that of the weakly positive control.

REFERENCES

Avrameas, D., and Ternynck, T. (1971). *Immunochemistry* **8,** 1175–1179.

Bayer, E. A., and Wilcheck, R. (1977). *Methods Biochem. Anal.* **8,** 1–43.

Berg, R. A., Yolken, R. H., Rennard, S. I., Dolin, R., Murphy, B. R., and Strauss, S. E. (1980). *Lancet* **1,** 851–853.

Chanock, R. M., and Parrott, R. H. (1965). *Pediatrics* **36,** 21–39.

Clem, T. R., and Yolken, R. H. (1978). *J. Clin. Microbiol.* **7,** 55–58.

Collett-Cassart, D., Magnusson, C., Ratcliffe, J., Cambiaso, C., and Masson, P. (1981). *Clin. Chem.* **27,** 64–67.

Cremer, N. R., Hoffman, M., and Lennette, E. H. (1978). *J. Clin. Microbiol.* **8,** 160–165.

Decker, R. H., and Overby, L. R. (1977). *In* "Quality Control in Nuclear Medicine" (B. A. Rhodes, ed.), pp. 427–436. Mosby, St. Louis, Missouri.

Emmons, R. W., and Riggs, J. L. (1977). *In* "Methods in Virology" (K. Maramorosch and H. Kaprowski, eds.). Academic Press, New York.

Engvall, E., and Perlmann, P. J. (1972). *J. Immunol.* **109,** 129–135.

Ferber, J. P. (1978). *Adv. Clin. Chem.* **20,** 129–179.

Fox, J. P., and Kilbourne, E. (1973). *J. Infect. Dis.* **128,** 361–386.

Frankel, M. E., and Gerhard, W. (1979). *Mol. Immunol.* **16,** 101–106.

Gardner, P. S. (1977). *Experientia* **33,** 1674–1676.

Gardner, P. S., and McQuillian, J. (1968). *Br. Med. J.* **3,** 340–343.

Gardner, P. S., and McQuillian, J. (1974). "Rapid Diagnosis of Infectious Diseases," p. 255. Butterworth, London.

Gerhard, W., Croce, C. M., Lopes, D., and Koprowski, H. (1978). *Proc. Natl. Acad. Sci. U.S.A.* **75,** 1510–1514.

Glezen, W. P., and Denny, F. W. (1973). *N. Engl. J. Med.* **288,** 498–505.

Grange, J. (1977). *J. Immunol. Methods* **18,** 365–375.

Harmon, M. W., and Pawlick, K. M. (1982). *J. Clin. Microbiol.* **15,** 5–11.

Harmon, M. W., Russo, L. L., Wilson, S. Z. (1983). *J. Clin. Microbiol.* **17,** 305–311.

Hawkes, P. A. (1979). "Diagnostic Procedures for Viral and Related Diseases" (E. Lennette, ed.), 5th ed., pp. 3–78. APHA, Washington, D.C.

Hendry, R. M., and McIntosh, K. (1982). *J. Clin. Microbiol.* **16,** 324–328.

Hornsleth, A., Grauballe, P., Friis, B., Genner, J., and Pederson, I. R. (1981). *J. Clin. Microbiol.* **14,** 501–509.

Kohler, G., and Milstein, C. (1975). *Nature (London)* **256,** 495–497.

Konishi, E., Iwasa, S., Kondo, K., and Hori, M. (1980). *J. Clin. Microbiol.* **12,** 140–143.

Lazekas de St. Groth, S., and Scheidegger, D. (1980). *J. Immunol. Methods* **35,** 1–21.

Lehtonen, O. P., and Viljanen, M. K. (1980). *J. Immunol. Methods* **34,** 61–70.

McIntosh, K., Hendry, R. M., Fahnestock, M. L., and Pierkik, L. T. (1982). *J. Clin. Microbiol.* **16,** 329–333.

Monto, A. S. (1978). *Am. J. Epidemiol.* **97,** 338–348.

Neurath, A. R., and Strick, N. (1981). *J. Virol. Methods* **3,** 155–165.

Pepple, J., Moxon, E. R., and Yolken, R. H. (1980). *J. Pediatr. (St. Louis)* **97,** 233–237.

Pesce, A. J., Modesto, R. R., Ford, D. J., Sethi, K., Clyne, D. N., and Pollak, V. E. (1976). *In* "Immunoenzymatic Techniques" (G. Feldman, P. Druett, J. Bignon, and S. Avrameas, eds.), pp. 7–23. North-Holland Publ., Amsterdam.

Pesce, A. J., Ford, D. J., Gaizutis, M., and Pollak, V. E. (1977). *Biochem. Biophys. Acta* **492,** 399–407.

Pronovost, A. D. (1980). *J. Infect. Dis.* **142,** 793–802.

Purcell, R. H., Wong, D. C., Mortisugu, Y., Dienstag, J. L., Routenberg, J. A., and Boggs, J. D. (1976). *J. Immunol.* **116,** 349–356.

Rietz, B., and Guilbault, G. G. (1975). *Clin. Chem.* **21,** 1791–1794.

Robbins, J. B., Haimovich, J., and Sela, M. (1967). *Immunochemistry* **4,** 11–22.

Rubin, R. L., Hardtke, M. A., and Carr, R. I. (1980). *J. Immunol. Methods* **33,** 277–292.

Rytel, M. (1979). *In* "Rapid Diagnosis in Infectious Disease" (M. Rytel, ed.), pp. 7–16. CRC Press, Boca Raton, Florida.

Salonen, E. M., Vaheri, A., Suni, J., and Wager, O. (1980). *J. Infect. Dis.* **142,** 250–255.

Sarkkinen, H. K., Halonen, P. E., Arstilla, P. P., and Salmi, A. A. (1981a). *J. Clin. Microbiol.* **13,** 258–265.

Sarkkinen, H. K., Halonen, P. E., Arstilla, P. P., and Salmi, A. A. (1981b). *J. Gen. Virol.* **56,** 49–57.

Sarkkinen, H. K., Halonen, P. E., Arstilla, P. P., and Salmi, A. A. (1981c). *J. Med. Virol.* **7,** 213–220.

Segal, E., Berg, R. A., Pizzo, P. A., and Bennett, J. E. (1979). *J. Clin. Microbiol.* **10,** 116–118.

Soini, E., and Hemmila, I. (1979). *Clin. Chem.* **25,** 353–361.

Unsigned editorial (1976). *Lancet* **1,** 1391–1393.

Voller, A., Bidwell, D., Bartlett, A. (1980). *In* "Manual of Clinical Immunology" (N. Rose and H. Friedman, eds.), pp. 359–371. American Society for Microbiology, Washington, D.C.

Wanatabe, H., Gust, I. D., Holmes, I. H. (1978). *J. Clin. Microbiol.* **7,** 405–409.

Whitehead, T. P., Kircka, L. J., Carter, T. J. N., Thorpe, G. H. G. (1979). *Clin. Chem.* **25,** 1531–1546.

Wisdom, G. B. (1976). *Clin. Chem.* **22,** 1243–1255.

Yolken, R. H. (1980). *Yale J. Biol. Med.* **53,** 85–92.

Yolken, R. H., and Leister, F. J. (1981). *J. Immunol. Methods* **43,** 209–218.

Yolken, R. H., and Leister, F. J. (1982). *J. Infect. Dis.* **146,** 43–46.

Yolken, R. H., and Leister, F. J. (1983). *J. Immunol. Methods* (in press).

Yolken, R. H., and Stopa, P. J. (1979). *J. Clin. Microbiol.* **10,** 317–321.

Yolken, R. H., and Stopa, P. J. (1980). *J. Clin. Microbiol.* **11,** 546–551.

Yolken, R. H., Stopa, P. J., and Harris, C. C. (1980). *In* "Manual of Clinical Immunology" (N. Rose and H. Friedman, eds.), 2nd ed., pp. 692–699. Am. Soc. Microbiol., Washington, D.C.

Zurawski, V., Black, P., and Haber, E. (1980). *In* "Monoclonal Antibodies" (R. H. Kennett, T. McKearns, and K. Bechtol, eds.), pp. 19–33. Plenum, New York.

19

Diagnosis of Hepatitis B and Non-A, Non-B Hepatitis

EDWARD TABOR

Hepatitis Branch
Division of Blood and Blood Products
Bureau of Biologics
Food and Drug Administration
Bethesda, Maryland

I. Hepatitis B ... 241
 A. Epidemiology ... 242
 B. Hepatitis B Surface Antigen ... 243
 C. Antibody to Hepatitis B Core Antigen ... 248
 D. Antibody to Hepatitis B Surface Antigen ... 249
 E. Hepatitis B e Antigen ... 251
 F. DNA Polymerase ... 252
 G. Delta Antigen and Antibody ... 253
 H. Nonspecific Screening for Hepatitis B ... 254
II. Non-A, Non-B Hepatitis ... 254
 A. Introduction ... 254
 B. Epidemiology ... 255
 C. Experimental Serologic Tests ... 258
 References ... 261

I. HEPATITIS B

Rapid technological developments have led to a dramatic reduction in the number of cases of posttransfusion hepatitis caused by hepatitis B virus (HBV). Nevertheless, even the availability of sensitive tests has not eliminated hepatitis B because the amount of hepatitis B surface antigen (HBsAg) in some blood donors is too small to be detected. Severe long-term consequences of hepatitis B include chronic active hepatitis, cirrhosis, and hepatocellular carcinoma. For-

The Direct Detection of Microorganisms in Clinical Samples

ISBN 0-12-187780-9

tunately, the development of a safe and effective vaccine may eventually lead to the control of hepatitis B infections.

HBV is a 42-nm double-shelled virus found in the serum of acutely and chronically infected humans. The virus contains a 27-nm core with an incomplete double-stranded DNA. It is unique in that vast quantities of surface protein (22-nm HBsAg spheres and tubules) are produced during infection and released into the circulation. The ratio of 22-nm HBsAg to intact HBV in serum may range from 10 : 1 to 1000 : 1 (Hoofnagle, 1980). Hepatitis B virus has not been grown in cell culture. Transmission of this virus to chimpanzees, the only reliably susceptible nonhuman animal species, has provided an important model for studying the disease and testing assays for HBV serologic markers.

A. Epidemiology

Hepatitis B surface antigen is a marker of active HBV infection. It is detectable by sensitive assays in approximately 0.3% of healthy adults in the United States (Barker *et al.*, 1977), ranging from 0.1 to 0.5% in different geographic areas of the United States. A number of risk factors affect the prevalence of HBsAg, including low socioeconomic status, sexual promiscuity, residence in institutions, and employment in certain health care occupations (for example, surgeons, oral surgeons, dentists, and ward and laboratory personnel who handle specimens of body fluids from patients are susceptible). Transmission from HBsAg carriers to susceptible individuals may occur by parenteral means, including blood transfusion, parenteral drug abuse with shared needles, or transmission by hematophagous insects. Inapparent parenteral transmission may also occur through shared toothbrushes or razors or through cuts. Although oral transmission is not believed to play a significant role in the spread of HBV, it may occur in certain circumstances. The finding of HBsAg in a number of body fluids other than blood suggests that spread may sometimes occur through infected saliva, urine, breast milk, semen, and vaginal secretions, even in the absence of detectable blood (Villarejos *et al.*, 1974). Hepatitis B has been experimentally transmitted to gibbons by ingestion of infected saliva (Bancroft *et al.*, 1977) and to chimpanzees by inoculation of semen from human HBsAg carriers (Alter *et al.*, 1977). The spectrum of nonparenteral or "inapparent" parenteral routes of infection may explain why the vast majority of individuals with HBV serologic markers have no history of blood transfusion.

The transmission of HBV occurs between infected mothers and infants; this is the reason for the persistence of the reservoir of HBsAg carriers in many parts of the world (Gerety and Schweitzer, 1977). Infections during infancy result in a chronic HBsAg carrier state in 90% of cases (Gerety *et al.*, 1974; Tong *et al.*, 1981) compared with 5 to 10% of infections that begin later in life (Hoofnagle *et al.*, 1978a).

Before the introduction of serologic testing, as many as 60% of posttransfusion hepatitis cases were due to hepatitis B (Alter *et al.*, 1975a). As a result of federal regulations requiring testing for HBsAg by sensitive techniques in all blood banks in the United States, the prevalence of hepatitis B has been reduced to 11 to 13% of cases of posttransfusion hepatitis (Aach *et al.*, 1978; Alter *et al.*, 1975b), or about 1.7% of all blood recipients (Aach *et al.*, 1978). Since 1977 all blood for transfusion in the United States has been required by federal regulations to be labeled "paid" or "volunteer," according to its source (*Federal Register*, 1977). This is because more than twice as many recipients of HBsAg-negative units of blood from paid donors develop HBV infection than those receiving HBsAg-negative blood from volunteer donors (Goldfield *et al.*, 1975). Labeling of blood units is believed to have had a substantial impact on the incidence of hepatitis B following transfusions, augmenting the impact of screening for HBsAg.

The infusion of certain pooled plasma derivatives presents a high risk for the transmission of HBV. Each plasma pool used in their manufacture includes more than 1000 units of plasma and therefore has a statistical risk of including a unit of plasma with a level of HBV that is below the limit of radioimmunoassay (RIA) detectability for HBsAg. At present, no effective method has been developed to ensure the absence of HBV infectivity from these plasma derivatives, despite testing of donor plasma for HBsAg. Derivatives that are too labile to withstand inactivation by heating at 60°C for 10 hr, such as antihemophilic factor (factor VIII, AHF) and factor IX complex (factor IX) may transmit HBV infections despite the absence of detectable HBsAg. Such plasma derivatives are administered to thousands of hemophiliac patients each year. More than 20% of adult hemophiliacs have a history of having had clinically recognized hepatitis (Peterson *et al.*, 1973), and 60–90% can be shown by serologic tests to have had hepatitis B (Hoofnagle *et al.*, 1975). Newly diagnosed patients with hemophilia who receive AHF or factor IX for the first time are at a particularly high risk for acquiring HBV infection. Factor IX has also been used in some patients without a documented deficiency of this clotting factor, which has caused great concern among those familiar with its associated HBV risk.

B. Hepatitis B Surface Antigen

Hepatitis B surface antigen is found on HBV and on the 22-nm particles that are present in high concentrations in the serum of infected individuals. It can be detected in the serum of most patients at some time during HBV infection and in the serum of almost all patients with symptomatic disease (Hoofnagle *et al.*, 1978a). HBsAg usually appears from 1 to 2 months before the onset of symptoms (Figs. 1 and 2) and is frequently but not universally cleared from the circulation after hepatitis symptoms have resolved. Some patients clear HBsAg

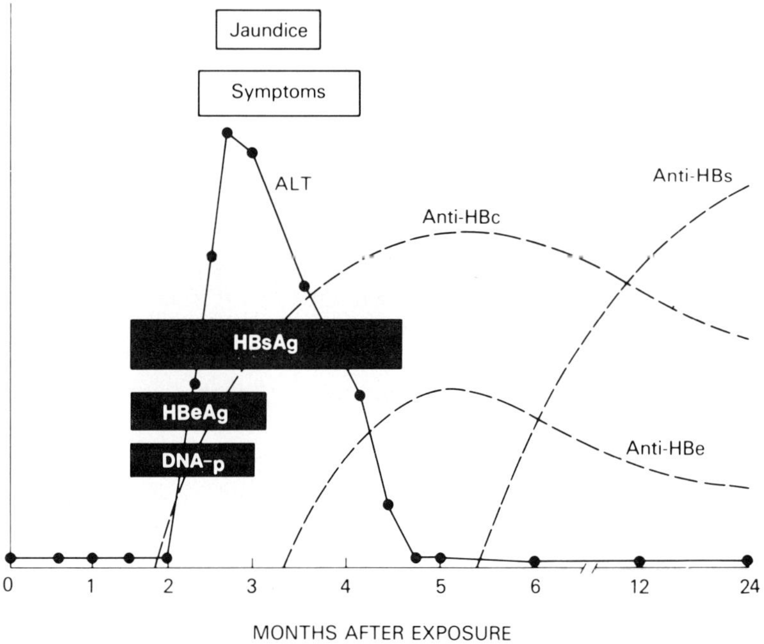

Fig. 1. Serologic pattern in a typical case of acute hepatitis B virus infection. ALT, alanine aminotransferase; SGPT, serum glutamic pyruvic transminase; HBsAg, hepatitis B surface antigen; HBeAg, hepatitis B e antigen; DNA-p, DNA polymerase; anti-HBc, antibody to hepatitis B core antigen; anti-HBs antibody to HBsAg; anti-HBe, antibody to HBeAg. From Hoofnagle (1980).

rapidly and are already HBsAg-negative at the time of onset of symptoms and jaundice. These patients may present a diagnostic dilemma requiring evaluation by other serologic tests and the testing of serial serum samples. In fact, only 92% of patients remain HBsAg-positive at the onset of symptoms due to hepatitis B (Hoofnagle *et al.,* 1978a). Even in patients whose HBV infections result in fulminant hepatic failure and coma, HBsAg may diminish (Tabor *et al.,* 1976a) or disappear (Hoofnagle *et al.,* 1978a; Tabor *et al.,* 1981a) by the time coma begins. The persistence of HBsAg for longer than 6 months indicates that a chronic carrier state has developed. This may occur in about 5 to 10% of infected individuals. Although HBsAg is usually detected in the presence of antibody to the hepatitis B core antigen (anti-HBc; see Section I,C), it may be found alone during the first 3 to 5 weeks of the infection (Table I). Theoretically, HBsAg without anti-HBc could also be found in an immunosuppressed patient or a patient with agammaglobulinemia. Just as the absence of HBsAg in a single serum sample does not rule out acute hepatitis B, so also the presence of HBsAg is not proof of acute hepatitis B. A blood recipient with clinical hepatitis may be an asymptomatic chronic carrier of HBsAg with superimposed acute hepatitis

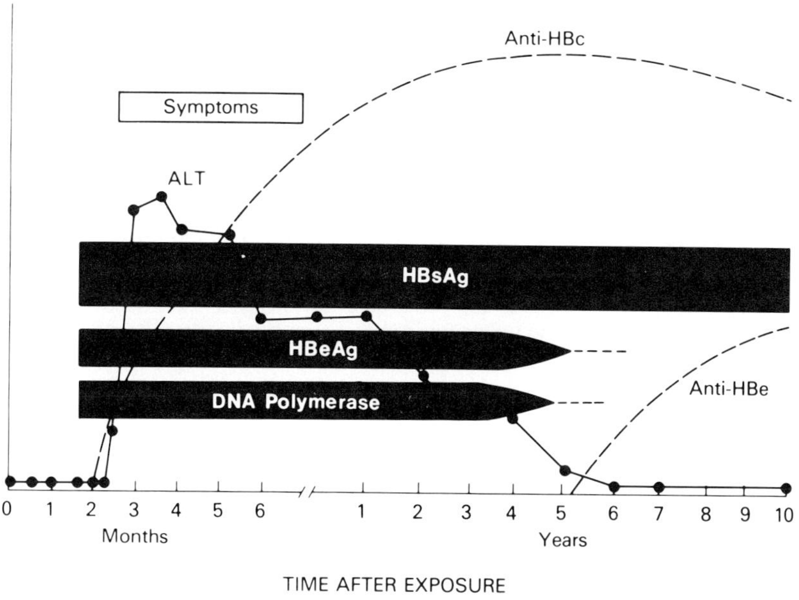

Fig. 2. Serologic pattern in a typical case of chronic hepatitis B virus infection. ALT, alanine aminotransferase; SGPT, serum glutamic pyruvic transaminase; HBsAg, hepatitis B surface antigen; HBeAg, hepatitis B e antigen; anti-HBc, antibody to hepatitis B core antigen; anti-HBe, antibody to HBeAg. From Hoofnagle (1980).

TABLE I

Serologic Patterns in Hepatitis B

HBsAg	Anti-HBs	Anti-HBc	Interpretation
+	−	−	Early acute
+	−	+	Acute or chronic
−	+	+	Convalescent
−	+	−	Late convalescence, seroconversion, or vaccination
−	−	+	IgM: early convalescence, chronic; IgG: Late convalescence
+	+	+	Chronic with heterotypic anti-HBs

due to a non-A, non-B hepatitis agent, hepatitus A virus (HAV) acquired in the community, halothane anesthesia, or another toxic agent. Testing for other HBV serologic markers in serial serum samples may be necessary in some cases to be completely certain of the diagnosis.

At present, only "third-generation" assays are of acceptable sensitivity for detecting HBsAg. The designation of third-generation sensitivity indicates the test's capacity to detect weakly positive HBsAg samples in the HBsAg Reference Panel distributed by the Bureau of Biologics, U.S. Food and Drug Administration. Such tests as agar gel diffusion (AGD) and counterelectrophoresis are used only for subtyping of HBsAg because their sensitivity is 10–1000 times lower than that of third-generation tests and is not acceptable in either blood banking or clinical practice (Table II).

TABLE II

Sensitivity of Assays for Hepatitis B Surface Antigen[a]

Designated generation of sensitivity	Test	Relative sensitivity	HBsAg prevalence among volunteer blood donors by this test (%)	Minimum HBsAg particles per ml serum required for detection
First generation	Agar gel diffusion	1	0.1	10^{13}
Second generation	Counterelectrophoresis	2–10	0.1–0.2	10^{12}
	Rheophoresis			
	Complement fixation			
	Reversed passive latex agglutination			
Third generation	Radioimmunoassay	100–10,000	0.2–0.4	10^{9}
	Enzyme-linked immunosorbent assay			
	Reversed passive hemagglutination			
	Reversed passive latex agglutination—T.G.[b]			
Infectivity	Chimpanzee inoculation	1,000,000–10,000,000	Not applicable	10^{6}

[a] From Gerety *et al.* (1978).

[b] T.G., third generation sensitivity of the indicated assay.

In the United States the solid-phase RIA is the most common method of detecting HBsAg. The basis of this method is a "solid phase," consisting of a plastic tube or bead, which is coated with antibody to hepatitis B surface antigen (anti-HBs). If HBsAg is present in the serum sample to be tested, it is bound to the solid phase and then binds a radiolabeled anti-HBs preparation. A ratio ("RIA ratio") of counts per minute to the mean counts per minute of negative control samples of ≥ 2.1 is a positive result. The advantages of RIA include sensitivity (Gerety *et al.*, 1978), infrequent false positive results (Reesink *et al.*, 1980), and an objective end point. The commercial tests also have a neutralization test, which can be used to confirm positive results. Although most RIAs require overnight incubation, tests requiring only 45 min are now available. Disadvantages of RIA include substantial costs of purchasing and maintaining gamma counters and the procedures needed to ensure employee safety when radioisotopes are being used. Another disadvantage is the relatively short shelf life of reactants (for most commercial tests 45 days; usually 30 days after delivery to the laboratory).

Reverse passive hemagglutination (RPHA), enzyme-linked immunosorbent assay (ELISA), and certain reverse passive latex agglutination (RPLA) assays also have third-generation sensitivity but are slightly less sensitive than RIA. Next to RIA, RPHA was until relatively recently the most sensitive commercial test (Gerety *et al.*, 1978; Reesink *et al.*, 1980). Improvements in the commercial ELISA test for HBsAg have made it more sensitive (Hepatitis Branch, Bureau of Biologics, unpublished data). In RPHA, erythrocytes coated with anti-HBs are agglutinated in the presence of HBsAg. In RPLA, latex particles are used. Both RPHA and RPLA are rapid, require only inexpensive equipment, and have a long shelf life (6 months). However, the reading of the results is subjective, and the accuracy varies with the reader's ability to detect weak reactions. In addition, false positive results are more common than with RIA.

The ELISA works on the same principle as the solid-phase RIA. Anti-HBs is bound to a solid phase; HBsAg in a test sample reacts with the anti-HBs and itself binds an enzyme-labeled anti-HBs. The ELISA has the advantage of combining a need for little equipment, a long shelf life (6–18 months, depending on the manufacturer), and objective reading of the results.

There are several subtypes of HBsAg that are of epidemiologic importance. HBsAg obtained from the serum of a given patient always contains the common **a** subtype determinant and either the **d** or **y** and the **w** or **r** determinants. Thus, a particular strain of HBV will be found to have one of the major subtypes of HBsAg, either **adw, adr, ayw,** or **ayr.** Rarely, infection by hybrid forms or dual infections may be responsible for the **adyw** and **adywr** subtypes. Numerous minor subtypes have also been described but have been poorly characterized. The HBsAg subtype permits one to trace the source of an infection or to confirm a point-source outbreak. In the United States and Western Europe, **adw** is the most common subtype among chronic carriers and hence among blood donors. However, **ayw** is not uncommon among descendants of individuals from Mediterranean countries (Sampliner *et al.*, 1981), and **adr** infections may be found among those of Oriental descent and servicemen returning from the Far East. In Eastern Europe and much of Africa, **ayw** is the most common subtype. In Japan and northern China, **adr** is common; in southern China, **adw** is most common. The **adr** subtype is frequently found among individuals whose ancestors emigrated from northern China to southern Japan or Taiwan (Yamashita *et al.*, 1975; Sung and Chen, 1978).

Recovery from infection with one subtype provides protection against infection by other subtypes (Purcell and Gerin, 1978; Gerety *et al.*, 1979b), and this forms the basis for vaccine production using only one or two subtypes; HBsAg subtypes and antibodies directed to them can be detected by a modification of the RIA test for HBsAg (Hoofnagle *et al.*, 1977a), rheophoresis (Jambazian and Holper, 1972), or AGD (Le Bouvier, 1971). Reagents for the detection of the subtypes are not available commercially. However, small volumes of reference reagents for use in the identification of donors whose sera are suitable as reagents can be obtained from several research laboratories.

C. Antibody to Hepatitis B Core Antigen

Antibody to the 27-nm core of HBV, anti-HBc, is detected in serum during acute and chronic hepatitis B and remains detectable for many years after recovery. It usually appears 2–4 weeks after HBsAg is first detected and 2–4 weeks before the onset of symptoms (Figs. 1 and 2). In unusual cases, anti-HBc may not be detected until after the onset of symptoms. The anti-HBc titer rises progressively during acute HBV infection; it remains at a constant high titer in chronic carriers. This can permit the differentiation between acute and chronic HBV infection in HBsAg-positive patients from whom paired serum samples are available.

Early in recovery, anti-HBc may be the only serologic marker of HBV detected for up to 20 weeks, during the "window period" between the disappearance of HBsAg and the appearance of anti-HBs (Table I). During this period, symptoms and infectivity may persist.

After recovery, anti-HBc persists at a lower titer. Usually, it coexists with anti-HBs for a few years; it then becomes undetectable, leaving anti-HBs as the only HBV marker. Occasionally, only anti-HBs may become undetectable. Such sera may be differentiated from those during the window period by their low titer of anti-HBc and the presence of only anti-HBc of the IgG class. However, at present, separate tests for anti-HBc of the IgM and IgG classes are available only in research laboratories.

It has been suggested that some individuals with anti-HBc in their serum and without detectable HBsAg and anti-HBs are chronic carriers of HBV, with HBsAg in such low titers that it cannot be detected (Hoofnagle *et al.*, 1978b). Although this hypothesis has not been confirmed, screening of blood donors for anti-HBc has been proposed as a means of preventing the cases of hepatitis B that occur despite testing of donor blood by the most sensitive tests for HBsAg. This is based on retrospective studies, which suggest that the presence of anti-HBc alone in transfused blood is associated with the transmission of hepatitis B to recipients (Hoofnagle *et al.*, 1978b; Rakela *et al.*, 1980; Seeff *et al.*, 1978). Clearly, most such blood does not contain HBV or the attack rate of hepatitis B in transfused patients would be much higher than is generally observed. It has been suggested that screening donors for IgM anti-HBc or high titers of anti-HBc would result in the identification of infected donors. However, the significance of anti-HBc alone of the IgM class is not known.

Several technical problems make it impractical to screen blood donors for anti-HBc. Each donor would have to be screened for HBsAg and anti-HBs as well as anti-HBc at markedly increased cost. Continued testing of HBsAg would be necessary to detect donors in the early stages of acute HBV infection, and testing for anti-HBs would be necessary in order not to exclude donors who have fully recovered from HBV infections. Whether tests for IgM anti-HBc are performed using principles of anti-μ, staphylococcal protein A, ultracentrifugation, or column chromatography, they are cumbersome and unstandardized. Tests for anti-HBc titer would further increase the cost. At present, the only commercially available RIA for anti-HBc is a competitive inhibition RIA. As such, it may have a higher false positive rate than would be expected with a "sandwich" assay, particularly in the testing of stored older serum samples. Although in general this assay is specific, it is not manufactured with a confirmatory second stage because of the relative lack of HBcAg for neutralization. The test is conducted by mixing equal volumes of test sample and radiolabeled anti-HBc; a plastic bead coated with HBcAg is then added. Greater than 50% inhibition of binding of the radiolabeled anti-HBc to the bead indicates that the test sample contains anti-HBc. Whenever anti-HBc is detected by RIA in the absence of all other HBV serologic markers, the test should be repeated to determine reproducibility.

D. Antibody to Hepatitis B Surface Antigen

Anti-HBs can be detected during recovery in most patients. It appears after clearance of HBsAg and remains detectable for life. In 8% of patients, HBsAg has disappeared and anti-HBs has appeared by the onset of symptoms (Hoofnagle *et al.*, 1978a); in these patients the diagnosis of hepatitis B can be made only by documenting rising titers of anti-HBc or anti-HBs. In 46% of patients there is a period of 1 to 20 weeks between the disappearance of HBsAg and the appearance of anti-HBs. In 3% of patients anti-HBs cannot be detected even after a year or more, despite the disappearance of HBsAg.

Four additional types of anti-HBs responses may be observed. The primary antibody response seen in about 22% of HBV infections is characterized by the appearance of anti-HBs as the first HBV serologic marker at 4 to 12 weeks, accompanied by anti-HBc. A secondary antibody response is characterized by a rise in anti-HBs titer in a person with preexisting anti-HBs in the absence of HBsAg or elevation of the anti-HBc titer. This response is believed to be due to immunization by HBsAg. A primary immunization response occurs after immunization with purified HBsAg or its immunogenic polypeptides in the form of a hepatitis B vaccine. Anti-HBs may appear any time after a first, second, or third dose of the vaccine and is not accompanied by HBsAg or anti-HBc. Heterotypic anti-HBs, which is usually found in low titer in the presence of coexisting HBsAg, is directed to a different HBsAg subtype determinant than that on the circulating HBsAg and is found in up to 15% of HBsAg chronic carriers (Bernier *et al.*, 1982). This may reflect frequent exposures to HBV of different HBsAg subtypes or exposure to more than one subtype at a single point in time. Whether it represents infection by two subtypes with recovery from one of them or immunization with the second subtype in a chronic carrier is not known (Tabor *et al.*, 1976b, 1977; Koziol *et al.*, 1976).

There is general agreement that the portion of anti-HBs directed to the group determinant of HBsAg, anti-**a**, is the protective antibody. This is based on the rarity of second episodes of HBV infection in individuals with natural or vaccine-induced anti-**a.** (Occasionally, reported reinfections have occurred after parenteral exposure to high titers of HBV and these exposures have resulted only in asymptomatic serologic evidence of reinfection.) In addition, it has been shown that chimpanzees that have recovered from HBV infection of one subtype are not susceptible to reinfection upon inoculation of a different subtype (Gerety *et al.*, 1979b).

Sensitive commercial assays for anti-HBs include RIA and passive hemagglutination (PHA). The design of the RIA for anti-HBs is the inverse of the RIA for HBsAg. The sample is added to a bead or tube coated with HBsAg, and radiolabeled HBsAg is subsequently added. A ratio of counts per minute to the mean counts per minute of the negative control samples of ≥2.1 is considered to be positive. Unlike the commercial RIAs for HBsAg, no confirmatory test is provided with the tests for anti-HBs, so results with RIA ratios of less than 20 should be repeated for confirmation. The PHA is performed by adding the test sample to red blood cells coated with HBsAg; agglutination indicates that anti-HBs is present. Considerations of equipment expense, shelf life, and objectivity are the same as those discussed above for RIA and RPHA for HBsAg. The RIA for anti-HBs detected 3–4% more positive samples in an unselected population than PHA, and in no case could anti-HBs be detected by PHA in samples in which it was not detectable by RIA (Gerety *et al.*, 1978).

E. Hepatitis B e Antigen

Hepatitis B e antigen (HBeAg) is distinct from the particulate hepatitis B core antigen, although it is believed to originate in the core of HBV because it can be released from the core by treatment with pronase and sodium dodecyl sulfate (Takahashi *et al.*, 1979). It is associated with two polypeptides of 19,000 and 45,000 daltons, respectively (Takahashi *et al.*, 1979), and three separate antigenic components: HBeAg/1, HBeAg/2, and HBeAg/3 (Murphy *et al.*, 1978). The detection of HBeAg in a patient's serum during HBV infection has been associated with progression of the disease to chronic persistent or chronic aggressive hepatitis (Magnius *et al.*, 1975; Eleftheriou *et al.*, 1975) and a greater likelihood of transmission of HBV to others (Magnius and Espmark, 1972; Nordenfelt and Kjellen, 1975; Takahashi *et al.*, 1976; Okada *et al.*, 1976). The HBeAg has been associated with the presence of intact HBV, specific DNA polymerase (Nordenfelt and Kjellen, 1975; Takahashi *et al.*, 1976), and an increased risk of transmission of HBV to newborn infants (Okada *et al.*, 1976; Tong *et al.*, 1981). Antibody to HBeAg (anti-HBe) has been associated with the asymptomatic chronic carrier state of HBsAg (Magnius and Espmark, 1972) and with less risk of transmitting HBV to others (Magnius and Espmark, 1972; Oda, 1978). Nevertheless, HBsAg-positive blood containing anti-HBe may still transmit hepatitis B (Oda, 1978).

With the development of sensitive RIAs for HBeAg and anti-HBe, it became necessary to redefine the interpretation of the detection of these markers. Earlier studies using the relatively insensitive AGD system detected either HBeAg or anti-HBe in only 14% of acute and 73% of asymptomatic chronic HBV infections (Eleftheriou *et al.*, 1975; Okada *et al.*, 1976). In contrast, RIA detects HBeAg or anti-HBe in 87 to 100% of HBV infections and has 500- to 1000-fold greater sensitivity than AGD (Mushahwar *et al.*, 1978; Frösner *et al.*, 1978; Aldershvile *et al.*, 1980). In addition, the appearance of HBeAg by RIA is not useful in predicting the outcome of HBV infections, because HBeAg may be detected by RIA early in the course of almost every HBV infection. The persistence of HBeAg for greater than 10 weeks with RIA, however, is associated with progression to chronicity (Aldershvile *et al.*, 1980). The detection of anti-HBe by RIA early in the course of HBV infection or conversion from HBeAg to anti-HBe may be associated with rapid clearance of HBsAg (Tabor *et al.*, 1980a). Two or more serial serum samples are usually needed to interpret HBeAg and anti-HBe results when the disease is of uncertain onset. The test may be extremely useful in confirming the existence of active HBV infection in a person whose only other HBV serologic marker is anti-HBc (Tabor *et al.*, 1980b, 1981a).

Occasionally, HBeAg and anti-HBe may be detected in the same serum sample. This has been reported using AGD (Tabor *et al.*, 1980e) and RIA (Ling *et al.*, 1979; Tabor *et al.*, 1980a). When detected by AGD, the components were shown to be HBeAg/1 and anti-HBe/2 (Tabor *et al.*, 1980e). The simultaneous presence of HBeAg and anti-HBe, particularly when detected at the time of transition from HBeAg to anti-HBe (Tabor *et al.*, 1980a), may be explained by the continuing circulation of one HBeAg component at a time when anti-HBe directed to another HBeAg component had already appeared. Other possible explanations include anti-HBe excess of one specificity in the presence of HBeAg of two or more specificities or the presence of immune complexes of HBeAg and anti-HBe of the same specificity. The current RIA, unlike AGD, cannot differentiate among HBeAg/1, HBeAg/2, and HBeAg/3.

The commercial RIA for HBeAg works on the same principle as the RIA for HBsAg. The test sample is added to a plastic bead coated with anti-HBe; if HBeAg is present, it binds to the bead and can then bind a radiolabeled anti-HBe in a "sandwich." An RIA ratio of ≥2.1 is a positive result. The RIA for anti-HBe is a competitive inhibition assay in which the test sample and a standardized sample of HBeAg are mixed and an anti-HBe-coated plastic bead is added. If anti-HBe is present in the sample, less of the standard HBeAg is bound to the bead; subsequently, radiolabeled anti-HBe is added, and less binds to the bead if anti-HBe was present in the test sample. A positive sample is one that results in ≥50% inhibition of the reactivity. A sensitive PHA technique for HBeAg and anti-HBe has also been described (Takekoshi *et al.*, 1979) but is not generally available.

F. DNA Polymerase

The HBV core contains a circular DNA, of which as much as 85% is double-stranded and the remainder single-stranded, although the percentage is variable (Robinson and Albin, 1980). The portion that is single-stranded is made double-stranded by the presence of a DNA polymerase molecule, which is specific for this virus. The core also contains an associated protein kinase activity (Robinson and Albin, 1980).

Often, DNA polymerase activity is present in serum early in HBV infection, when a large number of HBV particles are present, and persists in some chronically infected individuals. However, there are many HBsAg-positive sera that probably are infectious but do not contain detectable DNA polymerase. This enzyme may provide an indication of the level of infectivity of a given serum, but it is not as sensitive a marker as HBsAg.

DNA polymerase in serum is associated with HBeAg, high titers of HBsAg, a greater number of intact HBV particles, and increased infectivity (Nordenfelt and Kjellen, 1975; Takahashi *et al.*, 1976). Measuring DNA polymerase in serum is a time-consuming procedure that involves centifuging intact HBV particles, adding tritiated triphosphate, "spotting" the mixture on a paper disk, acid precipitating the bound ^3H, washing, and measuring the remaining radioactivity (Robinson, 1975). This assay is available in only a few laboratories.

G. Delta Antigen and Antibody

Delta antigen, or delta agent, was first thought to be an antigenic variant of the core antigen of HBV; however, it was soon recognized as an apparently separate virus found in the liver and blood of HBV-infected individuals. Delta agent is thought to be a defective virus that requires helper functions from HBV. This virus is most common in Italians and in individuals of Italian descent living in other countries; whether this represents vertical transmission or genetic susceptibility is not known. Non-Italians may also become infected, particularly if repeatedly exposed to blood or blood derivatives (Rizzetto *et al.*, 1979). For instance, antibody to the delta antigen (anti-delta) was found in 45 to 100% of HBsAg-positive hemophiliac patients from Italy, Germany, and the United States (Rizzetto *et al.*, 1980c). HBsAg-positive drug addicts in several countries have been shown to have had delta agent infections, a finding that is believed to reflect repeated exposure to blood (Rizzetto *et al.*, 1980c).

It seems likely that delta agent is transmitted primarily by blood transfusion and that the presence of HBV is required in order for infection to take place. It is almost certain, however, that "inapparent parenteral" routes of infection are also common because of the reservoir of delta agent in areas where it is endemic. Vertical transmission from an infected mother to her newborn infant also occurs (Smedile *et al.*, 1981). Delta agent is transmitted to the newborn simultaneously with or after transmission of HBV. Maternal HBeAg may be an important requirement for such transmission (Smedile *et al.*, 1981), as it is for the transmission of HBV to the newborn (Tong *et al.*, 1981).

Delta antigen consists of a protein of 68,000 daltons with a buoyant density of 1.25 to 1.28 g/cm^3 in CsCl (Rizzetto *et al.*, 1980b; Bonino *et al.*, 1981). It has been found in association with 35 to 37-nm particles that have HBsAg incorporated into their surface. Each particle contains an RNA molecule of approximately 5.5×10^5 daltons; the RNA has not been fully characterized (Bonino *et al.*, 1981). Although the delta agent is usually detected serologically only in the liver during infection, low titers of its associated antigen can also be detected in the serum following detergent treatment in rare cases (Bonino *et al.*, 1981).

However, the delta antigen is probably present transiently in serum in many acute cases, but it is not detected unless frequent serum samples are obtained (Rizzetto *et al.*, 1980a).

Sensitive serologic tests have been developed for the detection of anti-delta. A solid-phase RIA for anti-delta employs reagent delta antigen purified from human liver obtained at autopsy from a patient with high titers of anti-delta in his serum (Rizzetto *et al.*, 1980b). This method is more than 1000 times more sensitive than the frequently used immunofluorescence test for anti-delta (Rizzetto *et al.*, 1977, 1980b), and it is about 10 times more sensitive than ELISA (Crivelli *et al.*, 1981). However, it is not clear how many additional cases of delta agent infection are detected using RIA.

The delta antigen can be detected in infected liver tissue by direct immunofluorescence and in occasional detergent-treated serum samples by RIA. Detergent treatment of the serum is necessary to expose the delta antigen that is contained within the 35- to 37-nm particles in the circulation.

H. Nonspecific Screening for Hepatitis B

In an attempt to prevent the transmission of hepatitis by blood early federal regulations prohibited anyone from donating blood who had a history of viral hepatitis, had close contact within the previous 6 months with anyone with hepatitis, or had received a transfusion of blood or any of its derivatives within the previous 6 months (Code of Federal Regulations, 1975). These regulations remain in effect today, and studies have confirmed that individuals with a history of hepatitis are more likely to have serologic evidence of hepatitis B (Tabor *et al.*, 1979c). These regulations have been retained despite the use of sensitive screening tests for HBsAg and the recommendation of a WHO Expert Committee that blood donors be excluded only if they have had hepatitis in the preceding year (WHO Expert Committee, 1977). The value of excluding blood donors with a history of hepatitis may vary from country to country, depending on the prevalence of hepatitis in a country, and may be of minimal value in areas where blood and plasma donors are less likely to be infected with hepatitis B or non-A, non-B hepatitis. The primary reason for retaining these regulations in the United States is the prevention of some cases of non-A, non-B hepatitis.

II. NON-A, NON-B HEPATITIS

A. Introduction

The development of sensitive tests for serologic markers of HBV and HAV and the exclusion of HBsAg-positive blood donors have led to an awareness that most cases of hepatitis associated with transfusion are not caused by either these agents or other agents known to cause hepatitis. As many as 89% of cases of transfusion-associated hepatitis in the United States are due to the agent or agents of non-A, non-B hepatitis (Alter *et al.*, 1975b; Seeff *et al.*, 1977; Goldfield *et al.*, 1975; Prince *et al.*, 1974). Between 5 and 15% of recipients of 1 to 5 units of blood transfused in the United States develop non-A, non-B hepatitis (Aach *et al.*, 1978). Although the name *hepatitis C* has been proposed for this disease (Prince *et al.*, 1974), most clinicians and investigators have continued to refer to it as non-A, non-B hepatitis because circumstantial evidence suggests that more than one agent may be responsible for the disease. The diagnosis of non-A, non-B hepatitis is still based on ruling out, by serologic tests, known etiologic agents of hepatitis because a universally recognized specific serologic test has not yet been developed. However, in a research setting the infectivity of serum containing the human non-A, non-B hepatitis agent(s) can be documented by transmission to chimpanzees (Tabor *et al.*, 1978).

B. Epidemiology

Non-A, non-B hepatitis may be transmitted by the transfusion of blood (Prince *et al.*, 1974), blood products, or plasma derivatives (factor VIII, factor IX complex, and fibrinogen) (Craske *et al.*, 1975, 1978; Wyke *et al.*, 1979; Bradley *et al.*, 1979; Tabor, 1981; Yoshizawa *et al.*, 1980) or by accidental needlestick inoculation (Tabor *et al.*, 1978). Nonparenteral transmission must also occur in order to account for the high prevalence of this infection among healthy blood donors, most of whom have had no exposure to potential sources of hepatitis. Although clinically apparent infection does not appear to be transmitted easily to close contacts (Dienstag *et al.*, 1977), 27% of sporadic cases of clinically recognizable hepatitis have been shown to be caused by this agent (Dienstag *et al.*, 1977). In a study of sporadic acute hepatitis conducted by our laboratory, 19% of sporadic acute non-A, non-B hepatitis patients had had prior contact with a person with hepatitis compared with 1% of controls; this suggests that nonparenteral transmission may have been responsible for some of these cases. In addition, 40% of the sporadic cases of non-A, non-B hepatitis had no exposure to parenteral or nonparenteral risk factors associated with hepatitis transmission. In a study of donors whose blood had been implicated in the transmission of non-A, non-B hepatitis to recipients, 93% had never received a transfusion (Tabor *et al.*, 1979c). Apparent immunity to non-A, non-B hepatitis in 53 to 90% of inoculated volunteers suggests that they had immunity acquired by previous nonparenteral transmission (Hoofnagle *et al.*, 1977b).

Chronic infections are a common outcome of non-A, non-B hepatitis. Post-transfusion non-A, non-B hepatitis results in chronic hepatitis in 33 to 46% of cases, as indicated by a prolonged elevation of aminotransferase levels (Knodell *et al.*, 1977; Berman *et al.*, 1979; Rakela and Redeker, 1979) or prolonged abnormal thymol turbidity and cephalin flocculation tests (Hoofnagle *et al.*,

1977b). In three studies, biopsy evidence of chronic active or chronic persistent hepatitis was observed in 80 to 100% of those with chronic elevation of aminotransferase levels (Knodell *et al.*, 1977; Berman *et al.*, 1979; Rakela and Redeker, 1979). Furthermore, a high prevalence of chronic hepatitis has been observed in hemophiliacs, which may in part be due to non-A, non-B hepatitis (Gerety *et al.*, 1980). The prevalence of chronic infections after sporadic non-A, non-B hepatitis has been reported to range from 7 (Norkrans *et al.*, 1979) to 44% (unpublished data from our laboratory). The different rate of chronicity in the former series of sporadic cases compared with all reported series of posttransfusion cases, if valid, may reflect different infecting doses, different routes of exposure, or more than one agent of non-A, non-B hepatitis. The number of cases resulting in chronic infections may, in fact, be greater than reported; subclinical cases cannot be identified at present. Infectivity during chronic non-A, non-B hepatitis was documented by inoculation of chimpanzees with each of four serum or plasma samples obtained over a 6-year period from a patient, including two samples obtained at a time when the patient's aminotransferase levels had temporarily returned to normal or nearly normal (Fig. 3) (Tabor, *et al.*, 1978; Tabor, 1980c). Even in apparently acute, self-limited non-A, non-B hepatitis, the agent has been shown to be present in blood for at least 10 weeks (Tabor *et al.*, 1979a) and as early as 12 days before the onset of elevated aminotransferase levels (Hollinger *et al.*, 1978) in the chimpanzee model.

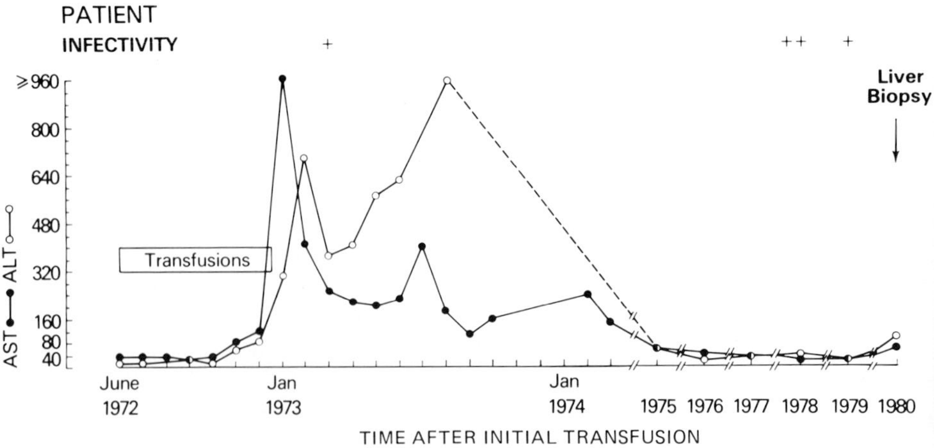

Fig. 3. Course of asymptomatic, chronic non-A, non-B hepatitis in a patient whose serum became inoculum I. "Transfusions" represent period of time when the patient was infused with 31 units of packed red blood cells and 43 units of platelets for aplastic anemia. Infectivity (shown by "+") indicates times when serum or plasma samples were shown to transmit non-A, non-B hepatitus. AST, ALT, aspartate and alanine aminotransferase (normal ≤ 40 IV/liter). Liver biopsy, chronic persistant hepatitus (Tabor *et al.*, 1980c).

Nevertheless, acquired immunity to non-A, non-B hepatitis appears to develop after apparent recovery in some cases. Immunity to reinfection followed apparent recovery in experimentally infected chimpanzees (Tabor *et al.*, 1979b; Alter, 1980; Tabor, 1981) and in human volunteers (Hoofnagle *et al.*, 1977b).

At present, most investigators believe that the majority of cases of non-A, non-B hepatitis following transfusion are caused by a single agent, whereas one or more additional agents may be responsible for the remaining cases. These conclusions are based in part on cross-challenge studies in chimpanzees using serum from humans with non-A, non-B hepatitis residing in different geographic areas (Tabor *et al.*, 1979b; Alter, 1980); in these studies inoculation of a second infectious serum did not result in detectable reinfections. However, accumulated evidence strongly suggests that a possible second agent is responsible for some cases, particularly those transmitted by derivatives of pooled plasma. Successive infections with non-A, non-B hepatitis have been reported in humans (Mosley *et al.*, 1977). A clinically milder form of non-A, non-B hepatitis has been observed among hemophiliacs, particularly when clotting factor concentrates (antihemophilic factor, or factor VIII; factor IX complex) manufactured from large pools of human plasma are first used in patients who previously had received only single-donor cryoprecipitate (Craske *et al.*, 1975). The likelihood that this is caused by a second agent is enhanced by the fact that it occurs in hemophiliacs whose prior exposure to many units of whole blood and single-donor cryoprecipitate makes it likely that they have had prior exposure and have immunity to agent(s) commonly transmitted by blood. The association with the introduction of plasma products made from pools of plasma from more than 1000 donors makes it more likely that a less common second agent could be transmitted. In addition, it is theoretically possible that the manufacturing process results in concentration of the second agent in these plasma products. The possibility that different preparations of factor VIII or factor IX complex may sometimes be infected with different agents of non-A, non-B hepatitis has been suggested on the basis of clinical observations (Craske *et al.*, 1978) and chimpanzee inoculations (Tsiquaye *et al.*, 1979). Apparently successful cross-challenge studies resulting in infection with a possible second agent of non-A, non-B hepatitis have been reported in chimpanzees using some lots of factor VIII, factor IX complex (Tsiquaye *et al.*, 1979), serum (Hollinger *et al.*, 1980), or fibrinogen (Tabor and Gerety, 1982; Tabor, 1981).

Four other facts have been cited in support of the existence of two separate agents, although the validity of these has been questioned. Higher rates of chronicity reported after transfusion compared with some studies of sporadic non-A, non-B hepatitis have already been discussed. There are different incubation periods (Aach *et al.*, 1978; Craske *et al.*, 1975; Alter *et al.*, 1978a; Wyke *et al.*, 1979; Bradley *et al.*, 1979; Prince *et al.*, 1978; Tabor *et al.*, 1978) with non-A, non-B hepatitis, most of which fall into categories of either 2 to 4 weeks or 8 to 12 weeks. However, in one study a single serum administered in a single dose by a single route in humans produced incubation periods ranging from 2 to 9 weeks (Hoofnagle *et al.*, 1977b). The elevation of serum aminotransferase levels in a monophasic or biphasic pattern has been suggested as a means of differentiating between two non-A, non-B hepatitis agents in infected humans, although a given inoculum has been shown to produce both monophasic and biphasic elevations of aminotransferase in chimpanzees (Tabor *et al.*, 1979a, 1980d). Ultrastructural differences in liver biopsies, originally thought to be associated with two different agents of non-A, non-B hepatitis (Shimizu *et al.*, 1979), have been found not to be exclusively associated with one inoculum or another because both changes have been seen after infection by one inoculum (Burk *et al.*, 1981). However, the characteristic tubular structures in infected liver tissue (Jackson *et al.*, 1979) appear to be specifically associated with non-A, non-B hepatitis.

Serologic studies have shown the agent(s) of non-A, non-B hepatitis to be unrelated to HAV, HBV, cytomegalovirus, Epstein–Barr virus, varicella-zoster virus, or herpes simplex virus (Gerety and Tabor, 1979; Tabor *et al.*, 1978). Non-A, non-B hepatitis has also been shown to be unrelated to the GB hepatitis agent (Tabor *et al.*, 1979d, 1980d) and to the delta agent (E. Tabor, unpublished data). The fact that the agent(s) has not yet been successfully grown in the common cell culture systems suggests that it is not one of the known common viral agents. However, the susceptibility of the agent to inactivation by formalin (Tabor and Gerety, 1980) and by heating at 60°C for 10 hr (Tabor and Gerety, 1982) suggests that its properties are similar to those of conventional infectious agents.

A variety of viruslike particles have been observed by electron microscopy in association with non-A, non-B hepatitis; no particle has been consistently detected. This probably reflects the difficulty of searching for an unknown agent in liver tissue by electron microscopy and of identifying particles in serum in a disease that may not be associated with high concentrations of virus in serum. Such viruslike particles have been reported in liver, serum, or plasma derivatives by 14 laboratories and range in diameter from 4 to 140 nm. Most, however, fall in a range of 20 to 35 nm or 60 to 80 nm in diameter.

C. Experimental Serologic Tests

The diagnosis of non-A, non-B hepatitis is still based on ruling out, by serologic tests, known etiologic agents of hepatitis and ruling out toxic causes of hepatitis. Extensive efforts have been made in many laboratories throughout the world to develop a serologic test to detect non-A, non-B hepatitis agents in blood donors. Candidate serologic assays, discussed in this section, have been the subject of 17 reports. However, there is little agreement in the scientific community on the specificity of these tests or whether the tests detect one or more antigens.

Using either direct or indirect immunofluorescence, four laboratories identified a nuclear antigen in hepatocytes during non-A, non-B hepatitis: one in liver biopsies from experimentally infected chimpanzees (Kabiri *et al.*, 1979) and three in liver biopsies from humans with acute or chronic non-A, non-B hepatitis (Trepo *et al.*, 1980; Arnold *et al.*, 1980; Alberti *et al.*, 1980). The specificity of these tests was documented by the absence of the antigen in uninfected or preinoculation control liver biopsies as well as biopsies obtained during either hepatitis A or hepatitis B infection and the appearance of antibodies only after acute illness.

Thirteen reports have appeared concerning antigens and antibodies detected by immunoprecipitation systems. Only one of these (Tabor *et al.*, 1979e) described detection by counterelectrophoresis, a system in which there is less difficulty in distinguishing the antigen-containing serum from the antibody-containing serum in an unknown system, because the antigen migrates toward the anode and the antibody toward the cathode at the alkaline pH used. Other laboratories have described antigen–antibody systems detected by immunodiffusion (Shirachi *et al.*, 1978; Prince *et al.*, 1980; Vitvitski *et al.*, 1979; Renger *et al.*, 1980; Overby *et al.*, 1980; Arnold *et al.*, 1980; Cossart *et al.*, 1975; Marciano-Cabral *et al.*, 1980; Williams *et al.*, 1980; Bamber *et al.*, 1980; Mori *et al.*, 1981; Hopkins *et al.*, 1981). Another study has suggested that some systems using immunodiffusion might be detecting the presence of immune complexes (Suh *et al.*, 1981).

Although relatively little is known about the nature of the antigen–antibody systems, it is clear that one or more antigens have been detected in the serum and/or liver of experimentally infected chimpanzees as well as patients with acute and chronic non-A, non-B hepatitis. The limited specificity testing that has been done suggests that the majority so far described in serum are associated only with non-A, non-B hepatitis. Specific antibodies to these antigens frequently develop. Some of these antibodies (perhaps analogous to antibodies to hepatitis B core antigen) may be present in either recovery from or chronic infection with a non-A, non-B agent (Tabor *et al.*, 1980a). Nonspecific antibodies and autoantibodies may also develop during or after hepatitis (Storch and Hagert, 1980) and may coexist with specific antibodies (Tabor *et al.*, 1979e). Given the complexity of the soluble and particulate antigens associated with HBV and their corresponding antibodies, the likelihood of there being more than one non-A, non-B hepatitis agent, and the possibility of misdiagnosed cases of non-A, non-B hepatitis, it is not surprising that there is neither clarity nor agreement as to the identity or specificity of the antigen–antibody systems associated with the agents of non-A, non-B hepatitis.

In the absence of specific serologic tests to detect blood donors who may transmit non-A, non-B hepatitis, many nonspecific measures have been suggested. Several of these originally had been used to prevent hepatitis B before the development of serologic tests to detect donors who transmit that disease. The exclusion of blood donors who are not in good health is one of these methods. The exclusion of prospective blood donors who have had clinically recognizable hepatitis (Code of Federal Regulations, 1975) has been continued in the United States in part because this may theoretically reduce the number of blood donors with chronic non-A, non-B hepatitis (Tabor *et al.*, 1979c). Blood from paid donors has been shown to be more likely to transmit both hepatitis B and non-A, non-B hepatitis than is blood from volunteer donors (Goldfield *et al.*, 1975), and this has resulted in the requirement that all blood for transfusion be labeled according to its source as either volunteer or paid (*Federal Register,* 1977). Similar requirements have not been applied to the collection of plasma. No similar increased risk has been documented for plasma; the pooling of thousands of units of plasma for the manufacture of plasma products makes it impossible to eliminate hepatitis completely from the pool by such labeling. In addition, several of the derivatives of these plasma pools, albumin, plasma protein fraction, and immune globulin, probably do not transmit non-A, non-B hepatitis.

Serum aminotransferase levels have been suggested as a means of identifying blood that may transmit non-A, non-B hepatitis (Hollinger and Alter, 1978; Aach *et al.*, 1981). However, the problems of nonspecific false positive results in donors with other causes of aminotransferase elevation and the prevalence of elevated aminotransferase levels among otherwise healthy adult blood donors make such screening difficult to interpret (Holland *et al.*, 1981). Data from a large multicentered prospective study group known as the Transfusion Transmitted Viruses (TTV) Study Group suggest that receiving multiunit blood transfusions, of which one unit is from a donor with elevated aminotransferase levels, may be associated with a greater likelihood of transmission of non-A, non-B hepatitis (Aach *et al.*, 1981). In that study, non-A, non-B hepatitis occurred in 6% of recipients transfused only with units with low normal aminotransferase levels [alanine aminotransferase (ALT, SGPT) $\leq$ 29 IU/liter) compared with 45% of recipients of multiple units of which at least one unit had elevated levels (ALT $\geq$ 60 IU/liter). However, 11% of non-A, non-B hepatitis cases in this study occurred among recipients of blood, all of whom had low normal aminotransferase levels (ALT $\leq$ 29 IU/liter). In another study of blood donors implicated in the transmission of non-A, non-B hepatitis to recipients of one or two units of blood, 80% of implicated donors had normal aminotransferase levels when tested retrospectively (Tabor *et al.*, 1979c). Furthermore, in a 5-year period following the introduction of ALT screening of blood donors in Hannover, West Germany, the incidence of non-A, non-B hepatitis actually increased, for reasons probably unrelated to the testing (Müller, 1982). Standardization of the ALT tests to ensure uniform application of the tests in blood banks is a technological difficulty that has not been resolved. It appears that additional study is needed before aminotransferase testing is routinely applied to the screening of blood for non-A, non-B hepatitis.

The use of tests for carcinoembryonic antigen (CEA) to screen donors has also been suggested on the basis of a single report that blood donors with elevated CEA levels may cause elevated CEA levels in recipients of their blood and may be more likely to transmit non-A, non-B hepatitis (Molnar and Gitnick, 1977). Until these findings are confirmed, the use of CEA as a screening test would be premature. Elevated CEA levels may be found in 11% of asymptomatic adults (Fuks *et al.*, 1974) and in 5% of asymptomatic adolescents (Tabor *et al.*, 1981b). These findings may represent nonspecificity of the test or may perhaps reflect the presence of asymptomatic carriers of non-A, non-B hepatitis in some cases. The greater prevalence of elevated CEA levels among apparently healthy tobacco smokers (Fuks *et al.*, 1974) suggests that a significant degree of nonspecificity would be found if this test were routinely used to screen blood donors.

The use of frozen-deglycerolized red blood cells for transfusion has been suggested as a means of preventing the transmission of viral hepatitis (Haynes *et al.*, 1960). The failure of this method to remove HBV has been documented in chimpanzees (Alter *et al.*, 1978b). The large-scale use of frozen or washed red blood cells in one hospital did not result in the elimination of clinically apparent non-A, non-B hepatitis (Haugen, 1979). Studies conducted with chimpanzees showed that the use of frozen blood cannot be relied on to prevent the transmission of human non-A, non-B hepatitis (personal communication, Dr. H. Alter, Clinical Center Blood Bank, National Institutes of Health).

REFERENCES

Aach, R. D., Lander, J. J., Sherman, L. A., Miller, W. V., Kahn, R. A., Gitnick, G. L., Hollinger, F. B., Werch, J., Szmuness, W., Stevens, C. E., Kellner, A., Weiner, J. M., and Mosley, J. W. (1978). *In* "Viral Hepatitis" (G. N. Vyas, S. N. Cohen, and R. Schmid, eds.), pp. 383–396. Franklin Inst. Press, Philadelphia, Pennsylvania.

Aach, R. D., Szmuness, W., Mosley, J. W., Hollinger, F. B., Kahn, R. A., Stevens, C. E., Edwards, V. M., and Werch, J. (1981). *N. Engl. J. Med.* **304,** 989–994.

Alberti, A., Realdi, G., Bortolotti, F., Busachi, C. A., and Badiali, L. (1980). Presented at the International Symposium on Non-A, Non-B Hepatitis, Vienna, June 16–18, 1980.

Aldershvile, J., Frösner, G. G., Nielsen, J. O., Hardt, F., Deinhardt, F., and Skinhøj, P. (1980). *J. Infect. Dis.* **141,** 293–298.

Alter, H. J. (1980). Presented at the International Workshop on Non-A, Non-B Hepatitis, Vienna, Austria, June 16–18, 1980.

Alter, H. J., Holland, P. V., and Purcell, R. H. (1975a). *Am. J. Med. Sci.* **270,** 329–334.

Alter, H. J., Purcell, R. H., Holland, P. V., Feinstone, S. M., Morrow, A. G., and Moritsugu, Y. (1975b). *Lancet* **2,** 838–841.

Alter, H. J., Purcell, R. H., Gerin, J. L., London, W. T., Kaplan, P. M., McAuliffe, V. J., Wagner, J., and Holland, P. V. (1977). *Infect. Immun.* **16,** 928–933.

Alter, H. J., Purcell, R. H., Holland, P. V., and Popper, H. (1978a). *Lancet* **1,** 459–463.

Alter, H. J., Tabor, E., Meryman, H. T., Hoofnagle, J. H., Kahn, R. A., Holland, P. V., Gerety, R. J., and Barker, L. F. (1978b). *N. Engl. J. Med.* **298,** 637–642.

Arnold, W. Reiter, H. J., Martini, G. A., and Meyer zum Büschenfelde, K. H. (1980). Presented at the International Symposium on Non-A, Non-B Hepatitis, Vienna, June 16–18, 1980.

Bamber, M., Murray, A., Thomas, H. C., Schever, P. J., Kernoff, P., Weller, I., Morelli, A., and Sherlock, S. (1980). *Gastroenterology* **79,** 1098. (Abstr.)

Bancroft, W. H., Snitbhan, R., Scott, R. H., Tingpalapong, M., Watson, W. T., Tanticharoenyos, P., Karwacki, J. J., and Srimarut, S. (1977). *J. Infect. Dis.* **135,** 79–85.

Barker, L. F., Gerety, R. J., and Tabor, E. (1977). *Adv. Intern. Med.* **23,** 327–351.

Berman, M., Alter, H. J., Ishak, K. G., Purcell, R. H., and Jones, E. A. (1979). *Ann. Intern. Med.* **91,** 1–6.

Bernier, R. H., Sampliner, R., Gerety, R., Tabor, E., Hamilton, F., and Nathanson, N. (1982). *Am. J. Epidemiol.* **116,** 199–211.

Bonino, F., Hoyer, B., Ford, E., Shih, J. W., Purcell, R. H., and Gerin, J. L. (1981). *Hepatology* **1,** 127–131.

Bradley, D. W., Cook, E. H., Maynard, J. E., McCaustland, K. A., Ebert, J. W., Dolana, G. H., Petzel, R. A., Kantor, R. J., Heilbrunn, A., Fields, H. A., and Murphy, B. L. (1979). *J. Med. Virol.* **3,** 253–269.

Burk, K. H., Cabral, G. A., Dreesman, G. R., Peters, R. L., and Alter, H. J. (1981). *J. Med. Virol.* **7,** 1–19.

Code of Federal Regulations (1975). 21: Food and Drugs, Section 640.3, p. 80.

Cossart, Y. E., Field, A. M., Cant, B., and Widdows, D. (1975). *Lancet* **1,** 72–73.

Craske, J., Dilling, N., and Stern, D. (1975). *Lancet* **2,** 221–223.

Craske, J., Spooner, R. J. D., and Vandervelde, E. M. (1978). *Lancet* **2,** 1051–1052 (letter).

Crivelli, O., Rizzetto, M., Lavarini, C., Smedile, A., and Gerin, J. L. (1981). *J. Clin. Microbiol.* **14,** 173–177.

Dienstag, J. L., Alaama, A., Mosley, J. W., Redeker, A. G., and Purcell, R. H. (1977). *Ann. Intern. Med.* **87,** 1–6.

Eleftheriou, N., Thomas, H. C., Heathcote, J., and Sherlock, S. (1975). *Lancet* **2,** 1171–1173.

Fed. Regist. (1977). **43,** 2142.

Frösner, G. G., Brodersen, M., Papaevangelou, G., Sugg, U., Haas, H., Mushahwar, I. K., Ling, C. M., Overby, L. R., and Deinhardt, F. (1978). *J. Med. Virol.* **3,** 67–76.

Fuks, A., Banjo, C., Shuster, J., Freedman, S. O., and Gold, P. (1974). *Biochim. Biophys. Acta* **417,** 123–152.

Gerety, R. J., and Schweitzer, I. L. (1977). *J. Pediatr. (St. Louis)* **90,** 368–374.

Gerety, R. J., and Tabor, E. (1979). *Infection (Munich)* **7,** 208.

Gerety, R. J., Hoofnagle, J. H., Markenson, J. A., and Barker, L. F. (1974). *J. Pediatr. (St. Louis)* **84,** 661–665.

Gerety, R. J., Tabor, E., Hoofnagle, J. H., Mitchell, F. D., and Barker, L. F. (1978). *In* "Viral Hepatitis" (G. N. Vyas, S. N. Cohen, and R. Schmid, eds.), pp. 121–138. Franklin Inst. Press, Philadelphia, Pennsylvania.

Gerety, R. J., Tabor, E., Purcell, R. H., and Tyeryar, F. J. (1979). *J. Infect. Dis.* **140,** 642–648.

Gerety, R. J., Eyster, M. E., Tabor, E., Drucker, J. A., Lusch, C. T., Proger, D., and Rice, S. A. (1980). *J. Med. Virol.* **6,** 111–118.

Goldfield, M., Black, H. C., Bill, J., Srihongse, S., and Pizzuti, W. (1975). *Am. J. Med. Sci.* **270,** 335–342.

Haugen, R. K. (1979). *N. Engl. J. Med.* **301,** 393–395.

Haynes, L. L., Tullis, J. L., Pyle, H. M., Sproul, M. T., Wallach, S., and Turville, W. C. (1960). *JAMA, J. Am. Med. Assoc.* **173,** 1657–1663.

Holland, P. V., Bancroft, W., and Zimmerman, H. (1981). *N. Engl. J. Med.* **304,** 1033–1034.

Hollinger, F. B., and Alter, H. J. (1978). *In* "Viral Hepatitis" (G. N. Vyas, S. N. Cohen, and R. Schmid, eds.), pp. 697–702. Franklin Inst. Press, Philadelphia, Pennsylvania.

Hollinger, F. B., Gitnick, G. L. Aach, R. D., Szmuness, W., Mosley, J. W., Stevens, C. E., Peters, R. L., Weiner, J. M., Werch, J. B., and Lander, J. J. (1978). *Intervirology* **10,** 60–68.

Hollinger, F. B., Mosley, J. W., Szmuness, W., Aach, R. D., Peters, R. L., and Stevens, C. (1980). *J. Infect. Dis.* **142,** 400–407.

Hoofnagle, J. H. (1980). *In* "Virus and the Liver" (L. Bianchi, W. Gerok, K. Sickinger, and G. A. Stalder, eds.), pp. 27–37. MTP Press, Lancaster, England.

Hoofnagle, J. H., Aronson, D., and Roberts, H. (1975). *Thromb. Diath. Haemorrh.* **33,** 606–609.

Hoofnagle, J. H., Gerety, R. J., Smallwood, L. A., and Barker, L. F. (1977a). *Gastroenterology* **72,** 290–296.

Hoofnagle, J. H., Gerety, R. J., Tabor, E., Feinstone, S. M., Barker, L. F., and Purcell, R. H. (1977b). *Ann. Intern. Med.* **87,** 14–20.

Hoofnagle, J. H., Seeff, L. B., Bales, Z. B., Gerety, R. J., Tabor, E., (1978a). *In* "Viral Hepatitis" (G. N. Vyas, S. N. Cohen, and R. Schmid, eds.), pp. 219–242. Franklin Inst. Press, Philadelphia, Pennsylvania.

Hoofnagle, J. H., Seeff, L. B., Bales, Z. B., and Zimmerman, H. J. (1978b). *N. Engl. J. Med.* **298,** 1379–1383.

Hopkins, R., Robertson, A. E., Haase, G., Brettle, R., Green, D., and Brown, T. (1981). *Lancet* **1,** 946–947 (letter).

Jackson, D., Tabor, E., and Gerety, R. J. (1979). *Lancet* **1,** 1249–1250 (letter).

Jambazian, A., and Holper, J. C. (1972). *Proc. Soc. Exp. Biol. Med.* **140,** 560–564.

Kabiri, M., Tabor, E., and Gerety, R. J. (1979). *Lancet* **2,** 221–224.

Knodell, R. G., Conrad, M. E., and Ishak, K. G. (1977). *Gastroenterology* **72,** 902–909.

Koziol, D. E., Alter, H. J., Kirchner, J. P., and Holland, P. V. (1976). *J. Immunol.* **117,** 2260–2262.

Le Bouvier, G. L. (1971). *J. Infect. Dis.* **123,** 671–675.

Ling, C. M., Mushahwar, I. K., Overby, L. R., Berquist, K. R., and Maynard, J. E. (1979). *Infect. Immun.* **24,** 352–356.

Magnius, L. O., and Espmark, J. A. (1972). *J. Immunol.* **109,** 1017–1021.

Magnius, L., Lindholm, A., Lundin, P., and Iwarson, S. (1975). *JAMA, J. Am. Med. Assoc.* **231,** 356–359.

Marciano-Cabral, F., Rublee, K. L., Carithers, R. L., Galen, E. A., and Cabral, G. A. (1980). *Gastroenterology* **79,** 1036. (Abstr.)

Molnar, I. G., and Gitnick, G. L. (1977). *Gastroenterology* **73,** 1235. (Abstr.)

Mori, Y., Ogata, S., Ata, S., and Nakano, Y. (1981). *Lancet* **2,** 98–99 (letter).

Mosley, J. W., Redeker, A. G., Feinstone, S. M., and Purcell, R. H. (1977). *N. Engl. J. Med.* **296,** 75–78.

Müller, R. (1983). *Vox Sang.* **44,** 48–64.

Murphy, B., Tabor, E., McAuliffe, V., Williams, A., Maynard, J., Gerety, R., and Purcell, R. (1978). *J. Clin. Microbiol.* **8,** 349–350.

Mushahwar, I. K., Overby, L. R., Frösner, G., Deinhardt, F., and Ling, C. M. (1978). *J. Med. Virol.* **2,** 77–87.

Nordenfelt, E., and Kjellen, L. (1975). *Intervirology* **5,** 225–232.

Norkrans, G., Frösner, G., Hermodsson, S., and Iwarson, S. (1979). *Scand. J. Infect. Dis.* **11,** 259–264.

Oda, T. (1978). *In* "Hepatitis Viruses" (T. Oda, ed.), pp. 285–291. Univ. Park Press, Baltimore, Maryland.

Okada, K., Kamiyama, I., Inomata, M., Imai, M., Miyakawa, Y., and Mayumi, M. (1976). *N. Engl. J. Med.* **294,** 746–749.

Overby, L., Chairez, R., Ling, C., and Gitnick, G. L. (1980). Presented at the International Symposium on Non-A, Non-B Hepatitis, Vienna, June 16–18, 1980.

Peterson, M. P., Barker, L. F., and Schade, D. S. (1973). *Vox Sang.* **24,** 66–75.

Prince, A. M., Brotman, B., Grady, G. F., Kuhns, W. J., Hazzi, C., Levine, R. W., and Millian, S. J. (1974). *Lancet* **2,** 241–255.

Prince, A. M., Brotman, B., van den Ende, M. C., Richardson, L., and Kellner, A. (1978). *In* "Viral Hepatitis" (G. N. Vyas, S. N. Cohen, and R. Schmid, eds.), pp. 633–640. Franklin Inst. Press, Philadelphia, Pennsylvania.

Prince, A. M., Trepo, C., Vitvitski, L., Brotman, B., Richardson, L., Huang, C. Y., van den Ende, M. C., and Hantz, O. (1980). Presented at the International Symposium on Non-A, Non-B Hepatitis, Vienna, June 16–18, 1980.

Purcell, R. H., and Gerin, J. L. (1978). *In* "Viral Hepatitis" (G. N. Vyas, S. N. Cohen, and R. Schmid, eds.), pp. 491–505. Franklin Inst. Press, Philadelphia, Pennsylvania.

Rakela, J., and Redeker, A. G. (1979). *Gastroenterology* **77,** 1200–1202.

Rakela, J., Mosley, J. W., Aach, R. D., Gitnick, G. L., Hollinger, F. G., Stevens, C. E., and Szmuness, W. (1980). *Gastroenterology* **78,** 1318. (Abstr.)

Reesink, H. W., Lafeber-Schut, L. J., Aaij, C., and Reerink-Brongers, E. E. (1980). *Vox Sang.* **39,** 61–72.

Renger, V. F., Frank, K. H., Porst, H., and Hinkley, K. (1980). Presented at the International Symposium on Non-A, Non-B Hepatitis, Vienna, June 16–18, 1980.

Rizzetto, M., Canese, M. G., Arico, S., Crivelli, O., Trepo, C., Bonino, F., and Verme, G. (1977). *Gut* **18,** 997–1003.

Rizzetto, M., Shih, J. W., Gocke, D. J., Purcell, R. H., Verme, G., and Gerin, J. L. (1979). *Lancet* **2,** 986–990.

Rizzetto, M., Canese, M. G., Gerin, J. L., London, W. T., Sly, D. L., and Purcell, R. H. (1980a). *J. Infect. Dis.* **141,** 590–602.

Rizzetto, M., Shih, J. W., and Gerin, J. L. (1980b). *J. Immunol.* **125,** 318–324.

Rizzetto, M., Purcell, R. H., and Gerin, J. L. (1980c). *Lancet* **1,** 1215–1219.

Robinson, W. S. (1975). *Am. J. Med. Sci.* **270,** 151–159.

Robinson, W. S., and Albin, C. (1980). *In* "Virus and the Liver" (L. Bianchi, W. Gerok, K. Sickinger, and G. A. Stalder, eds.), pp. 39–47. MTP Press, Lancaster, England.

Sampliner, R. E., Loevinger, B. L., Tabor, E., and Gerety, R. J. (1981). *Am. J. Epidemiol.* **113,** 50–54.

Seeff, L. B., Zimmerman, H. J., Wright, E. C., Finkelstein, J. D., Garcia-Pont, P., Greenlee, H. B., Dietz, A. A., Leevy, C. M., Tamburro, C. H., Schiff, E. R., Schimmel, E. M., Zemel, R., Zimmon, D. S., and McCollum, R. W. (1977). *Gastroenterology* **72,** 111–121.

Seeff, L. B., Wright, E. C., Zimmerman, H. J., Hoofnagle, J. H., Dietz, A. A., Felsher, B. F., Garcia-Pont, P. H., Gerety, R. J., Greenlee, H. B., Kiernan, T., Leevy, C. M., Nath, N., Schiff, E. R., Schwartz, C., Tabor, E., Tamburro, C., Vlahcevic, Z., Zemel, R., and Zimmon, D. S. (1978). *In* "Viral Hepatitis" (G. N. Vyas, S. N. Cohen, R. Schmid, eds.), pp. 371–381. Franklin Inst. Press, Philadelphia, Pennsylvania.

Shimizu, Y. K., Feinstone, S. M., Purcell, R. H., Alter, H. J., and London, W. T. (1979). *Science* **205,** 197–200.

Shirachi, R., Shiraishi, H., Tateda, A., Kikuchi, K., and Ishida, N. (1978). *Lancet* **2,** 853–856.

Smedile, A., Dentico, P., Zanetti, A., Sagnelli, E., Nordenfelt, E., Actis, G. C., and Rizzetto, M. (1981). *Gastroenterology* **81,** 992–997.

Storch, W., and Hagert, M. (1980). Presented at the International Symposium on Non-A, Non-B Hepatitis, Vienna, June 16–18, 1980.

Suh, D. J., White, Y., Eddleston, A. L., Amini, S., Tsiquaye, K., Zuckerman, A. J., and Williams, R. (1981). *Lancet* **1,** 178–180.

Sung, J. L., and Chen, D. S. (1978). *J. Formosan Med. Assoc.* **77,** 263–270.
Tabor, E. (1981). *In* "Non-A, Non-B Hepatitis" (R. J. Gerety, ed.), pp. 189–206. Academic Press, New York.
Tabor, E., and Gerety, R. J. (1980). *J. Infect. Dis.* **142,** 767–770.
Tabor, E., and Gerety, R. J. (1982). *In* "Viral Hepatitis: 1981 International Symposium" (W. Szmuness, H. J. Alter, J. E. Maynard, eds.), pp. 305–317. Franklin Inst. Press, Philadelphia, Pennsylvania.
Tabor, E., Gerety, R. J., Hoofnagle, J. H., and Barker, L. F. (1976a). *Gastroenterology* **71,** 635–640.
Tabor, E., Gerety, R. J., Smallwood, L. A., and Barker, L. F. (1976b). *J. Immunol.* **117,** 2038–2040.
Tabor, E., Gerety, R. J., Smallwood, L. A., and Barker, L. F. (1977). *J. Immunol.* **118,** 369–370.
Tabor, E., Gerety, R. J., Drucker, J. A., Seeff, L. B., Hoofnagle, J. H., Jackson, D. R., April, M., Barker, L. F., and Pineda-Tamondong, G. (1978). *Lancet* **1,** 463–466.
Tabor, E., April, M., Seeff, L. B., and Gerety, R. J. (1979a). *Gastroenterology* **76,** 680–684.
Tabor, E., April, M., Seeff, L. B., and Gerety, R. J. (1979b). *J. Infect. Dis.* **140,** 789–793.
Tabor, E., Hoofnagle, J. H., Smallwood, L. A., Drucker, J. A., Pineda-Tamondong, G. C., Ni, L. Y., Greenwalt, T. J., Barker, L. F., and Gerety, R. J. (1979c). *Transfusion (Philadelphia)* **19,** 725–731.
Tabor, E., Seeff, L. B., and Gerety, R. J. (1979d). *J. Infect. Dis.* **140,** 794–797.
Tabor, E., Mitchell, F. D., Goudeau, A. M., and Gerety, R. J. (1979e). *J. Med. Virol.* **4,** 161–169.
Tabor, E., Frösner, G., Deinhardt, F., and Gerety, R. J. (1980a). *J. Med. Virol.* **6,** 91–99.
Tabor, E., Ziegler, J. L., and Gerety, R. J. (1980b). *J. Infect. Dis.* **141,** 289–292.
Tabor, E., Seeff, L. B., and Gerety, R. J. (1980c). *N. Engl. J. Med.* **303,** 139–143.
Tabor, E., Peterson, D. A., April, M., Seeff, L. B., and Gerety, R. J. (1980d). *J. Med. Virol.* **5,** 103–108.
Tabor, E., Goldfield, M., Black, H. C., and Gerety, R. J. (1980e). *Transfusion (Philadelphia)* **20,** 192–198.
Tabor, E., Krugman, S., Weiss, E. C., and Gerety, R. J. (1981a). *J. Med. Virol.* **8,** 277–282.
Tabor, E., Gerety, R. J., Needy, C. F., Elisberg, B. L., Colon, A. R., and Jones, R. (1981b). *Eur. J. Cancer* **17,** 257–258 (letter).
Takahashi, K., Imai, M., Tsuda, F., Takahashi, T., Miyakawa, Y., and Mayumi, M. (1976). *J. Immunol.* **117,** 102–105.
Takahashi, K., Akahane, Y., Gotanda, T., Mishiro, T., Imai, M., Miyakawa, Y., and Mayumi, M. (1979). *J. Immunol.* **122,** 275–279.
Takekoshi, Y., Tanaka, M., Miyakawa, Y., Yoshizawa, H., Takahashi, K., and Mayumi, M. (1979). *N. Engl. J. Med.* **300,** 814–819.
Tong, M. J., Thursby, M., Rakela, J., McPeak, C., Edwards, V. M., and Mosley, J. W. (1981). *Gastroenterology* **80,** 999–1004.
Trepo, C., Vitvitski, L., Hantz, O., Chevallier, P., Grimaud, J. A., Pichoud, C., and Sepetjan, M. (1980). *Gastroenterology* **79,** 1060. (Abstr.)
Tsiquaye, K. N., and Zuckerman, A. J. (1979). *Lancet* **1,** 1135–1136 (letter).
Villarejos, V. M., Visoná, K. A., Gutiérrez, A., Rodríguez, A. (1974). *N. Engl. J. Med.* **291,** 1375–1378.
Vitvitski, L., Trepo, C., Prince, A. M., and Brotman, B. (1979). *Lancet* **2,** 1263–1267.
WHO Expert Committee (1977). "Advances in Viral Hepatitis," p. 59. World Health Organization, Geneva.
Williams, A. E., Wright, J. D., and Miller, D. J. (1980). *Gastroenterology* **79,** 1066. (Abstr.)
Wyke, R. J., Tsiquaye, K. N., Thornton, A., White, Y., Portmann, B., Das, P. K., Zuckerman, A. J., and Williams, R. (1979). *Lancet* **1,** 520–524.

Yamashita, Y., Kurashina, S., Miyakawa, Y., and Mayumi, M. (1975). *J. Infect. Dis.* **131,** 567–569.

Yoshizawa, H., Akahane, Y., Itoh, Y., Iwakiri, S., Kitajima, K., Morita, M., Tanaka, A., Nojiri, T., Shimizu, M., Miyakawa, Y., and Mayumi, M. (1980). *Gastroenterology* **79,** 512–520.

20

Immunoassays for the Diagnosis of Rotavirus and Norwalk Virus Infections

GEORGE CUKOR, SARAH H. CHEESEMAN, AND NEIL R. BLACKLOW

University of Massachusetts Medical School
Worcester, Massachusetts

I. Introduction 267
II. Rotavirus 268
A. Epidemiology and Virology 268
B. Considerations in Establishing Rotavirus Solid-Phase Immunoassays 269
C. Commerical Tests 274
III. Norwalk Virus 274
References 277

I. INTRODUCTION

Acute viral gastroenteritis (reviewed by Blacklow and Cukor, 1981) is an extremely common illness that affects all age groups throughout the world and occurs in both epidemic and endemic forms. Among illnesses affecting U.S. families, it is second in frequency only to the common cold. The disease is usually self-limited, although it can be lethal to the malnourished, elderly, debilitated, or infant patient. The two viral pathogens known to have medically important etiologic roles in human gastroenteritis, rotavirus and Norwalk virus, were discovered during the 1970s by electron microscopy (Bishop *et al.*, 1973; Kapikian *et al.*, 1972). The development of rapid diagnostic immunoassays for these viruses has led to an understanding of their epidemiology and medical importance.

The Direct Detection of Microorganisms in Clinical Samples

ISBN 0-12-187780-9

II. ROTAVIRUS

A. Epidemiology and Virology

Rotavirus infection is the most important cause of acute diarrheal illness requiring hospitalization of infants and young children and accounts for 40 to 60% of such hospitalization in temperate climates. The virus also appears to have a major role in gastroenteritis in infants and young children not requiring hospitilization. Infection occurs throughout the world, usually in sporadic but occasionally in epidemic forms. In temperate climates the peak incidence of rotaviral disease occurs during the cooler months, whereas in tropical areas cases occur year round. Asymptomatic rotavirus infections occur frequently in adults who have had contact with ill children. Symptomatic adult rotavirus infections have been reported among parents of ill children and also among adults who lack contact with children and in travelers. Rotavirus is transmitted from person to person by the fecal–oral route. The incubation period is 1–3 days. The virus is detectable in feces during acute illness and is commonly shed for up to 8 days after the onset of symptoms and occasionally longer. Nosocomial spread of infection among hospitalized pediatric patients and medical personnel is well documented. Prevalence studies indicate rapid acquisition of serum antibody, with most individuals possessing antirotavirus antibody by 3 years of age.

Serologic assays have been useful for epidemiologic studies but, because of the virtually universal presence of preexisting serum antibody, the diagnosis of an episode of illness depends on the detection of rotavirus in diarrheal stool specimens. A rapid diagnosis of rotavirus illness provides the physician with useful information for formulating a prognosis, may prevent the unnecessary use of antibiotic therapy and prolonged hospitalization, and allows for the institution of appropriate control measures.

Some human rotaviruses cannot be directly cultivated in cell culture. A human rotavirus strain was adapted to replication in cell culture after 11 passages in gnotobiotic pigs (Wyatt *et al.*, 1980). Other strains have been induced to replicate *in vitro* after treatment with trypsin, concentration by ultracentrifugation, and approximately six blind passages in cell culture (Urasawa *et al.*, 1981; Sato *et al.*, 1981). Even with the use of these special and time-consuming techniques, approximately 50% of the isolates currently remain noncultivatable. Therefore, at present, tissue culture isolation of human rotavirus remains an exciting research tool but not a practical diagnostic technique.

The characteristic morphology of rotaviruses is readily recognized in negatively stained stool preparations under the electron microscope. Electron micrograph studies resulted in the original identification of rotavirus (Bishop *et al.*, 1973). This technique has been used extensively for rotavirus diagnosis and remains the standard by which other assays must be judged. However, the usefulness of electron microscopy is limited by the requirement for highly specialized equipment and personnel as well as by the difficulty of handling a large number of specimens.

The diagnostic test of choice for most situations is the solid-phase immunoassay. It offers the possibility of same-day or next-day diagnosis of rotavirus infection and can be performed on a large scale with routinely available equipment and personnel. In this assay the stool specimen to be tested is exposed to a solid phase, which has been coated with antirotavirus antibody. After an absorption period and washing, a labeled virus-specific detection antibody is used to demonstrate the presence of rotavirus bound to the solid phase.

B. Considerations in Establishing Rotavirus Solid-Phase Immunoassays

1. Choice of Serum Reagents

At least three serotypes of human rotavirus have been identified on the basis of virus neutralization assays (Wyatt *et al.*, 1982). In addition, two different antigenic specificities, designated as subgroups, have been identified by differential reactivity of human rotaviruses in immunoassays (Kapikian *et al.*, 1981). All human rotaviruses seem to exhibit the same clinical and epidemiologic patterns, and the grouping or typing of isolates would serve no practical diagnostic purpose, although it is useful for following epidemiologic trends of infection.

There have been several approaches to the production of high-titered antirotavirus serum for use in immunoassay techniques. Convalescent human sera have not been found to be of sufficiently high titer for practical use, and therefore hyperimmune animal sera have been used. In one approach human rotavirus is purified directly from human diarrheal stool specimens and is used as antigen for the immunization of laboratory animals (Kjeldsberg and Mortensson-Egnurd, 1982; Middleton *et al.*, 1977; Grauballe *et al.*, 1981; Obert *et al.*, 1981; Sarkkinen *et al.*, 1979). The major drawback of this approach is that the hyperimmune serum produced may contain antibody reacting with human stool components that copurified with the virus. This would result in higher background reactions and perhaps false positive reactions with certain stools that contain high concentrations of the reacting antigen. To circumvent this problem, a second approach has been taken, in which human rotavirus is used to cause illness in gnotobiotic calves. The virus is then purified from calf stool; thus, contamination with human stool components is avoided and the virus becomes a cleaner immunogen (Kalica *et al.*, 1977; Yolken *et al.*, 1977a). The third approach exploits the presence of a group-specific antigen shared by mammalian rotaviruses. Several of the animal rotaviruses, such as the simian SA-11 and bovine NCDV, are readily propagated in tissue culture and can be purified as antigen for the produc-

TABLE I

Selection of Hyperimmune Serum[a]

Serum	Complement fixation titer	RIA for SA-11 (P/N)	Accuracy of RIA for stool (%)
Guinea pig A	1 : 8,192	37.0	100
Guinea pig B	1 : 16,384	5.6	50
Guinea pig C	1 : 4,096	10.1	50
Guinea pig D	1 : 8,192	16.3	50
Guinea pig E	1 : 8,192	21.5	100

[a] Sera from five hyperimmunized guinea pigs were evaluated in three ways (*1*) the titer of CF antibody to SA-11 virus; (*2*) P/N ratios obtained by reaction in RIA with cell-culture-grown SA-11 virus; (*3*) the capacity to distinguish by RIA between four electron microscope-positive and four electron microscope-negative human stool specimens.

tion of hyperimmune sera (Cukor *et al.*, 1978; Rubinstein and Miller, 1982; Hammond *et al.*, 1982). Because gnotobiotic animals are not readily available, the third approach is preferable for most laboratories.

In order to avoid additional problems associated with nonspecific reactivity of serum reagents, it is preferable that the coating and detection antibodies for immunoassays be prepared in different animal species. In this way, undesired reactivity present in only one of the reagent sera would not interfere with the test. Goats, rabbits, and guinea pigs have been used successfully for antirotavirus serum production. Not all individuals of a given species produce satisfactory hyperimmune serum. Therefore, several animals should be immunized and their sera separately tested in the immunoassay. Other assays, such as complement fixation, are not adequate in our experience for the screening of sera for use in solid-phase immunoassays (Table I).

2. *Choice of Solid Phase*

Flexible 96-well polyvinyl chloride microtiter plates have been used with success when goat serum has been employed as a coating antibody. Hard polystyrene plates may also be used with other animal sera. Individual polystyrene beads offer an advantage when small or variable numbers of specimens are to be tested, but they are slightly more laborious to handle. The flexible plates may also be cut up to obtain the desired number of wells for each test.

3. *Sample Preparation*

Only diarrheal stools from the acute phase of illness are likely to contain virus. Stool samples may be stored at −70°C. When stools are processed for study, no serum should be used in the diluent unless it has been shown to be negative for rotavirus antibody. In our laboratory an approximate 2–10% suspension of stool is prepared in veal infusion broth supplemented with 0.5% bovine serum albumin, mixed on a Vortex apparatus for 1 min, and clarified by centrifugation at 200 *g* for 15 min (Blacklow and Cukor, 1980). Rectal swabs have been used with somewhat less success than stool specimens (Brandt *et al.*, 1981). The incorporation of EDTA into the suspension medium (0.025 *M*) has been reported to increase the sensitivity of the test for some specimens, presumably by exposure of additional latent group-specific antigen on the surface of the virus particles in the presence of the chelating agent (Hammond *et al.*, 1982; Beards and Bryden, 1981).

4. Choice of Test Conditions

A wide variety of test conditions have been used in different laboratories. Incubations have usually been performed at either room temperature or 37°C for periods ranging from 1 to 12 hr. Some studies indicate that incubation times of only 5 to 10 min may be sufficient (Yolken and Leister, 1982) because when high-titered reagents are used at optimal concentrations (as predetermined by checkerboard titration) the incubation conditions appear not to be critical and depend largely on the convenience of the operator.

5. Choice of an Indicator System

Although the use of radioiodinated detection antibody is convenient in a research situation in which a planned number of specimens are to be tested and iodination facilities are available, radioimmunoassay (RIA) is less practical in a clinical diagnostic setting. The reason for this is the short shelf life of radioactive antibody and the need for special facilities for handling and disposing of radioactive materials. The enzyme immunoassay (EIA) uses a stable enzyme–antibody conjugate as a detection system. The addition of a chromogenic substrate allows for the convenient quantification of bound antibody with the use of a spectrophotometer. The availability of specially modified spectrophotometers (EIA readers) makes the EIA system the most practical for clinical laboratories.

The most commonly used enzymes are alkaline phosphatase and horseradish peroxidase. Alkaline phosphatase has the disadvantage of being expensive; also, some clinical material may contain endogenous enzyme activity, which could contribute to higher background values. Peroxidase is inexpensive and readily available; however, the enzyme is inhibited by metal ions, bacterial contamination, and antibacterial agents such as methanol and sodium azide. β-Galactosidase has also been used in some EIA systems. It offers the advantage of being absent from clinical specimens, but because of the low specific activity of available preparations color development is much slower than it is for other enzymes. The various enzymes and substrates available for EIA systems have been reviewed (Yolken, 1982).

6. Choice of Direct or Indirect Assay

Thus far our discussion has been confined to the direct immunoassay for rotavirus. In its indirect modification (Yolken *et al.*, 1977b), the antibody-coated solid phase is reacted with the test specimen as in the direct test. Next, an unlabeled antirotavirus serum (derived from a species other than the source of the coating antibody) is added. After the incubation and washing steps, a labeled antiglobulin directed against the animal species of the second antibody is utilized as the detection antibody. With the inclusion of appropriate controls (see Section 7) the indirect assay system has been used successfully for the detection of rotavirus (Sarkkinen *et al.*, 1979; Brandt *et al.*, 1981; Yolken *et al.*, 1980). By analogy with fluorescent antibody tests, the indirect EIA includes an attenuation step and therefore provides the theoretical advantage of increased sensitivity. In addition, it precludes the necessity of labeling each detection antibody preparation by allowing the use of a single labeled antiglobulin for many different types of immunoassays.

When direct and indirect EIA tests for rotavirus using the same reagents were compared (Hammond *et al.*, 1982), the theoretical advantage of increased sensitivity with the indirect test was not realized. Rotavirus is excreted by ill children in copious amounts (Vesikari *et al.*, 1981); the amount of viral antigen has been estimated to be close to 1 mg per milliliter of diarrheal stool. In light of this, increased sensitivity is probably not an important consideration in the routine clinical diagnostic situation. The indirect test is somewhat less convenient than the direct assay because extra incubation and washing steps are required. However, it appears that the major drawback of the indirect test is an increased incidence of nonspecific reactions, which has been repeatedly observed (see Section 7; Hammond *et al.*, 1982; Brandt *et al.*, 1981; Yolken *et al.*, 1977b) and high background levels of reaction (Sarkkinen *et al.*, 1979; Hammond *et al.*, 1982), reportedly resulting in a test of low sensitivity and specificity (Hammond *et al.*, 1982; Brandt *et al.*, 1981).

7. Nonspecific Reactions and Selection of Controls

In the original design of the solid-phase immunoassay for rotavirus (Kalica *et al.*, 1977; Yolken *et al.*, 1977a), the reactivity of test specimens was compared with the reactivity of a panel of known negative stool samples. A confirmatory test was recommended for determining the status of weakly positive specimens. The confirmatory test, consisting of exogenous blocking (neutralization) of sample reactivity with a high-titered antirotavirus serum, was laborious and thought to be unnecessary for routine use with each specimen tested. However, after several years of experience with this test, it became apparent that the frequency of nonspecific reactions was a serious problem (Brandt *et al.*, 1981; Yolken *et al.*, 1977b; Yolken and Stopa, 1979). Nonspecific reactions to goat serum were first recognized in stools from patients living in certain geographic areas, especially the Far East, and thought to be related to their diet. Other stools contained nonspecific antiglobulin activity related to an IgM molecule present in the specimen. To circumvent these problems normal goat serum and a reducing agent were added to all stool specimens (Yolken and Stopa, 1979). Others have attempted to eliminate nonspecific reactions by the addition of EDTA or bovine serum albumin or by neutralization of the pH of the stool specimen (Middleton *et al.*, 1977; Hammond *et al.*, 1982). Some nonspecific reactions could be eliminated by these measures. In an extensive study reported in 1981 from the Children's Hospital National Medical Center in Washington, D.C. (Brandt *et al.*, 1981) it was found that, among 5626 fecal specimens tested, 1344 gave positive results with the indirect EIA test, which could not be confirmed by electron microscopic visualization of rotavirus particles, blocking EIA, confirmatory EIA, or a combination of these methods. Thus, 73% of the 1834 presumptively positive EIA tests were in fact not positive. False positive EIA reactions were especially common with specimens from hospitalized young infants in a tertiary care nursery. It was postulated that nonspecific binding of rotavirus antibody to intestinal bacteria or their products such as staphylococcal protein A may have been the basis of some of the false positive reactions.

On the basis of this experience it is apparent that measures such as the addition of reducing agent are inadequate because they prevent only some kinds of nonspecific reactions from taking place. Reliance on a panel of known negative stools as controls is also inadequate. Furthermore, not only borderline specimens but all potential positive reactions must be confirmed. As in the system originally employed for the detection of Norwalk virus (Greenberg *et al.*, 1978), the reaction of each stool specimen in the well coated with rotavirus-positive (postimmunization) serum is compared with the reaction of the same specimen in a control well coated with rotavirus-negative (preimmunization) serum (Yolken *et al.*, 1980; Kapikian *et al.*, 1979). When such a confirmatory test was included in the indirect EIA at Children's Hospital in Washington (Brandt *et al.*, 1981), false positive results were not encountered in 400 consecutive tests.

8. *Calculation of Cutoff Value for Positive Results*

A variety of calculations have been used to determine the cutoff value for positive specimens. In an RIA system, positive/negative (P/N) values are often calculated by dividing the counts bound to the experimental well (P) by the counts bound to the control well (N); P/N values of ≥ 2 are usually considered to be positive (Kalica *et al.*, 1977; Cukor *et al.*, 1978).

In an EIA system in which control absorbance values may be lower and more variable, a P minus N calculation may be used (Yolken *et al.*, 1977a; Rubenstein and Miller, 1982). It is clear that there is not a single method for the calculation of results that applies to all of the different immunoassays. Rather, for each new assay developed the test and calculation method must be validated by testing a large number of positive and negative specimens and comparing the results with a standard test, preferably electron microscopy (Kjeldsberg and Mortensson-Egnund, 1982; Middleton *et al.*, 1977; Grauballe *et al.*, 1981; Obert *et al.*, 1981; Sarkkinen *et al.*, 1979; Kalica *et al.*, 1977; Yolken *et al.*, 1977a; Cukor *et al.*, 1978; Rubenstein and Miller, 1982; Hammond *et al.*, 1982; Brandt *et al.*, 1981; Beards and Bryden, 1981; Wall *et al.*, 1982).

C. Commercial Tests

As of the middle of 1982 one rotavirus EIA kit was licensed for diagnostic use. (Rotazyme, Abbott Laboratories, North Chicago, Illinois), and two others were available for investigational use (Dakopatts, Acurate Chemical Corp., Westbury, New York, and Enzygnost, Calbiochem. Corp., LaJolla, California). Table II compares the design of these three tests with the current versions of the indirect EIA used at the Johns Hopkins University (Yolken and Leister, 1981) and the RIA test used at the University of Massachusetts Medical School (Blacklow and Cukor, 1980). A large number of specimens have been tested by the Rotazyme assay in several laboratories (Rubenstein and Miller, 1982; Hammond *et al.*, 1982; Beards and Bryden, 1981; Wall *et al.*, 1982; Yolken and Leister, 1981) with favorable reports as to the sensitivity and specificity of the test. The test appears to be somewhat less sensitive than most electron microscopic or indirect EIA assays. Despite the fact that the manufacturer has not provided a confirmatory test, problems of specificity have not been reported for the Rotazyme test. Apparently, the direct test design and special formulations of sample diluent and wash solutions are sufficient to eliminate most false positive results. Laboratories may elect to use an exogenous confirmatory blocking test with any high-titered antirotavirus serum in conjunction with the Rotazyme test. Only limited experience with the other two commercial assays has been reported (Kjeldsberg and Mortensson-Egnund, 1982; Grauballe *et al.*, 1981; Yolken and Leister, 1981). Both tests suffer from the theoretical disadvantage of using serum from the same animal species as both coating and detection antibodies.

III. NORWALK VIRUS

Norwalk virus is responsible for a high proportion of outbreaks of acute gastroenteritis among older children and adults. This virus has been associated with 42% of 74 outbreaks of acute nonbacterial gastroenteritis in the United States studied epidemiologically by the Centers for Disease Control (Kaplan *et al.*, 1982). These outbreaks occur during all seasons and in a variety of locations, including private homes, elementary schools and colleges, nursing homes, cruise

TABLE II

Immunoassays for Rotavirus[a]

Feature	Assay				
	Rotazyme	Dakopatts	Enzygnost	Johns Hopkins	University of Massachusetts
Solid phase	Polystyrene beads	Polystyrene microplates	Polystyrene microplates	Polyvinyl microplates	Polyvinyl microplates
Capture antibody	Guinea pig anti-SA-11 (tissue culture-derived)	Rabbit anti-HRV (human stool-derived) or control normal rabbit serum	Rabbit anti-SA-11 (tissue culture-derived)	Goat anti-HRV (gnotobiotic calf stool-derived) or control normal goat serum	Goat anti-SA-11 (tissue culture-derived) or control normal goat serum
Second antibody	—	—	—	Guinea pig anti-HRV	—
Detection antibody	Rabbit anti-SA-11–peroxidase	Rabbit anti-HRV–peroxidase	Rabbit anti-NCDV–alkaline phosphatase (tissue culture-derived)	Rabbit anti-guinea pig immunoglobulin–alkaline phosphatase	Guinea pig ^{125}I-labeled anti-SA-11
Confirmatory test	Not available	Included	Not available	Included	Included

[a] HRV, Human rotavirus.

ships, swimming pools, recreational camps, and restaurants. The virus is usually transmitted by person-to-person contact or contaminated water and occasionally food.

It would clearly be desirable for state laboratories and other public health authorities to have available a diagnostic test for Norwalk virus. This is difficult for several reasons: the fact that a relatively small amount of the virus is shed in feces, the fact that the virus cannot yet relicate *in vitro* or in laboratory animals, and the failure of current technology to prepare highly purified virus from human stool samples. Because of its small size, the virus can be visualized only by the laborious and technically difficult immune electron microscopic (IEM) procedure (Kapikian *et al.*, 1972). The lack of purified viral antigen precludes the possibility of preparing hyperimmune animal serum, which would be useful in an immunoassay.

Virtually all that we know about Norwalk virus is based on the one model system available for study of the pathogenesis of the infection: human volunteers. When these individuals orally ingest the virus, they experience an illness clinically indistinguishable from the naturally occurring disease. Stool and serum specimens generated by volunteers have provided the only laboratory reagents currently available to identify the virus and are used for RIA and IEM tests. These human reagents also permit the quantification of antibody responses in sera and intestinal secretions. Unfortunately, because the reagents are specimens from human volunteers, the assay for Norwalk virus is available in only a few research laboratories.

The RIA test for Norwalk virus (Greenberg *et al.*, 1978; Blacklow *et al.*, 1979) is similar in principle to the confirmatory RIA test described for rotavirus. In brief, wells of microtiter plates are coated with acute- and convalescent-phase sera from a selected volunteer who has experienced experimentally induced Norwalk virus illness. The stool samples to be tested are then added, and finally convalescent-phase serum from a different volunteer is used as the source of radioiodine-labeled detection antibody. It should be noted that the serologic assay for Norwalk virus is more sensitive than the antigen test and that most Norwalk outbreaks have been identified primarily by the findings of seroconversions.

Circumventing the use of human-derived reagents for Norwalk virus diagnosis and developing RIA tests for other 27-nm Norwalk-like viruses that may be important in human disease (Blacklow and Cukor, 1981) are two major goals of research in the field of viral gastroenteritis.

ACKNOWLEDGMENTS

The work in the authors' laboratory was supported by cooperative agreement CR808801010 from the U.S. Environmental Protection Agency, and the World Health Organization.

REFERENCES

Beards, G. M., and Bryden, A. S. (1981). *J. Clin. Pathol.* **34,** 1388–1391.

Bishop, R. F., Davidson, G. P., Holmes, I. H., and Ruck, B. J. (1973). *Lancet* **ii** 1281–1283.

Blacklow, N. R., and Cukor, G. (1980). *In* "Manual of Clinical Microbiology" (E. H. Lennette, A. Balows, W. Hausler, and J. Truant, eds.), 3rd ed., pp. 891–898. Am. Soc. Microbiol., Washington, D.C.

Blacklow, N. R., and Cukor, G. (1981). *N. Engl. J. Med.* **304,** 397–406.

Blacklow, N. R., Cukor, G., Bedigian, M. K., Echeverria, P., Greenberg, H. B., Schreiber, D. S., and Trier, J. S. (1979). *J. Clin. Microbiol.* **10,** 903–909.

Brandt, C. D., Kim, H. W., Rodriguez, W. J., Thomas, L., Yolken, R. H., Arrobio, J. O., Kapikian, A. Z., Parrott, R. H., and Chanock, R. M. (1981). *J. Clin. Microbiol.* **13,** 976–981.

Cukor, G., Berry, M. K., and Blacklow, N. R. (1978). *J. Infect. Dis.* **138,** 906–910.

Grauballe, P. C., Vestergaard, B. F., Meyling, A., and Genner, J. (1981). *J. Med. Virol.* **7,** 29–40.

Greenberg, H. B., Wyatt, R. G., Valdesuso, J., Kalica, A. R., London, W. T., Chanock, R. M., and Kapikian, A. Z. (1978). *J. Med. Virol.* **2,** 97–108.

Hammond, G. W., Ahluwalia, G. S., Barker, F. G., Horsman, G., and Hazelton, P. R. (1982). *J. Clin. Microbiol.* **16,** 53–59.

Kalica, A. R., Purcell, R. H., Sereno, M. M., Wyatt, R. G., Kim, H. W., Chanock, R. M., and Kapikian, A. Z. (1977). *J. Immunol.* **118,** 1275–1279.

Kapikian, A. Z., Wyatt, R. G., Dolin, R., Thornhill, T. S., Kalica, A. R., and Chanock, R. M. (1972). *J. Virol.* **10,** 1075–1081.

Kapikian, A. Z., Yolken, R. H., Greenberg, H. B., Wyatt, R. G., Kalica, A. R., Chanock, R. M., Kim, H. W. (1979). *In* "Diagnostic Procedures for Viral, Rickettsial and Chlamydial Infection" (E. H. Lennette and N. J. Schmidt, eds.), pp. 927–955. Am. Public Health Assoc., Washington, D.C.

Kapikian, A. Z., Cline, W. L., Greenberg, H. B., Wyatt, R. G., Kalica, A. R., Banks, C. E., James, H. D., Flores, J., Chanock, R. M. (1981). *Infect. Immun.* **33,** 415–425.

Kaplan, J. E., Gary, G. W., Baron, R. C., Singh, N., Schonberger, L. B., Feldman, R., and Greenberg, H. B. (1982). *Ann. Int. Med.* **96,** 756–761.

Kjeldsberg, E., and Mortensson-Egnund, K. (1982). *J. Virol. Methods* **4,** 45–53.

Middleton, P. J., Holdaway, M. D., Petric, M., Szymanski, M. T., and Tam, J. S. (1977). *Infect. Immun.* **16,** 439–444.

Obert, G., Gloeckler, R., Burckard, J., and Regenmortel, M. H. V. (1981). *J. Virol. Methods* **3,** 99–107.

Rubenstein, A. S., and Miller, M. F. (1982). *J. Clin. Microbiol.* **15,** 938–944.

Sarkkinen, H. K., Halonen, P. E., and Arstila, P. P. (1979). *J. Med. Virol.* **4,** 255–260.

Sato, K., Inaba, Y., Shinozaki, T., Fujil, R., and Matumoto, M. (1981). *Arch. Virol.* **69,** 155–160.

Urasawa, T., Urasawa, S., and Taniguchi, K. (1981). *Microbiol. Immunol.* **25,** 1025–1035.

Vesikari, T., Sarkkinen, H. K., and Maki, M. (1981). *Acta Paediatr. Scand.* **70,** 717–721.

Wall, R. A., Mellars, B. J., Luton, P., and Boulding, S. (1982). *J. Clin. Pathol.* **35,** 104–106.

Wyatt, R. G., James, W. D., Bohl, E. H., Theil, K. W., Saif, L. J., Kalica, A. R., Greenberg, H. B., Kapikian, A. Z., and Chanock, R. M. (1980). *Science* **207,** 189–191.

Wyatt, R. G., Greenberg, H. B., James, W. D., Pittman, A. C., Kalica, A. R., Flores, J., Chanock, R. M., and Kapikian, A. Z. (1982). *Infect. Immun.* **37,** 110–115.

Yolken, R. H. (1982). *Rev. Infect. Dis.* **4,** 35–68.

Yolken, R. H., and Stopa, P. J. (1979). *J. Clin. Microbiol.* **10,** 703–707.

Yolken, R. H., and Leister, F. J. (1981). *J. Infect. Dis.* **144,** 379.

Yolken, R. H., and Leister, F. (1982). *J. Infect. Dis.* **146,** 43–46.

Yolken, R. H., Kim, H. W., Clem, T., Wyatt, R. G., Kalica, A. R., Chanock, R. M., and Kapikian, A. Z. (1977a). *Lancet* **ii,** 263–267.

Yolken, R., Wyatt, R. G., and Kapikian, A. Z. (1977b). *Lancet* **ii,** 819.

Yolken, R. H., Stopa, P. J., and Harris, C. C. (1980). *In* "Manual of Clinical Immunology" (N. Rose and H. Friedman, eds.), 2nd ed., pp. 692–699. Am. Soc. Microbiol, Washington, D.C.

III

NONIMMUNOLOGIC DETECTION OF MICROBIAL PRODUCTS

Introduction

MARY JANE FERRARO

The chapters in this section describe nonimmunologic methods for the direct detection of microorganisms in clinical samples. Chapter 21 deals with the revival in the late 1970s and 1980s of the *Limulus* amoebocyte lysate test for the detection of endotoxin. Recent applications to a variety of body fluids have proved to be far more successful than the early application to the detection of gram-negative septicemia. Chapter 22 deals with toxin detection as a means of diagnosing antimicrobial-associated diarrhea. Particularly interesting is the discussion of the relative sensitivity and specificity of the cytotoxin and counterimmunoelectrophoresis assays as compared with improved culture sensitivity using highly selective media. Chapters 23–27 describe the use of instrumentation and automated equipment in the clinical microbiology laboratory. Specifically, Chapters 23 and 24 discuss the role of gas–liquid chromatographic analysis in the detection of microbial end products. Although much progress has been made in this area, as is detailed in Chapter 24, the techniques are still emerging and require standardization; an in-depth critique of the future role of gas–liquid chromatography in the clinical microbiology laboratory can be found in Chapter 23. The desire to improve in speed and quality the results emanating from the clinical microbiology laboratory has prompted clinical microbiologists to search for automation. The impact of increasing demands on the laboratory coupled with economic restraints on reimbursement have made this search even more compelling. Chapters 26 and 27 deal specifically with instrumentation or new methods for the detection of bacteremia or urinary tract infections, respectively. Although there has been much activity in this area since the early 1970s, relatively few instruments are available today to meet what will probably become an ever-increasing need in microbiology laboratories. A critique of progress to date and suggestions for future directions in this area can be found in Chapter 25.

21

Current Uses of the Limulus *Amoebocyte Lysate Test*

JAMES H. JORGENSEN

Department of Pathology
The University of Texas
Health Sciences Center
San Antonio, Texas

I.	Background and Mechanism of the *Limulus* Amoebocyte Lysate Test	283
II.	*Limulus* Amoebocyte Lysate Testing in the Pharmaceutical Industry	285
III.	Clinical Applications of the *Limulus* Amoebocyte Lysate Test	285
	A. Detection of Endotoxemia	285
	B. Diagnosis of Gram-Negative Bacterial Meningitis	286
	C. Detection of Gram-Negative Bacteriuria	287
	D. Presumptive Diagnosis of Gonorrhea	290
	E. Detection of Gram-Negative Bacteria in Ocular Infections	291
	F. Other Clinical Applications	292
	G. Summary	292
	References	293

I. BACKGROUND AND MECHANISM OF THE *LIMULUS* AMOEBOCYTE LYSATE TEST

The *Limulus* amoebocyte lysate (LAL) test was first described as a sensitive method for detecting endotoxins in the late 1960s by Levin and Bang (1968). The development of the test followed Bang's (1956) observation at the Woods Hole Oceanographic Institution that horseshoe crabs (*Limulus polyphemus*) exhibit a type of disseminated intravascular coagulation when seriously infected by marine gram-negative bacteria such as *Vibrio*. This initial observation was extended by the discovery that clotting of *Limulus* hemolymph could be produced *in vitro* by the addition of either viable gram-negative bacteria or purified gram-negative

The Direct Detection of Microorganisms in Clinical Samples

ISBN 0-12-187780-9

endotoxin (Levin and Bang, 1968). These workers also discovered that the amoebocyte, which is the only circulating cell in *Limulus* hemolymph, is the source of all of the factors necessary for coagulation. The coagulation components were further localized to the dense granules that pack the cytoplasm of the amoebocyte (Murer *et al.,* 1975). Thus, when tissue damage occurs, the amoebocytes aggregate and degranulate, and finally coagulation occurs (Levin and Bang, 1964).

A primitive immunologic function appears to be served by activation of the coagulation system through an alternate pathway when gram-negative bacteria or their endotoxins are present. In this manner invading bacteria can be localized and their spread limited by clotting of the hemolymph. The current use of the test as an *in vitro* indicator or endotoxin is based on the fact that physical disruption of amoebocytes that have been harvested from limuli yields a suspension (the amoebocyte lysate) that reacts only with bacterial endotoxin [because key membrane receptors for *in vivo* activation of the coagulation pathway are no longer present (Levin and Bang, 1968)]. This principle has been used to make coagulation of LAL the most sensitive method available (Rojas-Corona *et al.,* 1969) for detecting gram-negative bacterial lipopolysaccaride (LPS, endotoxin). Current commerical LAL preparations can reliably detect 0.05–0.25 ng/ml of purified *Escherichia coli* standard endotoxin.

It appears that LAL reacts to either bound or free LPS (Jorgensen and Smith, 1974) incorporated in the cell walls of virtually all gram-negative bacteria. Endotoxin has been demonstrated (Tai and Liu, 1977) to function as an activator of a proenzyme (MW 150,000) that is present in native LAL. Once the enzyme is activated in the presence of suitable divalent cations (such as Ca^{2+} or Mg^{2+}), it behaves as a serine protease in the hydrolysis of a substrate (coagulogen or clottable protein; MW 23,000–27,000) that is also present in the LAL. The activated enzyme cleaves the coagulogen at one specific site between the amino acids arginine and lysine (Tai *et al.,* 1977). This results in cleavage of the coagulogen into two fragments, the larger of which contains approximately 170 amino acid residues and which aggregate forming a secondary structure. The clottable protein forms a polymerized gel based on both hydrophilic and hydrophobic bonds between the peptide strands (Tai *et al.,* 1977). Electron microscopy of the final clot has revealed large aggregated strands whose structure resembles a single α-helix (Holme and Solum, 1973). The smaller peptide fragments (approximately 45 amino acid residues), which are released from the clottable protein during this process, do not appear to contribute to the integrity of the clot (Tai *et al.,* 1977). The visible end result of this cascade of reactions is usually a clot, which forms in a reaction tube and is recognized by a failure of the contents to flow when the test tube is inverted 180°. Reactions of lesser strength have been described as an increase in viscosity and often an associated increase in opacity (Jorgensen and Smith, 1973a). Weaker reactions are recognized by the production of starchlike granules, which adhere to the wall of a reaction tube. However, this subjectivity in the interpretation of end points of lesser integrity has been a criticism of the gelation method of performing the test.

Alternative methods for detecting a positive LAL test include recording the associated increase in optical density that occurs as a result of the clottable protein polymerization (Levin and Bang, 1968; Teller and Kelly, 1979) or, more recently, directly measuring the action of the enzyme on a synthetic chromogenic substrate (Nakamura *et al.*, 1977; Iwanaga *et al.*, 1978) in preference to the natural substrate, clottable protein. In either case the slowest and rate-limiting step in the reaction is the time required for activation of the proenzyme by endotoxin (Young *et al.*, 1972). Most LAL test methods include incubation periods of up to 1 hr, with at least 5 to 10 min required for the initial, rate-limiting step to begin. The majority of clinical studies of the LAL test have employed the traditional gelation end-point method of performing the test.

II. *LIMULUS* AMOEBOCYTE LYSATE TESTING IN THE PHARMACEUTICAL INDUSTRY

Several human clinical applications of the LAL test have been shown to be very promising and are discussed in this chapter. However, by far the largest user of the LAL test has been the pharmaceutical industry, which now performs hundreds of thousands of tests annually on a variety of large- and small-volume parenteral fluids, biologicals, and medical devices (Cooper and Neely, 1980). The use of the assay has been clearly successful for this purpose, because it is more sensitive, less expensive, and easier to perform than rabbit pyrogen tests (Cooper and Neely, 1980) and is amenable to automation (Jorgensen and Alexander, 1981). The LAL test is now commonly performed for both in-process and final-release testing of many injectable fluids and invasive medical devices.

III. CLINICAL APPLICATIONS OF THE *LIMULUS* AMOEBOCYTE LYSATE TEST

A. Detection of Endotoxemia

The first clinical application of the LAL assay to be described, the detection of gram-negative septicemia, has unfortunately been clouded by controversy regarding its diagnostic usefulness. The presence of poorly characterized "inhibitors" of the LAL test in human blood (Johnson *et al.*, 1977) requires that some procedure for the denaturation of proteins be performed on plasma before it can be tested using amoebocyte lysate. Although several methods have been proposed for this purpose (Levin *et al.*, 1970; Reinhold and Fine, 1971; DuBose *et al.*, 1980), none has proved to be universally successful. The results of LAL tests on the blood of patients suspected of having gram-negative sepsis have generally shown rather poor correlation with those obtained by conventional methods such as blood culture (Stumacher *et al.*, 1973; Elin *et al.*, 1975b). Such inconsistencies have raised difficult theoretical questions such as the possibility of bacteremia without endotoxemia or of endotoxemia without demonstrable bacteremia (Levin *et al.*, 1972; Sibbald, 1976). These theories have been debated without resolution for nearly a decade. It seems likely that the usefulness of this aspect of the LAL assay will require a completely different method of detecting endotoxin, to which the LAL test may be compared for ultimate clarification.

B. Diagnosis of Gram-Negative Bacterial Meningitis

Perhaps the most useful clinical application of the LAL test to date has been the rapid diagnosis of gram-negative bacterial meningitis. At least eight published studies (Nachum *et al.*, 1973; Ross *et al.*, 1975; Berman *et al.*, 1976; Butler *et al.*, 1976; Dyson and Cassady, 1976; McCracken and Sarff, 1976; Tuazon *et al.*, 1977; Jorgensen and Lee, 1978) involving almost 1300 patients with meningitis have shown encouraging results with the LAL test (Table I). This application was first described in 1973 (Nachum *et al.*, 1973) in a study involving 112 patients suspected of having acute bacterial meningitis. The LAL test was positive on all initial CSF specimens from 38 patients with culture-proven gram-negative bacterial meningitis. Negative tests were obtained in 74 patients with gram-positive bacterial, tuberculous, or aseptic meningitis and in patients without meningitis. Patients with meningitis in this study demonstrated growth of the microorganisms most commonly associated with acute bacterial meningitis: *Haemophilus influenzae, Neisseria meningitidis,* and *Streptococcus pneumoniae.* Nachum *et al.* (1973) and others have achieved almost 100% detection of *H. influenzae, N. meningitidis,* various Enterobacteriaceae, and pseudomonads. However, a major disadvantage of the LAL test that must be recognized is its inability to detect gram-positive bacteria such as *S. pneumoniae,* group B *Streptococcus,* and *Staphylococcus aureus.*

If the data from all eight published studies regarding meningitis are considered, well over 90% of patients with documented gram-negative meningitis were rapidly and objectively detected by the LAL test (Table I). This compares with a detection rate by the cerebrospinal fluid (CSF) Gram's stain of only about 70% in the same studies. Only the study of neonatal meningitis reported by McCracken and Sarff (1976) has indicated a lower rate of positivity of the LAL test on CSF.

The main predictive value of the LAL assay of CSF is its low percentage of false positive reactions (<2%). Thus, a patient with a positive LAL test of carefully collected CSF should be presumed to have a gram-negative bacterium

TABLE I

Accumulated Data on LAL Testing for the Diagnosis of Meningitis

	LAL assay results[a]		
No. of patients studied	Patients with meningitis (%)	Patients with gram-negative meningitis (%)	Reference
112	88 (38/43)	100 (38/38)	Nachum *et al.* (1973)
335	73 (37/51)	97 (37/38)	Ross *et al.* (1975)
232	80 (86/107)	100 (86/86)	Berman *et al.* (1976)
3	100 (3/3)	100 (3/3)	Butler *et al.* (1976)
145	60 (6/10)	100 (6/6)	Dyson and Cassady (1976)
84	64 (60/94)	71 (60/84)	McCracken and Sarff (1976)
81	57 (8/14)	100 (8/8)	Tuazon *et al.* (1977)
305	82 (61/74)	100 (61/61)	Jorgensen and Lee (1978)

[a] Results expressed as percentage of positive tests; numbers in parentheses indicate number of positive tests/total tests.

of some species in this CSF. However, a negative test does not rule out the possibility of meningitis due to a gram-positive microorganism.

The results of the LAL test can be effectively combined with the results of simultaneous counterimmunoelectrophoresis (CIE) or latex agglutination tests for bacterial antigens in CSF, which may detect organisms missed by LAL, for example, *S. pneumoniae* and group B *Streptococcus,* or which may clarify the species identity of a gram-negative bacterium associated with a positive LAL test, for example, *H. influenzae* or *N. meningitidis*. Neither method should be performed in lieu of a smear of the CSF, which might reveal staphylococci, *Listeria,* or other microorganisms that are not routinely detected by LAL or CIE methods.

An additional advantage of the LAL or immunologic methods, which do not require growth of viable bacteria, may be improved documentation of meningitis in patients who have been partially or inappropriately treated with antimicrobials, because bacterial antigens (and LPS) may persist for some time in the CSF (Jorgensen and Lee, 1978).

C. Detection of Gram-Negative Bacteriuria

The definitive diagnosis of urinary infections involves the cultivation of viable bacteria from a urine specimen, which may require 24–48 hr. Several more rapid techniques have been proposed which attempt to detect bacterial products in urine (Neter, 1965), provide a simple means for microscopic visualization of bacteria in urine (Brody *et al.*, 1968), or rapidly recognize microbial growth of photometric detection of broth cultures (Pezzlo *et al.*, 1982). An extensive discussion of some of these methods appears in Chapters 25 and 27 of this volume.

Several studies have demonstrated that the LAL test may also be used for this purpose (Table II). Jorgensen *et al.* (1973; also see Jorgensen and Jones, 1975) showed in two separate studies that gram-negative bacteria in significant numbers ($\geq 10^5$ per milliliter of urine) can be reliably detected using diluted urine and gelation LAL testing. It was necessary to dilute the urine specimens to differentiate significant bacteriuria from a smaller number of inadvertent contaminants encountered from collection procedures. These two studies of more than 1200 patients indicated that 99% of significant gram-negative bacteriurias were detected by the LAL test. However, as previously discussed, gram-positive bacteria and fungi cannot be detected using LAL. Approximately 87% of all significant bacteriurias (both gram-positive and gram-negative) were detected by the LAL test. The overall capacity of the test to distinguish between the presence or absence of bacteriuria was 95% (Jorgensen *et al.*, 1973; Jorgensen and Jones, 1975).

Nachum and Shanbrom (1981) used a different approach to LAL quantification for detecting significant bacteriuria; that is, the time required for LAL gelation to occur was changed, and undiluted urine was tested. All urines that contained $\geq 10^5$ gram-negative bacteria per milliliter yielded positive LAL tests within 15 min of incubation. This method of quantification correctly differentiated 93% of urines as containing either $<10^5$ or $\geq 10^5$ bacteria per milliliter of urine (Table II).

One attempt to apply LAL methodology to the detection of bacteriuria makes use of automated, spectrophotometric interpretation of LAL reactions (Jorgensen and Alexander, 1982). The Abbott MS-2 instrument was used to detect kinetic changes in optical density that occurred when urine samples (diluted 1 : 2000) containing $\geq 10^5$ gram-negative bacteria per milliliter were incubated with LAL in special MS-2 cuvettes. The total LAL test time used in this study was 20 min, with positive urines usually detected in the first 5–10 min of the test. The accuracy of the automated method was comparable to that of the three manual gelation studies cited in Table II. The advantages of the automated LAL test include increased speed of detection of positive reactions and a totally objective indication of a positive test (determined by kinetic changes in optical density that exceed a preestablished threshold value).

The usefulness of the LAL test for detecting bacteriuria lies in the predominance of gram-negative species that cause urinary tract infections; that is, almost 90% of all infections are due to gram-negative bacteria. The almost total detection of gram-negative bacteria but lack of recognition of gram-positive bacteria makes the overall rate of detection very similar to that of other screening methods

TABLE II

Data from Four Studies on the Use of the LAL Test for Detecting Significant Gram-Negative Bacteriuria

LAL test method	No. of patients or urines tested	LAL assay results (%)[a]			Reference
		$\geq 10^5$ organisms/ml	$\geq 10^5$ gram-negative bacteria/ml	Distinction between urines with $< 10^5$/ml or $\geq 10^5$/ml	
Gelation testing of diluted urine	209	92 (23/25)	100 (23/23)	99 (207/209)	Jorgensen *et al*. (1973)
Gelation testing of diluted urine	1077	86 (175/203)	99 (159/161)	94 (1016/1077)	Jorgensen and Jones (1975)
Time to gelation	190	84 (36/43)	100 (36/36)	93 (177/190)	Nachum and Shanbrom (1981)
Automated turbidometric	580	85 (78/92)	94 (78/83)	93 (537/580)	Jorgensen and Alexander (1982)

[a] Results expressed as percentage of positive tests; numbers in parentheses indicate number of positive tests/total tests.

(Pezzlo *et al.*, 1982). Thus, the LAL test may be a very useful and cost-effective technique for rapidly assessing the likelihood of bacteriuria in patients but cannot be considered to be as definitive as quantitative bacterial cultures.

D. Presumptive Diagnosis of Gonorrhea

Perhaps one of the most novel applications of LAL has been that of the diagnosis of gonococcal urethritis and cervicitis. In a study of gonococcal urethritis in men, Spagna *et al.* (1979) demonstrated that a dilution (1 : 200) of urethral exudate allowed the distinction to be made between gonococcal and nongonococcal urethritis with LAL. Dilution of the sample was necessary to prevent false positive LAL tests; it also seemed to exclude positive LAL tests due to infections caused by microorganisms such as *Chlamydia* and *Mycoplasma*. A subsequent communication (Prior and Spagna, 1981a) suggested that LPS from gonococci was either more reactive with LAL or possibly was more abundantly available at the cell surface than LPS in other genera. Therefore, fewer cells of *N. gonorrhoeae* were required to cause LAL gelatin than some other gram-negative bacteria in urethral or cervical secretions (Prior and Spagna, 1981a).

This principle was applied successfully by Prior, Spagna, and colleagues (Table III) to the examination of female cervical secretions in two separate studies (1980, 1981). In the first study 94% of patients with gonorrhea had positive LAL tests, and all patients without gonorrhea were negative. However, a similar study performed by Young *et al.* (1981) did not provide such convincing data. Only 71% of all women with gonorrhea yielded a positive LAL test at a relatively low dilution of the sample (1 : 50), whereas 7% of patients without any history of gonorrhea also gave positive LAL reactions. However, the lysate preparation used by these authors was different from the nonorganic-solvent-extracted lysate found to be more sensitive to gonococcal endotoxin by Prior and Spagna (1981a).

It appears that the LAL test is more easily employed for the examination of male urethral exudate than for use on endocervical secretions. Prior and Spagna (1981b) described the use of a "kit" for examining male urethral fluids in a group of 550 men. Their test device included a syringe for sample collection, a dilution reservoir to provide a 1 : 400 dilution, and a single test vial of LAL. With this procedure and a 30-min LAL incubation period, a sensitivity of 99.2% and a specificity of 96.4% were achieved. Concomitant Gram's stain examination of the same fluids provided lower sensitivity (96.4%) but greater specificity (99.5%). However, there was no statistically significant difference in the capacity of the LAL test and of Gram's stain examination to predict gonococcal culture results.

The diagnosis of gonorrhea by LAL is a promising and novel application of the test. It should probably be considered a more objective means of establishing a

TABLE III

Studies on the Use of the LAL Test for Diagnosing Gonorrhea

Specimen examined	Specimen dilution	LAL assay results[a]: Patients with gonorrhea[b] (%)	LAL assay results[a]: Patients without gonorrhea[b] (%)	Reference
Male urethral exudates	1 : 200	100 (73/73)	4 (1/27)	Spagna *et al.* (1979)
Cervical secretions	1 : 800	94 (17/18)	0 (0/22)	Spagna *et al.* (1980)
Male urethral exudates	1 : 1600	100 (61/61)	4 (2/55)	Prior and Spagna (1981a)
Cervical secretions	1 : 800	100 (3/3)	—	Spagna and Prior (1981)
Cervical secretions	1 : 50	71 (17/24)	7 (3/41)	Young *et al.* (1981)
Male urethral exudates	1 : 400	99 (363/366)	3 (6/184)	Prior and Spagna (1981b)

[a] Results expressed as percentage of positive tests; numbers in parentheses indicate number of positive tests/total tests.

[b] Presence or absence of *N. gonorrhoeae* documented by culture.

diagnosis of gonorrhea than Gram's stain examination, but it cannot replace the specificity of a good culture.

E. Detection of Gram-Negative Bacteria in Ocular Infections

The possibility of applying the LAL assay to the diagnosis of ocular diseases was first noted by Poirier and Jorgensen (1977) when they detected the endotoxin of *Pseudomonas aeruginosa* in the ulcer bed of a patient's infected cornea. A minute sample of corneal ulcer material obtained with a spatula was emulsified directly in a tube of LAL. The corneal scraping caused abrupt gelatin of the lysate within 30 min of collection. Since then, the use of the LAL test for recognizing gram-negative bacteria in corneal ulcers has proved to be a novel application (Wolters *et al.*, 1979).

It is also possible to test minute volumes of intraocular fluid for the presence of gram-negative bacteria by this method (Ellison, 1978; McBeath *et al.*, 1978). The principal advantages of the LAL test for this purpose are speed (because gram-negative ocular infections may lead to irreversible damage within a few hours) and objectivity (because stained smears of ocular fluids may be difficult to interpret). In this regard, the LAL test should be considered a rapid adjunctive technique to provide useful information earlier than culture results. Suitable means for detecting gram-positive bacteria, fungi, and viruses must also be provided to establish the presence of microorganisms that do not produce endotoxin.

F. Other Clinical Applications

Virtually any infectious process in which it is important to distinguish gram-positive from gram-negative bacteria or septic from aseptic conditions provides potential uses of the test. Bernstein *et al.* (1980) found the LAL test to be useful for detecting gram-negative bacteria in middle ear fluids in otitis media. The prominence of *Haemophilus influenzae* and possibly *Branhamella catarrhalis* in otitis media in children makes this another plausible application.

The potential of LAL testing of joint fluids is more difficult to evaluate. The etiology of pyogenic arthritis is often difficult to establish because bacterial cultures are sometimes negative and smears can be difficult to interpret. However, it is the lack of reliable diagnostic standards that makes it difficult to assess LAL testing of joint fluids. In the series of Tuazon *et al.* (1977) and of Elin *et al.* (1978), all patients with septic arthritis due to gram-negative bacteria had positive LAL tests of joint fluids. It is not clear, however, whether positive LAL tests on joint fluids with negative cultures represented false positive tests or simply reflected the problem of making the diagnosis by conventional means. All patients with positive LAL assays had an increased number of leukocytes in their joint fluids and a concomitantly depressed glucose level, two findings strongly suggestive of bacterial infection, despite negative culture results. Apparently, LAL is especially sensitive to gonococcal endotoxin (Prior and Spagna, 1981a), and *N. gonorrhoeae* is a prominent cause of septic arthritis in some patient groups (Holmes *et al.*, 1971).

Another application of the LAL test is in epidemiology. The concept of LAL testing of parenteral fluids for pyrogens can be readily applied to in-use contamination of fluids in a hospital setting. Jorgensen and Smith (1973b) used the assay for the rapid detection of an intravenous fluid preparation that had been inadvertently contaminated with *Pseudomonas aeruginosa* during processing in the hospital. Elin *et al.* (1975a) made similar use of the test in recognizing a contaminated bottle of sodium chloride solution used for washing frozen red blood cells. The LAL testing of solutions and source waters used in hemodialysis has also been useful in recognizing gram-negative bacterial contamination (Hindman *et al.*, 1975).

G. Summary

Numerous successful uses of the LAL test have been described since the early 1970s. Applications depend on the speed, sensitivity, and high specificity of LAL for bacterial endotoxin, whether in bound or free form. None of these applications, however, has completely replaced conventional methods. Thus, the LAL test is a useful but adjunctive means of detecting gram-negative bacteria in a variety of clinical circumstances. The lack of an FDA approved and licensed product for clinical laboratory use in the test has limited its usefulness.

REFERENCES

Bang, F. B. (1956). *Bull. Johns Hopkins Hosp.* **98,** 325–351.

Berman, N. S., Siegel, S. E., Nachum, R., Lipsey, A., and Leedom, J. (1976). *J. Pediatr.* (*St. Louis*) **88,** 553–556.

Bernstein, J. M., Praino, M. D., and Neter, E. (1980). *Can. J. Microbiol.* **26,** 546–548.

Brody, L. M., Webster, C., and Kark, R. M. (1968). *J. Am. Med. Assoc.* **206,** 1777–1781.

Butler, T., Levin, J., Linh, N. N., Chau, D. M., Adickman, M., and Arnold, K. (1976). *J. Infect. Dis.* **133,** 493–499.

Cooper, J. F., and Neely, M. E. (1980). *Pharm. Technol.* **4,** 72–79.

DuBose, D. A., Lemaire, M., Basamania, K., and Rowlands, J. (1980). *J. Clin. Microbiol.* **11,** 68–72.

Dyson, D., and Cassady, G. (1976). *Pediatrics* **58,** 105–109.

Elin, R. J., Lundberg, W. B., and Schmidt, P. J. (1975a). *Transfusion* (*Philadelphia*) **15,** 260–265.

Elin, R. J., Robinson, R. A., Levine, A. S., and Wolff, S. M. (1975b). *N. Engl. J. Med.* **293,** 521–524.

Elin, R. J., Knowles, R., Barth, V. F., and Wolff, S. M. (1978). *J. Infect. Dis.* **137,** 507–513.

Ellison, A. C. (1978). *Arch. Ophthalmol.* (*Chicago*) **96,** 1268–1271.

Hindman, S. J., Carson, L. A., Favero, M. S., Peterson, N. J., Schonberger, L. B., and Solano, J. T. (1975). *Lancet.* **ii,** 732–734.

Holme, R., and Solum, N. O. (1973). *J. Ultrastruct. Res.* **44,** 329–338.

Holmes, K. K., Wiesner, P. J., and Pedersen, A. H. B. (1971). *Ann. Intern. Med.* **75,** 470–471.

Iwanaga, S., Morita, T., Harada, T., Nakamura, S., Makoto, N., Takada, K., Kimura, T., and Sakakibara, S. (1978). *Haemostasis* **7,** 183–188.

Johnson, K. J., Ward, P. A., Goralnick, S., and Osborn, M. J. (1977). *Am. J. Pathol.* **88,** 559–574.

Jorgensen, J. H., and Alexander, G. A. (1981). *Appl. Environ. Microbiol.* **41,** 1316–1320.

Jorgensen, J. H., and Alexander, G. A. (1982). *J. Clin. Microbiol.* **16,** 587–589.

Jorgensen, J. H., and Jones, P. M. (1975). *Am. J. Clin. Pathol.* **63,** 142–148.

Jorgensen, J. H., and Lee, J. C. (1978). *J. Clin. Microbiol.* **7,** 12–17.

Jorgensen, J. H., and Smith, R. F. (1973a). *Appl. Microbiol.* **26,** 43–48.

Jorgensen, J. H., and Smith, R. F. (1973b). *Appl. Microbiol.* **26,** 521–524.

Jorgensen, J. H., and Smith, R. F. (1974). *Proc. Soc. Exp. Biol. Med.* **146,** 1024–1031.

Jorgensen, J. H., Carvajal, H. F., Chipps, B. E., and Smith, R. F. (1973). *Appl. Microbiol.* **26,** 38–42.

Levin, J., and Bang, F. B. (1964). *Bull. Johns Hopkins Hosp.* **115,** 337–345.

Levin, J., and Bang, F. B. (1968). *Thromb. Diath. Haemorrph.* **19,** 186–197.

Levin, J., Poore, T. E., Zauber, N. P., and Oser, R. S. (1970). *N. Engl. J. Med.* **283,** 1313–1316.

Levin, J., Poore, T. E., Young, N. S., Margolis, S., Zauber, N. P., Townes, A. S., and Bell, W. R. (1972). *Ann. Intern. Med.* **76,** 1–7.

McBeath, J., Forster, R. K., and Rebell, G. (1978). *Arch. Ophthalmol.* (*Chicago*) **96,** 1265–1267.

McCracken, G. H., Jr., and Sarff, L. D. (1976). *JAMA, J. Am. Med. Assoc.* **235,** 617–620.

Murer, E. H., Levin, J., and Holme, R. (1975). *J. Cell Physiol.* **86,** 533–539.

Nachum, R., and Shanbrom, E. (1981). *J. Clin. Microbiol.* **13,** 158–162.

Nachum, R., Lipsey, A., and Siegel, S. E. (1973). *New Engl. J. Med.* **289,** 931–934.

Nakamura, S., Morita, T., Iwanaga, S., Niwa, M., and Takahashi, K. (1977). *J. Biochem.* (*Tokyo*) **81,** 1567–1569.

Neter, E. (1965). *J. Am. Med. Assoc.* **142,** 769.

Pezzlo, M. T., Tan, G. L., Peterson, E. M., and de la Maza, L. M. (1982). *J. Clin. Microbiol.* **15,** 468–474.

Poirier, R. H., and Jorgensen, J. H. (1977). *Lancet.* **ii,** 85–86.

Prior, R. B., and Spagna, V. A. (1981a). *J. Clin. Microbiol.* **13,** 167–170.

Prior, R. B., and Spagna, V. A. (1981b). *J. Clin. Microbiol.* **14,** 256–260.

Reinhold, R. B., and Fine, J. (1971). *Proc. Soc. Exp. Biol. Med.* **137,** 334–340.

Rojas-Corona, R. R., Skarnes, R., Tamakuma, S., and Fine, J. (1969). *Proc. Soc. Exp. Biol. Med.* **132,** 599–601.

Ross, S., Rodriquez, W., Controni, G., Korengold, G., Watson, S., and Khan, W. (1975). *J. Am. Med. Assoc.* **233,** 1366–1369.

Sibbald, W. J. (1976). *Heart Lung* **5,** 765–771.

Spagna, V. A., and Prior, R. B. (1981). *Sex. Trans. Dis.* **8,** 18–20.

Spagna, V. A., Prior, R. B., and Perkins, R. L. (1979). *Br. J. Vener. Dis.* **55,** 179–182.

Spagna, V. A., Prior, R. B., and Perkins, R. L. (1980). *Am. J. Obstet. Gynecol.* **137,** 595–599.

Stumacher, R. J., Kovnat, M. J., and McCabe, W. R. (1973). *N. Engl. J. Med.* **288,** 1261–1264.

Tai, J. V., and Liu, T. (1977). *J. Biol. Chem.* **252,** 2178–2181.

Tai, J. Y., Seid, R. C., Jr., Huhn, R. D., and Liu, T. (1977). *J. Biol. Chem.* **252,** 4773–4776.

Teller, J. D., and Kelley, K. M. (1979). *In* "Biomedical Applications of the Horseshoe Crab (Limulidae)" (E. Cohen, ed.), pp. 423–433. Alan R. Liss, Inc., New York.

Tuazon, C. U., Perez, A. A., Elin, R. J., and Sheagren, J. N. (1977). *Arch. Intern. Med.* **137,** 55–56.

Wolters, R. W., Jorgensen, J. H., Calzada, E., and Poirier, R. H. (1979). *Arch. Ophthalmol.* (*Chicago*) **97,** 875–877.

Young, N. S., Levin, J., and Prendergast, R. A. (1972). *J. Clin. Invest.* **51,** 1790–1797.

Young, H., Sarafian, S. K., and McMillan, A. (1981). *Br. J. Vener. Dis.* **57,** 200–203.

22

Laboratory Diagnosis of Antimicrobial-Associated Diarrhea

JON E. ROSENBLATT

Mayo Clinic and Mayo Foundation
Rochester, Minnesota

I. Introduction ... 295
II. *Clostridium difficile* and Its Toxins ... 296
III. Laboratory Procedures ... 299
A. Culture ... 299
B. Cytotoxicity Assay for *Clostridium difficile* Toxin in Stool ... 300
References ... 301

I. INTRODUCTION

Experimental and clinical studies (Bartlett, 1979; Lusk *et al.*, 1978; Rifkin *et al.*, 1977) have established *Clostridium difficile* as a significant cause of diarrhea in patients who have been treated with antimicrobials. *Clostridium difficile* is resistant to many antimicrobials and is probably selected as more susceptible organisms are eliminated during antimicrobial therapy. This bacterium elaborates an enterotoxin that is responsible for the diarrhea and can produce a wide spectrum of disease from a mild gastroenteritis, which resolves with cessation of antimicrobial therapy, to a fulminating and potentially fatal pseudomembranous colitis (PMC).

Patients with PMC are often very toxic and have high fever, leukocytosis, abdominal pain, and severe diarrhea. The large bowel wall is friable, with extensive erosions leaving only islands of normal mucosa; this results in a "cobblestone" or "pseudomembrane" appearance on sigmoidoscopy. This appearance is virtually pathognomic of antimicrobial-associated diarrhea (AAD) due to *C. difficile*. Pseudomembrane colitis has rarely occurred in patients not treated with antimicrobials and was recognized in the preantibiotic era. Many patients with AAD, however, have only mild inflammation or normal mucosa

The Direct Detection of Microorganisms in Clinical Samples

ISBN 0-12-187780-9

seen through the sigmoidscope. It has also been reported that colonoscopy may detect focal areas of PMC in regions of the colon that are beyond the reach of the sigmoidoscope. Fecal leukocytes are often present but are neither a sensitive nor a specific indicator of AAD; a stool Gram's stain is not particularly useful either.

Clindamycin was the first antimicrobial to be definitely associated with AAD in patients. Subsequently, virtually every antimicrobial in use has been implicated as a precipitator of AAD. Currently, the use of penicillins and cephalosporins is associated with the greatest number of cases of AAD. There is no clear association between the development of AAD and such antimicrobial factors as dosage, route, or duration of administration. The AAD may develop within 1 to 2 days of initiation of antimicrobial therapy or may not occur until several weeks after cessation. Relapses occur in as many as one-third of adequately treated patients even in the absence of further treatment with antimicrobials. Such relapses are thought to be due to the persistence of *C. difficile* spores, which are resistant to even those antimicrobials, such as vancomycin, that are capable of eliminating the vegetative forms.

Trnka and LaMont have implicated *C. difficile* as a factor in symptomatic relapse of chronic inflammatory bowel disease (IBD). They found *C. difficile* toxin in the stools of 19% of patients with IBD and showed a strong correlation with activity of the IBD. The toxin was present in 9 of 15 patients with severe flares of their disease, including 4 who had received no antimicrobials. However, Meyers and associates could find no such specific association between the presence of *C. difficile* toxin in stools and severity of IBD. In their patients, toxin appeared only in those who had been exposed to antimicrobials within the preceding 2 months. Obviously, the exact role that *C. difficile* plays in the course of IBD has yet to be definitely determined.

II. *CLOSTRIDIUM DIFFICILE* AND ITS TOXINS

Clostridium difficile is an anaerobic spore-forming gram-positive bacillus originally described by Hall and O'Toole (1935). On blood agar, it produces colonies 2–3 mm in diameter that are slightly raised, white semitranslucent, glossy, and nonhemolytic. George *et al.* (1979) described a selective and differential medium for the isolation of *C. difficile* from stools. The medium consists of an egg yolk–fructose base to which cycloserine and cefoxitin have been added (CCFA). Most of the usual components of the bowel flora, except for *C. difficile,* are inhibited on this medium. After 48 hr of incubation, colonies are 7–8 mm in diameter, yellow, ground-glass appearing, circular with slightly filamentous edge, and flat to low umbonate in profile. Lecithinase and lipase reactions, which can be seen on media containing egg yolk, are negative. The initial orange color of the CCFA medium often changes to yellow for 2 to 3 mm around the colony. When examined under long-wave ultraviolet light, colonies fluoresce yellow after 1 to 5 days of incubation. Cultures growing *C. difficile* have the distinctive odor of *p*-cresol. Gas–liquid chromatography (GLC) has been used to detect *p*-cresol in cultures of *C. difficile* as a rapid means of presumptively identifying this organism (Phillips and Rogers, 1981). Other characteristics of *C. difficile* include fermentation of fructose, glucose, mannitol, and mannose, hydrolysis of esculin and gelatin, sluggish motility, and production of acetic, isobutyric, butyric, isovaleric, valeric, and isocaproic acids from fermentation of glucose as detected by GLC (Holdeman *et al.*, 1977).

Clostridium difficile has rarely been isolated from soil and water and is not usually a component of the flora of animals. It has occasionally been recovered in humans from such general sources as wounds and blood and in cases of gas gangrene. It can be isolated from most patients with classical AAD–PMC (58–96%, Table I) but is cultured from far fewer patients who are thought to have AAD but do not have typical PMC (55–58%).

The organism is not commonly recovered from stools of healthy adults (0–2%) or patients with diarrhea unrelated to antimicrobial therapy (0–12%). However, its value in the diagnosis of AAD has been questioned because it can be found in 21 to 46% of patients who have received antimicrobials but do not have diarrhea. Nevertheless, the presence of *C. difficile* in the stool of patients with a clinical picture typical of AAD (with or without PMC) in the absence of other known enteric pathogens strongly suggests that this organism is the etiologic agent. The cytotoxicity assay has been cited as a more specific diagnos-

TABLE I

***Clostridium difficile* Stool Culture or Cytotoxin Assay Results from Three Studies of Patients with or without Antimicrobial-Associated Diarrhea**[a]

	Culture positive[b]			Cytotoxin positive[b]		
Patient	(1)	(2)	(3)	(1)	(2)	(3)
Antimicrobial-associated PMC[c]	89[c]	~96	58	83	96	42
Antimicrobial-associated, not PMC[c]	58	—	55	21	27	40
Diarrhea, no antimicrobials	12	—	0	0	2	0
Antimicrobials, no diarrhea	46	21	—	25	2	—
Healthy adults	0	0	—	0	0	—
Healthy children (3–24 months)	—	9	—	—	4	—
Healthy neonates	—	29	—	—	27	—

[a] Results expressed as percentage of positive tests.

[b] (1) George *et al.* (1982); (2) Viscidi *et al.* (1981); (3) E. A. McVey and J. E. Rosenblatt (unpublished data from the Mayo Clinic).

[c] PMC, Pseudomembranous colitis as diagnosed by sigmoidoscopy.

tic aid, as described later. However, a review of Mayo Clinic patients with characteristic features of AAD (diarrhea, fever, abdominal pain, leukocytosis, abnormal sigmoidoscopic results, and presence of fecal leukocytes) showed them to be present as often in those with positive cultures as in those who were cytotoxin-positive. At present, there is adequate rationale for continuing to use culture of stools for *C. difficile* in the diagnosis of AAD.

The earliest indication of a specific etiology for AAD was the demonstration that clindamycin-associated colitis in hamsters was due to a toxin produced by a *Clostridium*, later identified as *C. difficile* (Bartlett *et al.*, 1977). The enterocolitis could be produced by a cell-free supernatant of a broth culture, and this effect could be neutralized by incubation first with gas gangrene antitoxin and later with *Clostridium sordellii* antitoxin. This cross-reactivity of *C. difficile* toxin with *C. sordellii* antitoxin initially led to the erroneous conclusion that *C. sordellii* was the offending organism. Subsequently, *C. difficile* was isolated from animals and patients with AAD (*C. sordellii* was not) and shown to be capable of producing a toxin that is neutralizable by gas gangrene, *C. sordellii,* and specific *C. difficile* antitoxin (Bartlett, 1979).

Studies in this area have been facilitated by the development of a cytotoxicity assay (Chang *et al.*, 1979). Broth culture supernatants and stools containing *C. difficile* contain a cytotoxin that produces a cytopathic effect on a variety of tissue culture cell lines including primary human amnion and human fibroblast (WI-38), which can be neutralized by the aforementioned antitoxins. This cytotoxicity assay has become the method of testing for *C. difficile* toxin in diagnostic and research laboratories.

The cytotoxin is present in up to 96% of cases of AAD–PMC but in only 21 to 40% of cases of AAD without PMC (Table I). Although rarely found in healthy adults and children, it has been detected in up to 25 to 27% of neonates and patients given antimicrobials who do not have AAD. In fact, an overall review of the data in Table I (granting their significant variability) suggests that positive cytotoxicity assays and *C. difficile* cultures occur with approximately equal frequency. The cytotoxicity assay is positive somewhat less often in AAD without PMC, but whether this indicates a lack of assay sensitivity or an excess of false positive cultures is unclear. Likewise, the assay is less often positive in patients given antibiotics who do not have diarrhea. Although the cytotoxin assay has been proposed as the more specific diagnostic test, it may simply be a somewhat insensitive indicator of the presence of *C. difficile* in stools, especially because most clinical isolates are capable of producing the cytotoxin *in vitro.*

The latter hypothesis is bolstered by the finding that *C. difficile* elaborates two toxins, designated toxin A and toxin B (Donta *et al.*, 1982). Toxin A has been described as an enterotoxin because it causes a fluid response in rabbit ileal loops. Toxin B has minimal enterotoxic activity but is the primary cytotoxin, its activity being approximately 1000-fold greater than that of toxin A. In cytotoxicity assays, toxin B masks the cytotoxic activity of toxin A. This information indicates that the currently used cytotoxicity assay probably detects the presence of toxin B and not toxin A, and a positive reaction is not necessarily an indicator of enterotoxin activity.

Counterimmunoelectrophoresis (which has been proposed as a rapid means of toxin assay in stools) is not sufficiently sensitive to detect toxin A and can give false positive results by detecting unspecified clostridial antigens other than the toxins (West and Wilkins, 1982).

This information leaves the diagnostic value of both the stool culture for *C. difficile* and cytotoxicity assay somewhat in doubt. Both tests detect the presence of the organism, and whether the presence of toxin B (positive cytotoxicity assay) is a more specific indicator of etiology is questionable. Because the assay may actually be a less sensitive test, it is probably best at present to perform both tests and realize that their positivity only helps to confirm the *C. difficile* etiology of clinically diagnosed AAD. Perhaps, the development of a specific assay for the presence of toxin A in stools will provide the specific diagnostic test that is so greatly needed (Lyerly *et al.*, 1983).

III. LABORATORY PROCEDURES

A single, freshly passed stool specimen (no cathartic or enema) is the specimen of choice. The toxin is thermolabile, and the organism is resistant to freezing, so the stool specimens should be sent to the laboratory frozen on dry ice. At least 10–15 cm^3 of a liquid stool or a similar amount of solid or semisolid stool is required for the two analyses. Holding a specimen at room temperature for 3 to 6 hr or in the refrigerator for up to 24 hr before freezing will not significantly affect the results. Undoubtedly, if there are delays greater than those mentioned before the specimen can be frozen, a replacement should be collected. Specimens should be tightly sealed because the carbon dioxide released from dry ice may acidify the specimen, resulting in toxin inactivation.

A. Culture

Upon arrival in the microbiology laboratory, culture plates (CCFA) are inoculated and incubated for 24 to 48 hr. The presence of suspicious colonies may be noted within 24 to 48 hr, but confirmatory identification by biochemical and chromatographic techniques may require 3–4 days of laboratory time.

***Preparation of Cycloserine, Cefoxitin, Fructose, Egg Yolk, Agar (CCFA) (George** et al.*, 1979)

Egg yolk–fructose–agar base	
Proteose peptone no. 2	40 g

Na_2HPO_4	5 g
KH_2PO_4	1 g
NaCl	2 g
$MgSO_4$, anhydrous	0.1 g
Fructose	6 g
Agar	20 g
Neutral red, 1% solution in ethyl alcohol	3 ml
Distilled water	1000 ml
Final pH, 7.28	

Dispense 100-ml amounts in screw-capped bottle. Sterilize at 121°C for 15 min. Store at 4°C until needed. Melt basal medium and maintain at 50°C. Aseptically add the following to each bottle:

Cycloserine base, final concentration 500 μg/ml
Cefoxitin base, final concentration 16 μg/ml
5 ml egg yolk (50% suspension in saline)

Mix well and dispense 20 ml per plate. This is a selective medium for *Clostridium difficile.*

B. Cytotoxicity Assay for *Clostridium difficile* Toxin in Stool

The cytopathic toxin produced by *C. difficile* in the feces of patients with antibiotic-associated PMC is neutralized *in vitro* by polyvalent gas gangrene or by *C. sordellii* or by *C. difficile* antitoxin (available from T. D. Wilkins, V. P. I. Anaerobe Laboratory, Blacksburg, Virginia). The assay is that developed by Chang *et al. (1979).*

1. Prepare a 1 : 2 suspension of feces in phosphate buffered saline (pH 7.2).
2. Mix on mechanical stirrer or shaker.
3. Centrifuge at 3000 *g* for 30 min or 10,000 *g* for 10 mins.
4. Aspirate supernatant and filter sterilize (0.45-μm membrane filter).
5. Add 0.5-ml aliquots of the filtrate and serial 10-fold dilutions, to give final dilutions of 1 : 20 to 1 : 20,000, to tubes containing human fibroblast (WI-38) cell cultures containing 0.9 ml of medium.
6. Add 0.5 ml of filtrate to one tube with WI-38 cell culture containing 0.9 ml of medium and 0.1 ml of *C. sordellii* or *C. difficile* or gas gangrene antitoxin diluted 1 : 10.
7. Incubate at 35°C for 18 to 24 hr.
8. Report the highest dilution demonstrating cytopathic effect with rounded cells involving at least 10% of the cell culture monolayer. Specimens with high titers of toxin can show CPE within 4 hr.

There is a minimal decrease in the titer of toxin in feces stored at 6°C for 5 days, whereas no decrease has been observed in specimens stored at −70°C for 5 days.

REFERENCES

Bartlett, J. G. (1979). *Rev. Infect. Dis.* **1,** 530–538.

Bartlett, J. G., Onderdonk, A. B., Cisneros, R. L., and Kasper, D. L. (1977). *J. Infect. Dis.* **136,** 701–705.

Chang, T.-W., Lauermann, M., and Bartlett, J. G. (1979). *J. Infect. Dis.* **140,** 765–770.

Donta, S. T., Sullivan, N., and Wilkens, T. D. (1982). *J. Clin. Microbiol.* **15,** 1157–1158.

George, W. L., Sutter, V. L., Citron, D., and Finegold, S. M. (1979). *J. Clin. Microbiol.* **9,** 214–219.

George, W. L., Rolfe, R. D., and Finegold, S. M. (1982). *J. Clin. Microbiol.* **15,** 1049–1053.

Hall, I., and O'Toole, E. (1935). *Am. J. Dis. Child.* **49,** 390.

Holdeman, L. V., Cato, E. P., and Moore, W. E. C. (1977). "Anaerobic Laboratory Manual," 4th ed. Virginia Polytechnic Inst. and State Univ., Blacksburg, Virginia.

Lusk, R. H., Fekety, R., Silva, J., Browne, R. A., Ringler, D. H., and Abrams, G. D. (1978). *J. Infect. Dis.* **137,** 464–474.

Lyerly, D. M., Sullivan, N. M., and Wilkins, T. D. (1983). *J. Clin. Microbiol.* **17,** 72–78.

Meyers, S., Mayer, L., Bottone, E., Desmond, E., and Janowitz, H. D. (1981). *Gastroenterology* **80,** 697–700.

Phillips, K. D., and Rogers, P. A. (1981). *J. Clin. Pathol.* **34,** 642–644.

Rifkin, G. D., Fekety, F. R., Silva, J., Jr., and Sack, R. B. (1977). *Lancet* **II,** 1103–1106.

Trnka, Y. M., and LaMont, J. T. (1981). *Gastroenterology* **80,** 693–696.

Viscidi, R., Willey, S., and Bartlett, R. G. (1981). *Gastroenterology* **81,** 5–9.

West, S. E. H., and Wilkins, T. D. (1982). *J. Clin. Microbiol.* **15,** 347–349.

23

Perspective on the Current and Future Role of Gas–Liquid Chromatographic Analysis

JAMES W. MAYHEW AND SHERWOOD L. GORBACH

Infectious Disease Service
New England Medical Center
Tufts University School of Medicine
Boston, Massachusetts

I.	Introduction	303
II.	Methodology	304
	Preparative Techniques	305
III.	Gas–Liquid Chromatography	305
IV.	Goals of Analytical Techniques	307
V.	Clinical Correlations	307
VI.	Pitfalls in Analysis	308
VII.	Future Prospects	309
	References	310

I. INTRODUCTION

Rapid diagnosis by chemical–chromatographic techniques is based on the hypothesis that certain detectable alterations in the chemical environment of the host are characteristic of specific diseases. Application to infectious diseases was systematically fostered by the early efforts of Mitruka and co-workers (Mitruka, 1975). Current rapid procedures rely on the detection of a limited number of specific substances in body fluids. These diagnostic markers are, in general, relatively small molecules such as short-chain fatty acids. The appearance of unique compounds as well as quantitative changes in the relationship between components in a limited metabolite profile may be correlated with specific infectious diseases. Metabolic profiles have been defined as multicomponent gas–liquid chromatography (GLC) analyses that define or describe chemical patterns for a group of metabolically or analytically related substances (Horning

The Direct Detection of Microorganisms in Clinical Samples

ISBN 0-12-187780-9

and Horning, 1971; Rehman *et al.*, 1982). Ideally, a total profile of all body chemicals could be correlated with disease states. The technology for practically conducting such an analysis awaits future development.

Particular emphasis has been placed on organic acids recovered from urine, cerebrospinal fluid, serum, purulent material, and tissue biopsy. These compounds usually are intrinsically volatile substances or can be made volatile by relatively simple and standardized chemical modifications. Gas–liquid chromatography is the preferred method for fractionation of the biological extracts and resolution of the markers of interest. The separating capacity and efficiency of GLC are exceeded by no other method. The instruments are reasonably priced and are frequently available in chemistry laboratories.

II. METHODOLOGY

The techniques of rapid diagnostic procedures can be categorized into a few essential elements. Fast and consistent handling is required at each step. The essential prelude to analysis is a thorough documentation of the specimen source and condition. A preliminary diagnosis of the patient, a list of drugs, general dietary features, and the mode of specimen collection and preservation are crucial to the analyst's choice of internal standard, preparative procedures, and analytical conditions.

An appropriate internal standard is added to the specimen before any manipulation or processing. A known quantity of a properly chosen internal standard minimizes errors related to extraction yield, derivatization efficiency, and instrumental variability. The standards are necessary for the localization of the diagnostic markers via relative retention times and for quantification using either ratio-recovery curves or calculations based on molar-response factors. The quantitative internal standard has special value when concentration changes in normal metabolites are in fact characteristic of an infectious disease. The internal standards are similar in solubility, concentration, and detector response to the diagnostic markers but should neither be present in the original specimen nor react with specimen components. The standard should elute near the diagnostic marker in analytical chromatography while remaining completely resolved.

Marker compound recovery and preliminary purification from the biological matrix can be coupled to a separate destructive or nondestructive deproteinization step. The former can be variously accomplished by heat, acidification, or treatment with organic solvents that can simultaneously deproteinize and favor partition of the marker into the organic phase. The nondestructive procedures, not suitable for recovering protein-bound diagnostic markers, are accomplished by membrane filtration, gel permeation chromatography, or dialysis. These methods are particularly useful for the recovery of amino acids and carbohydrates (Jellum *et al.*, 1976). Extraction of the diagnostic marker into an organic solvent is done at physiologic pH or after alteration to acidic or alkaline conditions.

Preparative Techniques

Preparative chromatographic processes have been useful for specimen workups. Anion-exchange chromatography is very efficient, although time-consuming, for the quantitative isolation of polar compounds, particularly organic acids, from the aqueous matrix (Horning and Horning, 1971; Rehman *et al.*, 1982; Thompson and Markey, 1975). Preparative cartridges packed with normal-phase silica, reversed-phase C_{18} resins, or diatomaceous earth are also available for rapid preliminary purifications. Reversed-phase resins are used to collect relatively apolar solutes from a polar milieu, whereas normal phase tends to retard the elution of a relatively polar substance from an aqueous specimen such as urine that has been absorbed onto the finely divided surface of the packing. Preparative high-performance liquid chromatography (HPLC) on reversed or normal phase can be a valuable tool for preliminary fractionations. The judicious use of preparative chromatographic procedures can simplify the interpretation of the analytical chromatography by limiting the number of extraneous eluates.

After the preliminary purification, nonvolatile compounds are prepared for GLC by simple chemical modifications. Silyl ethers, esters, or amides, methyl esters, and oximes are frequently formed. Electronegative (halogenated) substituents may be added for specialized detector analysis. Several excellent comprehensive tests contain derivatization procedures for the full range of organic molecules of biological origin (Pierce, 1977; Blau and King, 1978; Knapp, 1979).

III. GAS–LIQUID CHROMATOGRAPHY

Analytical chromatography is done by GLC. Electron capture and flame ionization detectors are usually preferred, but selective detectors such as thermionic sensitive detectors can be useful, for example, in the case of nitrogen-bearing compounds. Open tubular capillary columns have become the fractionating surfaces of choice. These columns have an extremely high number of theoretical plates; when coupled with a mass-sensitive detector such as the flame ionization detector, they offer a wide dynamic range with a high picogram column load detection limit.

To anyone who has used these systems, the number of chromatographic bands resolved from derivatizated biological preparations is astounding. As a result, the interpretation of these chromatograms has increased in complexity. Jellum and co-workers (Jellum *et al.*, 1976) have demonstrated that, whereas a packed column resolves 40–80 bands from normal urine, the capillary column can fractionate nearly 300! Approximately 30% of those bands have been positively identified, and most of the remaining peaks probably represent multicomponent eluates. Absolute retention times in this setting are unreliable for identification. Retention times relative to an internal standard can provide a presumptive or tentative identification but not a confirmation. The coelution of unknown chromatographic bands at the appropriate relative retention time for the diagnostic marker may still be encountered. Additional analytical options may be required.

The ideal combination for rapid diagnosis is probably the more formidable array of capillary GLC interfaced with a mass spectrometer. The relatively new HPLC–mass spectrometer combination will offer exciting alternatives to bypass GLC requirements for volatility. The spectroscopic technique of mass fragmentography (selected ion monitoring) offers precise identity and quantification well into the picogram range (Sweeley *et al.*, 1966; Hammar *et al.*, 1968). Compounds that are not cleanly resolved by capillary GLC can nevertheless be distinguished by characteristic ion patterns in mass fragmentography. The general technique of selected ion monitoring can provide precise measurement of a limited number of chromotographic bands in a mixed component analysis (Gates and Sweeley, 1978). The specialized computerized technique of "mass chromatography" is an alternative of less precision and sensitivity but of much more generalized usefulness for profile studies (Hites and Biemann, 1970).

Unfortunately, this level of sophisticated technology is often overlaid on the background of an uncontrolled population of patients. Misleading analytical information due to diet, pharmaceuticals and their metabolites, and chemical breakdown products related to extraction, derivatization, and GLC are regularly encountered hazards. An additional limitation of these procedures is that ion patterns of closely related isomers may not be distinguishable by mass fragmentography and must be resolved by the capillary column (Takahashi *et al.*, 1979). Computer attachment to the mass spectrometer offers several advantages, including automatic repetitive scanning, storage and documentation of data, spectrum manipulations, and ion pattern matching for multicomponent profile analysis in which a correlation can be made with a specific diagnosis (Liebich, 1978).

In summary, an internally standardized analytical protocol for diagnostic markers can variously include deproteinization extraction with non-water-miscible organic solvents and preparative chromatography. Chemical modification or derivatization precedes analytical GLC. Although certain diagnostic marker compounds or patterns can be tentatively identified by relative elution characteristics on GLC, mass fragmentography provides the highest level of precision for the identification and quantification of these substances.

IV. GOALS OF ANALYTICAL TECHNIQUES

As procedures for the rapid diagnosis of infectious diseases evolve to the level of the clinical chemistry laboratory, many factors should be carefully scrutinized. The first goal is positive identification of the diagnostic chemical markers that are uniquely correlated with each disease. The second goal is standardization of procedures for extraction, chromatography, and mass spectroscopy of the specific compounds. It is necessary to determine which analyses lend themselves to capillary GLC alone, which lend themselves to direct mass fragmentography after simple extraction (Liebich and Wöll, 1977), and which require GLC–mass spectroscopy (Lawson, 1975).

The importance of positive chemical identifications has been clearly demonstrated in studies of metabolic disorders. Significant increases in urinary levels of 1,4-butanedicarboxylic and 1,6-hexanedicarboxylic acids have been found to be characteristic of diabetes-linked ketoacidosis (Pettersen *et al.*, 1972; Pettersen, 1974; Niwa *et al.*, 1981). The urinary excretion of 2-hydroxybutyric acid has been positively correlated with a diagnosis of lactic acidosis syndrome (Pettersen *et al.*, 1973; Landass and Pettersen, 1975). The determination of methylmalonic acid in urine and amniotic fluid has been used to diagnose prenatal methylmalonic acidemia (Nakamuva *et al.*, 1976). Urinary excretion of a wide variety of phenylacetic, phenyllactic, and pyruvic aromatic organic acids has been classically associated with phenylketonuria (Chalmers and Watts, 1974; Issachar and Yinon, 1976; Koepp and Hoffman, 1975). Approximately half of the 200 metabolic disorders currently recognized can be studied by gas-phase analytical methods (Jellum, 1977). The techniques have been extremely valuable for confirming diagnoses.

As the chemical markers for infectious diseases are identified, more specific information on solubilities, reactivity with derivitizing reagents, chromatographic retention characteristics, and fragmentation patterns will be obtained. The likelihood of recovering the specific compound from the normal, noninfected biological milieu must be carefully evaluated.

V. CLINICAL CORRELATIONS

Many examples of specific chemical identifications have been reported for infectious diseases. In a GLC study of 98 clinical specimens, succinic, isobutyric, and butyric acids were recovered from 19 of 20 purulent exudates that harbored *Bacteroides* and *Fusobacterium.* Significant four-carbon acids were recovered from none of 18 infections without gram-negative anaerobes nor from 60 sterile noninfected specimens of body fluids (Gorbach *et al.*, 1976). Correlation between specific organic acids with infections caused by gram-negative anaerobes was confirmed by subsequent studies of pus from 10 intraabdominal abscesses, from 44 mixed infected sites, and from 52 infected pleural effusions (Nord, 1977; Phillips *et al.*, 1976; Thadepalli and Gangopadhyaya, 1980). In another study of 129 cerebrospinal fluid specimens, high levels of lactic acid were recovered from eight cases of bacterial meningitis, whereas levels in 15 cases of nonbacterial meningitis could not be distinguished from 106 normal specimens (Ferguson and Tearle, 1977). 3-(2-Ketohexyl)indoline was recovered in a study of tuberculous meningitis (Brooks *et al.*, 1977). With the use of selected ion monitoring, tuberculostearic acid was recovered from the sputum of eight patients with pulmonary tuberculosis but not from six other pneumonia patients. Strikingly high levels of mannose and arabinitol have been correlated with *Candida* infections (Monson *et al.*, 1978; Kiehn *et al.*, 1979; Roboz *et al.*, 1980). However, these results have been complicated by the suggestion that renal insufficiency in the absence of mycotic infection is also related to high serum arabinitol levels (Roboz *et al.*, 1980; Eng *et al.*, 1981).

The foregoing citations of analytical procedures to diagnose bacterial and mycotic diseases provide a sense of the variety of classes of chemicals that may be of value: short- and long-chain organic acids, an indole-based compound, and monosaccharides. Specimen management would be facilitated by a preliminary clinical evaluation as well as a reasonable estimate for which markers are being sought. The pathways from choice of internal standard to isolation, derivitization, and chromatography are quite distinct for each of the compounds cited.

Alternatives to dealing with specifically identified marker compounds have been reported. Gas–liquid chromatographic "fingerprinting," the detection of patterns composed of multiple chromatographic bands (frequently representing unknown eluates), has produced some promising results (Brooks *et al.*, 1974, 1981). In the absence of precise identifications of the relevant compounds, however, this procedure leaves a vast area for variations among individual analysts and their techniques. An analogous mass spectrophotometric technique, pattern recognition or ion profile analysis by mass fragmentography, can lead to more precise correlations between mixed chemical composition and specific infectious diseases (Liebich, 1978). Multicomponent profile analyses of normal and diseased body fluids coupled with computer matching will more effectively differentiate the chemical anomalies of disease. An important drawback is that much more material is required for spectroscopy in the scan mode than for selected ion monitoring.

VI. PITFALLS IN ANALYSIS

In the performance of rapid analytical procedures for probing the uncontrolled biological matrix, one must be aware of certain pitfalls. The analytical background may contain pharmaceuticals and their metabolites, normal serum constituents, and serum constituents related to physiologic changes during disease but not specific to a single disease (Gates and Sweeley, 1978). The GLC technique itself can contribute its own share of misleading background and artifacts. Methylmalonic acid can be decarboxylated to propionic acid (Duran *et al.*, 1973), 3-ethylcrotonic acid can arise from 3-hydroxyvaleric acid (Faull *et al.*, 1976), phenylpyruvic can be decarboxylated to phenylacetic acid (Thompson *et al.*, 1975), and crotonic acid can be formed from *p*-hydroxybutyric acid (Gompertz, 1971). Diet, for example, can contribute furan-containing organic acids to the urine (Pettersen and Jellum, 1972). Anhydrides of compounds such as lactic acid may be formed. Multiple chromatographic bands from derivatized monosaccharides are routinely encountered. Subtle anomalies in organic acid metabolism between normal individuals can contribute to significant variation in the metabolite proportions found on analysis of biological fluids (Rehman *et al.*, 1982). As in the case of arabinitol (Roboz *et al.*, 1980; Eng *et al.*, 1981), questions may arise as to the specificity of the diagnostic marker: Is it specifically correlated with infection (*Candida*), or are high levels related to renal insufficiency?

Preparative and chromatographic procedures can be designed to reduce background by selectively isolating the chemical groups of the marker. Internal standards can be carefully chosen for precise calculations of relative retention time. Heavy isotopes (deuterated) of the marker that do not chromatographically resolve from the native compound can be included in analyses when mass spectrometry is used. Classical procedures such as the use of two different derivatives (e.g., trifluoroacetic acid and trimethylsilane) can be examined. The relative retention time for the suspected marker should correlate with those of authentic compounds for both derivatives.

VII. FUTURE PROSPECTS

It is clear that rapid diagnostic procedures are becoming a valuable tool in the diagnosis of bacterial and mycotic infections. However, it is critically important that standardization of procedures continue to be developed. Important areas where progress can be made include the following:

1. Identification of the diagnostic markers. Is the presence of single unique compounds or significant alterations in the balance of normally present metabolites the most precise marker for infectious disease?
2. Establishment of a clear connection between the diagnostic markers and the disease process; assignment of clinical relevance.
3. Definition and management of the biological background with special attention to extraction and preparative chromatography procedures.
4. Selection of appropriate and commonly accepted internal standards.

5. Use of standardized methods with capillary columns and consideration of the use of more than a single chemical.
6. Investigation of the use of the HPLC–mass spectrometer combination as an alternative to bypass derivatization steps required for GLC–mass spectroscopy.
7. Definition of the role of mass spectroscopy and a clear understanding of which problems can be solved by GLC alone, mass spectroscopy alone, or by combinations of GLC and mass spectroscopy.
8. Careful investigation of the value of GLC fingerprinting and multiple ion profiles by mass fragmentography or mass chromatography.
9. Application of data management, sorting, and comparison techniques to electronically evaluate analytical results.

As the use of capillary columns spreads and as mass spectroscopy becomes more commonly available in laboratories and reference centers, rapid diagnostic procedures should naturally evolve into vital diagnostic tools in the management of infectious diseases.

REFERENCES

Björkhem, I., Blomstrand, R., Lantto, O., Svensson, L., and Ohman, G. (1976). *Clin. Chem.* **22,** 1789–1801.

Blau, K., and King, G. (1978). ''Handbook of Derivatives for Chromatography.'' Heyden, London.

Brooks, J. B., Kellogg, D. S., Alley, C. C., Short, H. B., Handsfield, H. H., and Huff, B. (1974). *J. Infect. Dis.* **129,** 660–668.

Brooks, J. B., Choudhary, G., Craven, R. B., Alley, C. C., Liddle, J. A., Edman, D. C., and Converse, J. D. (1977). *J. Clin. Microbiol.* **5,** 625–628.

Brooks, J. B., McDade, J. E., and Alley, C. C. (1981). *J. Clin. Microbiol.* **14,** 165–172.

Chalmers, R. A., and Watts, R. W. (1974). *Clin. Chim. Acta* **55,** 281–294.

Duran, M., Ketting, D., Wadman, S. K., Trijbels, J. M., Bakkeren, J. A., and Waelkens, J. J. (1973). *Clin. Chim. Acta* **49,** 177–179.

Eng, R. H., Chmel, H., and Buse, M. (1981). *J. Infect. Dis.* **143,** 677–683.

Faull, K. F., Bolton, P. D., Halpern, B., Hammond, J., and Danks, D. M. (1976). *Clin. Chim. Acta* **73,** 553–559.

Ferguson, I. R., and Tearle, P. V. (1977). *J. Clin. Pathol.* **30,** 1163–1167.

Gates, S. C., and Sweeley, C. C. (1978). *Clin. Chem.* (*Winston-Salem, N.C.*) **24,** 1663–1673.

Gompertz, D. (1971). *Clin. Chim. Acta* **33,** 457.

Gorbach, S. L., Mayhew, J. W., Bartlett, J. G., Thadepalli, H., and Onderdonk, A. B. (1976). *J. Clin. Invest.* **57,** 478–484.

Hammar, C. G., Holmstedt, B., and Ryhage, R. (1968). *Anal. Biochem.* **25,** 532–548.

Hites, R. A., and Biemann, K. (1970). *Anal. Chem.* **42,** 855–863.

Horning, E. C., and Horning, M. G. (1971). *Clin. Chem.* **17,** 802–809.

Issachar, D., and Yinon, J. (1976). *Clin. Chim. Acta* **73,** 307–314.

Jellum, E. (1977). *J. Chromatogr.* **143,** 427–462.

Jellum, E., Storseth, P., Alexander, J., Helland, P., Stokke, P., and Teig, E. (1976). *J. Chromatogr.* **126,** 487–493.

Kiehn, T. E., Bernard, E. M., Gold, J. W., and Armstrong, D. (1979). *Science* **206,** 577–580.
Knapp, D. R. (1979). "Handbook of Analytical Derivatization Reactions." Wiley, New York.
Koepp, P., and Hoffmann, B. (1975). *Clin. Chim. Acta* **58,** 215–221.
Landaas, S., and Pettersen, J. E. (1975). *Scand. J. Clin. Lab. Invest.* **35,** 259–266.
Lawson, A. M. (1975). *Clin. Chem. (Winston-Salem, N.C.)* **21,** 803–824.
Liebich, H. M. (1978). *J. Chromatogr.* **146,** 185–196.
Liebich, H. M., and Wöll, J. (1977). *J. Chromatogr.* **142,** 505–516.
Mitruka, B. M. (1975). "Gas Chromatographic Applications in Microbiology and Medicine." Wiley, New York.
Monson, T. P., Wilkinson, K. P., and Borer, W. Z. (1978). *Clin. Res.* **26,** 402A.
Nakamura, E., Rosenberg, L. E., and Tanaka, K. (1976). *Clin. Chim. Acta* **68,** 127–140.
Niwa, T., Maeda, K., Ohki, T., Saito, A., and Tsuchida, I. (1981). *J. Chromatogr.* **225,** 1–8.
Nord, C. E. (1977). *Acta Pathol. Microbiol. Scand.* No. 259, pp. 55–59.
Pettersen, J. E. (1974). *Diabetes* **23,** 16–20.
Pettersen, J. E., and Jellum, E. (1972). *Clin. Chim. Acta* **41,** 199–207.
Pettersen, J. E., Jellum, E., and Eldjarn, L. (1972). *Clin. Chim. Acta* **38,** 17–24.
Pettersen, J. E., Landaas, S., and Eldjarn, L. (1973). *Clin. Chim. Acta* **48,** 213–219.
Phillips, K. D., Tearle, P. V., and Willis, A. T. (1976). *J. Clin. Pathol.* **29,** 428–432.
Pierce, A. E. (1977). "Silylation of Organic Compounds." Pierce Chemical Company, Rockford, Illinois.
Rehman, A., Gates, S. C., and Webb, J. W. (1982). *J. Chromatogr.* **228,** 103–112.
Roboz, J., Suzuki, R., and Holland, J. F. (1980). *J. Clin. Microbiol.* **12,** 594–602.
Sweeley, C. C., Elliot, W. H., Fries, I., and Ryhage, R. (1966). *Anal. Chem.* **38,** 1549–1553.
Takahashi, G., Kinoshita, K., Hashimoto, K., and Yasuhira, K. (1979). *Cancer Res.* **39,** 1814–1818.
Thadepalli, H., and Gangopadhyaya, P. K. (1980). *Chest* **77,** 507–513.
Thompson, J. A., and Markey, S. P. (1975). *Anal. Chem.* **47,** 1313–1321.
Thompson, R. M., Belanger, B. G., Wappner, R. S., and Brandt, I. K. (1975). *Clin. Chim. Acta* **61,** 367–374.

24

*Gas–Liquid Chromatography as an Aid in Rapid Diagnosis by Selective Detection of Chemical Changes in Body Fluids**

JOHN B. BROOKS

Centers for Disease Control
Atlanta, Georgia

I.	Introduction	313
II.	Practical Methods for Recovering Volatile Chemical Compounds from Body Fluids for Derivatization Purposes	314
III.	Selective and Sensitive High-Resolution Gas–Liquid Chromatography Systems for Body Fluid Analysis	315
IV.	Practical Derivatization Methods for Analysis by FPEC–GLC	318
V.	Application of FPEC–GLC to Detection of Chemical Changes in Spent Culture Media and Infected Tissue Culture	319
VI.	Application of FPEC–GLC to Detection of Disease-Specific Profiles in Body Fluids	319
	A. Lymphocytic Meningitis and Encephalitis	319
	B. Arthritis	323
	C. Pleural Effusions	323
	D. Rocky Mountain Spotted Fever and Viral Diseases	326
VII.	Identification of Unknown Peaks Detected by FPEC–GLC	330
VIII.	Interpretation of Data Obtained by FPEC–GLC	331
IX.	Possibilities for Automation and Computerization of Data Obtained from FPEC–GLC	332
X.	Summary	333
	References	333

I. INTRODUCTION

Problems in the rapid diagnosis of disease still exist today. Some of the troublesome areas include the lymphocytic meningitides and encephalitides

*The use of trade names is for identification only and does not imply endorsement by the Public Health Service or by the U.S. Department of Health and Human Services.

The Direct Detection of Microorganisms in Clinical Samples

ISBN 0-12-187780-9

caused by viruses, certain bacteria such as *Mycobacterium tuberculosis,* fungi, and parasitic infections. Dr. Robert Craven has written in *CRC Critical Reviews in Clinical Laboratory Sciences* (Edman *et al.*, 1981) about some of the problems involved in devising treatments for bacterial meningitis presenting with a granulocytic pleocytosis with a low cerebrospinal fluid (CSF) glucose level and no detectable organisms. He states:

> Most practitioners, however, would probably rather face that dilemma than the one posed when confronted with a central nervous system (CNS) infection which presents with lymphocytic pleocytosis and equivocal CSF glucose. The differential diagnosis of lymphocytic meningitis includes partially treated bacterial meningitis, tuberculosis, asceptic, viral encephalitis and fungal, parasitic meningitis. The dilemma is heightened by the fact that those potential causes of lymphocytic meningitis require different therapeutic approaches. The incorrect therapeutic decision can produce a fatal result.

Other areas that pose problems in rapid diagnosis are pneumonia, arthritis, "exotic" infections such as dengue fever, rabies, Rocky Mountain spotted fever, and a host of other viral, fungal, and parasitic diseases.

The basic problems surrounding the rapid diagnosis of most of the diseases that are difficult to diagnose involve (*a*) difficulty in characterizing the organism, (*b*) lack of growth or slow growth of the organism *in vitro,* (*c*) length of time required to form antibodies during infection, and (*d*) cross-reactions in immunologic tests for antibody or antigen.

Most investigators would agree that there are changes in body chemistry during infection. These changes do not necessarily require the presence of large numbers of an infectious agent. The changes may be due to (*a*) a response to the disease by the body's defense mechanisms, (*b*) metabolites from an infectious agent, (*c*) changes in cellular metabolism caused by intracellular agents such as viruses, or (*d*) a combination of any of these. If practical methods of sufficient selectivity and sensitivity to detect infection-related chemical changes were developed, etiologic identification of the disease would be possible. This would not depend on antibody–antigen interaction and could be accomplished through the analysis of body fluids without the isolation of a causative agent. The new diagnostic process might be affected by therapy, processing, multiple disease states, and the state of the illness. In this chapter attempts to develop a diagnostic procedure in the face of these variables are discussed.

II. PRACTICAL METHODS FOR RECOVERING VOLATILE CHEMICAL COMPOUNDS FROM BODY FLUIDS FOR DERIVATIZATION PURPOSES

Techniques and instrumentation that would detect the entire spectrum of chemical changes in the body during a disease state would be too time-consuming and expensive to employ as a general diagnostic procedure. The major goal must be to develop a system with sufficiently broad capabilities to find differences in certain disease states and at the same time be compatible with the time schedule and expected cost of analyses normally performed by hospitals and clinical laboratories. Naturally, some extra time and expense might be tolerated if the procedure reliably produced results unattainable by less expensive methods. The most logical types of compounds to look for would be those normally produced by infectious agents and metabolites for which test procedures have already been demonstrated to be valuable for differentiation between disease-causing agents. Metabolic products of microorganisms consisting of carboxylic acids, hydroxy acids, alcohols, and amines have been used for years to develop specific tests to distinguish among microorganisms. Most of these compounds are volatile, with a boiling range between 40 and 400°C, and as such they can be separated, quantified, and identified by gas–liquid chromatography (GLC) in a short time. Not all of these compounds are easy to remove from aqueous solutions, but complete recovery is not necessary. It is necessary only to obtain these compounds reproducibly in quantities sufficient to produce a profile with qualitative and quantitative differences.

In order to prepare derivatives of carboxylic acids, alcohols, and amines, it is desirable, and in most cases necessary, to retrieve these compounds from aqueous solutions. The most practical way to accomplish this is to take advantage of their solubilities under acidic and basic conditions and their extraction characteristics with organic solvents. Because there are no universally appropriate solvents, the best approach is to use a solvent that is midscale in polarity and possesses characteristics that permit easy extraction, concentration, derivatization, and sample washing after derivatization. Another factor to consider is the availability of ultrapure solvents that will be necessary for use in trace analysis. Chloroform is a solvent that meets all these requirements and the one we have selected for our studies. However, the short-chain hydroxy acids are so polar that they require a more polar solvent for the initial extraction. We have chosen diethyl ether for this purpose, but the ether is evaporated and chloroform is used for the solvent derivatization and sample cleanup. It must be realized that selectivity, functional group identification, and increased detectability through sample concentration begin with the extraction procedure. A schematic drawing outlining the extraction and separation procedures used in our laboratory is shown in Fig. 1.

III. SELECTIVE AND SENSITIVE HIGH-RESOLUTION GAS–LIQUID CHROMATOGRAPHY SYSTEMS FOR BODY FLUID ANALYSIS

Because GLC was introduced in the late 1950s, it has been used extensively to separate, quantify, and identify volatile chemical compounds or chemical com-

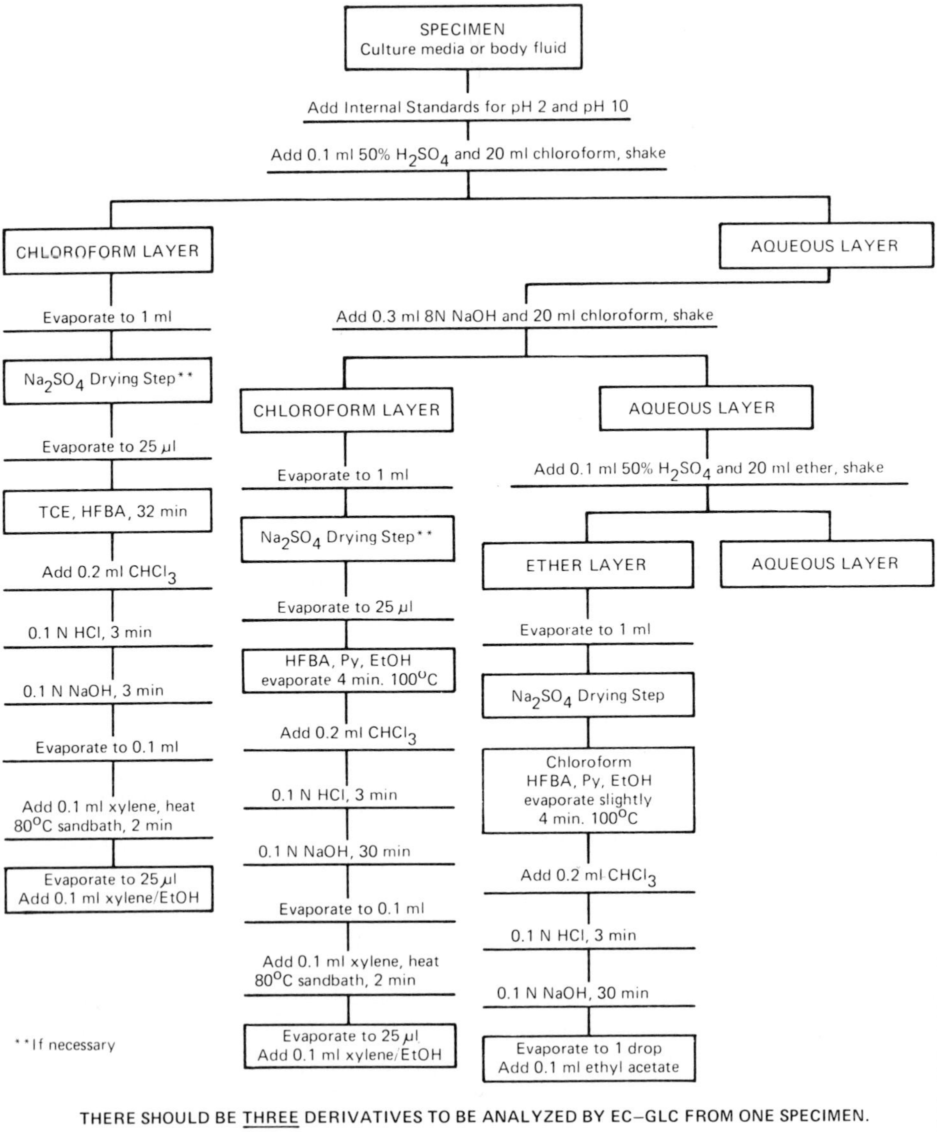

Fig. 1. Flow chart for preparing specimens for electron capture/gas–liquid chromatography (EC–GLC).

pounds that can be made volatile through derivatization. Chemical compounds with boiling ranges between 40 and 400°C are easily separated and tentatively identified. The reliability of the separation and identification depends on what is done to the sample before analysis, the specificity of the derivative formed, the column resolution, and the selectivity of the detector. In addition to selectivity, the detector must have the sensitivity necessary to detect compounds present in the nano- and pico-molar range. The thermoconductivity detector, the flame ionization detector, and the electron capture detector (ECD) have been used to detect metabolic products and cellular fatty acids of microorganisms, but the ECD alone possesses the sensitivity and, more importantly, the selectivity necessary for detecting most disease-induced changes in body fluids. The most dramatic improvement of the ECD came after the introduction of the frequency-pulsed electron capture (FPEC) detector by Maggs *et al.* (1971). The FPEC detector permits a much better degree of linearity than the older electron capture modes of operation, and in addition this mode of operation has greated improved capacity for recovery from overloading of the detector. The FPEC detector is also much less affected by contaminating agents than the older ECD.

The columns that have proved to be highly satisfactory for use in our laboratory are glass columns of 7.3-m length by 2-mm internal diameter packed with 3% OV-101 on Chromsorb W 80–100 mesh. These columns have excellent resolution characteristics, and they are very efficient for heptafluorobutyric anhydride (HFBA) and trichloroethanol (TCE) derivatives of carboxylic acids, hydroxy acids, alcohols, and amines. They last up to a year without repacking, and once conditioned they have a low bleed rate and can withstand high temperatures. The type of instrument used in the analysis must be of a quality that permits reproducible temperature programming between 90 and 265°C. The detector oven must be maintained with $\pm 1°$ tolerance. In addition to the packed column just listed, we have found that 50-m fused silica OV-101 columns operated in conjunction with splitless injectors are effective in this type of analysis. These columns also require programming but offer high resolution and are not easily broken.

In conjunction with either the packed or capillary column, the instrument should be equipped with makeup of flush gas. The carrier gas for the packed column is 95% argon–5% methane. The rate of flow through the column is 50 ml/min. Makeup gas is added to make the total flow of 68 ml/min through the detector at 125°C. Helium is used as the carrier gas through the capillary column at a flow rate of 2 ml/min (about 40 psi pressure), and makeup gas (95% argon–5% methane) is added to make the total flow through the detector 50 ml/min.

The sensitivity of the FPEC–GLC system is a very important function of the operation. The sensitivity of a newly foiled detector is low, and that of a very aged detector foil (usually more than a year old) is high. The sensitivity of the detector should be adjusted so that the internal standards heptanoic acid and di-*n*-butylamine, described by Brooks *et al.* (1980a,b), just exceed full scale with a 2-μl injection of sample. The standing current, attenuation, flush gas, and sample dilution can be adjusted to obtain the desired response. The standing current is the one most often used to make the sensitivity adjustment. Overattenuation can create problems with peak shape distortion and detector overloading.

IV. PRACTICAL DERIVATIZATION METHODS FOR ANALYSIS BY FPEC–GLC

In order to take maximum advantage of the ECD, one must ensure the formation of electron-capturing derivatives from the functional groups of carboxylic acids, hydroxy acids, alcohols, and amines. Derivatization makes the functional group detectable by the FPEC detector and, in addition, changes the volatility and polarity of the compound. Once esters of the acids and alcohols are formed, they become very insoluble in water and highly soluble in chloroform. The chloroform layer containing the derivatized extracts can be washed with acid and base to remove the derivatizing agent. In most cases the derivative is less volatile than the derivatizing agent, and thus in some cases the reagent can be removed by evaporation (Alley *et al.*, 1976, 1979). The functional groups of the excess reagent, which are free or sometimes halogenated by products of the reaction, elute slowly from the nonpolar column. They are highly electron absorbing and elute from the column in such a fashion that they will obscure other peaks of lesser intensity. In order to prevent interference, most of the excess reagent is removed either by washing with an acid and base solution or by evaporation in a solvent with a boiling temperature that is slightly higher than that of the reagent.

The derivatizing agent should be as specific for the functional group as practical methods will permit. Esters of acids from esterification with a halogenated alcohol and halogenated derivatives of alcohols and amines formed by reaction with halogenated anhydrides are specific and form highly electron-capturing derivatives, and the derivatization procedures are highly reproducible. Not only does derivatization with these reagents prepare electron-capturing derivatives, but, because they form specific derivatives, they also produce valuable information about the functional group. Details of the preparation of these derivatives have been described by Brooks *et al.* (1974b, 1980a) and Alley *et al.* (1976, 1979). After derivatization is complete, chloroform is removed to almost complete dryness by evaporation with clean, dry air. Next is added 0.1 ml of final solvent (xylene–ethanol, 1 : 1, for carboxylic acids, alcohols, and amines and ethyl acetate for diesters of hydroxy acids), and then 2 μl of the derivative is used for analysis by FPEC–GLC.

V. APPLICATION OF FPEC–GLC TO DETECTION OF CHEMICAL CHANGES IN SPENT CULTURE MEDIA AND INFECTED TISSUE CULTURE

FPEC–GLC analyses have been done on fatty acids in spent bacterial culture media and tissue culture media infected with viruses and whole bacterial cells after degradation with base. These analyses have been useful in the identification of anaerobes, *Neisseria,* the major causative agents of bacterial meningitis (Brooks *et al.*, 1980a,b), and dengue fever (unpublished data). A defined medium as described by Catlin (1974) is best when it can be used because chromatographic profiles can be reproduced from lot to lot. Undefined media can be used, but changes in media, metabolic products, and growth characteristics must be determined on a lot-to-lot basis.

As shown in Fig. 2 some of the major causative agents of bacterial meningitis can be differentiated by FPEC–GLC analysis of spent medium. Both the acid profiles (Brooks *et al.*, 1980a) and the amine profiles (Brooks *et al.*, 1980b) have been used. A Catlin-defined medium was used to culture the organisms. One of the disadvantages of this approach is that the organism must be cultured before analysis. In the development of FPEC–GLC methods to distinguish among certain organisms, it is advisable to conduct an initial study to find the best media and cultural conditions possible for the analysis. It is conceivable that infected spinal fluid could be mixed with culture media, incubated a few hours, and then analyzed by FPEC–GLC to obtain patterns useful for rapid differentiation. FPEC–GLC analysis could be very beneficial in the chemotoxonomy of microorganisms. In detailed studies of *Neisseria meningitidis* strains, Brooks *et al.* (1981a) showed the diversity of metabolism among strains of a single species. Differences in strains were detected even in a defined medium used with rigidly controlled culture conditions. Serologic methods have also demonstrated differences among strains of this genus. The FPEC–GLC information can be used either to create new species or to lump the organisms on the basis of peaks given increased importance by the investigator.

VI. APPLICATION OF FPEC–GLC TO DETECTION OF DISEASE-SPECIFIC PROFILES IN BODY FLUIDS

A. Lymphocytic Meningitis and Encephalitis

The diagnosis of disease through the use of FPEC–GLC patterns obtained from analysis of body fluids is one of the most exciting possible applications of this new tool. This is also an area with many potential pitfalls. The technique is

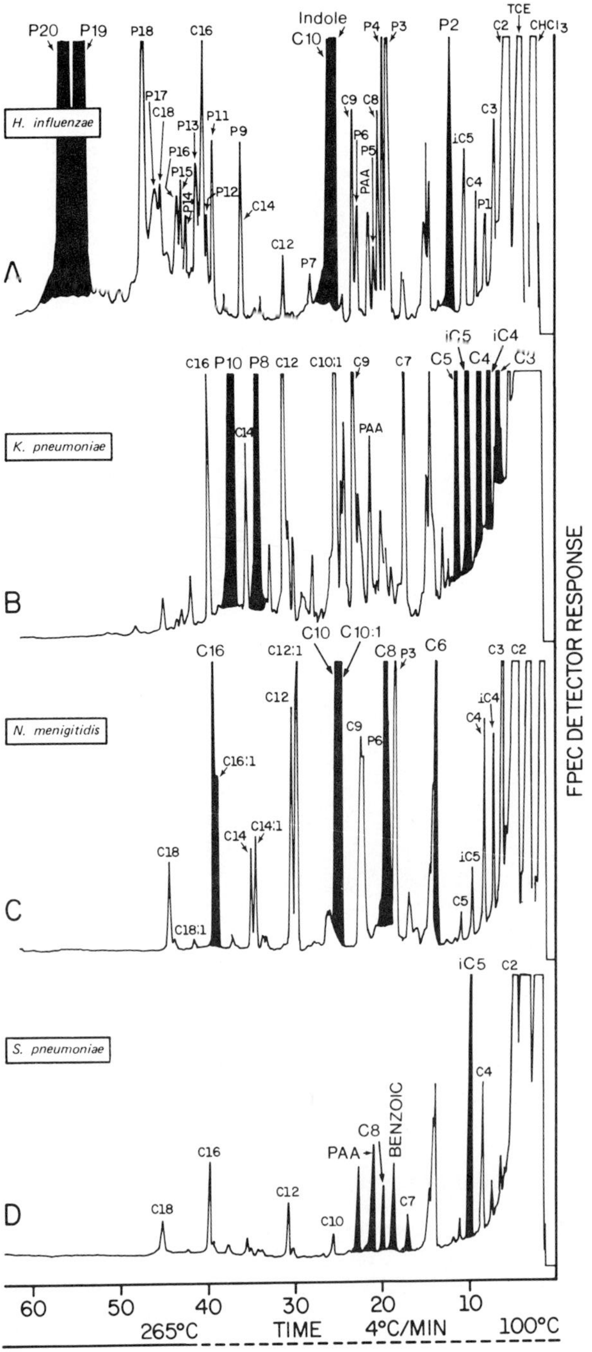
H. influenzae
K. pneumoniae
N. menigitidis
S. pneumoniae
A
B
C
D
FPEC DETECTOR RESPONSE
60 50 40 30 20 10
265°C TIME 4°C/MIN 100°C

new, and many investigators hesitate to embark into this area until the FPEC–GLC technique can be proven reliable and can be automated and computerized. For the reasons just mentioned it is logical that FPEC–GLC technology may initially find greater acceptance in areas where diagnosis is difficult or reliable rapid methods are lacking. As stated in the introduction, lymphocytic meningitis poses such a problem. We have been studying the possible application of FPEC–GLC to the rapid diagnosis of tuberculous meningitis for the past several years. When the specimens are taken properly before therapy, correct identification in our laboratory has exceeded 90% (Brooks *et al.*, 1980c). Figure 3 shows the type of FPEC–GLC profile normally obtained. The interpretation is made with the knowledge that (*a*) no growth was obtained after 24 hr of culturing for the major causative agent of meningitis, and (*b*) a lymphocytic type of meningitis is present. FPEC–GLC analysis of both CSF and serum is beneficial. Serum from patients with tuberculous bacteremia could potentially show an abnormal FPEC–GLC profile without the spinal fluid being affected, but we have not seen a case of tuberculous meningitis in which both the serum and CSF were not affected.

As previously stated lymphocytic meningitis encompasses a large area of study. We have only begun to meet some of the potential needs in this area. The earlier studies by Schlossberg *et al.* (1976) indicated that FPEC–GLC could be useful as an aid in the identification of cryptococcal meningitis. Schlossberg found differences in the amine patterns in CSF resulting from certain enteroviruses and those induced by cryptococci. He grew cryptococci in supplemented spinal fluid and found that some of the peaks of this material matched those found in the CSF of a patient with cryptococcal meningitis. In later studies (Brooks *et al.*, 1980c; Craven *et al.*, 1977) FPEC–GLC patterns obtained from CSF of patients with encephalitis caused by herpes simplex were different from those found in cases of tuberculous meningitis (Fig. 4). It must be pointed out, however, that the amount of research devoted to the rapid diagnosis of tuberculous meningitis has not yet been applied to herpes encephalitis. The reason for this lack of study is related to the lack of specimens from well-documented cases of herpes encephalitis available for study in our laboratory. We studied suspected cases of tuberculous meningitis that we recognized by FPEC–GLC as herpes,

---

Fig. 2. FPEC–GLC chromatograms of TCE–HFBA derivatives prepared from chloroform extracts of spent Catlin-defined medium. The letter "C" followed by a number indicates a saturated straight-chain carboxylic acid with the number of carbon atoms indicated by the number. The letter "i" indicates iso, and a colon between two numbers indicates unsaturation. The letter "P" next to a number indicates an unidentified compound. PAA denotes phenylactic acid; $CHCl_3$, residual chloroform; TCE, trichloroethanol; HFBA, heptafluorobutyric anhydride. Peaks have been blackened to indicate which peak we consider important for purposes of differentiation among organisms. From Brooks *et al.* (1980a, p. 47) with permission (American Society for Microbiology). All rights reserved.

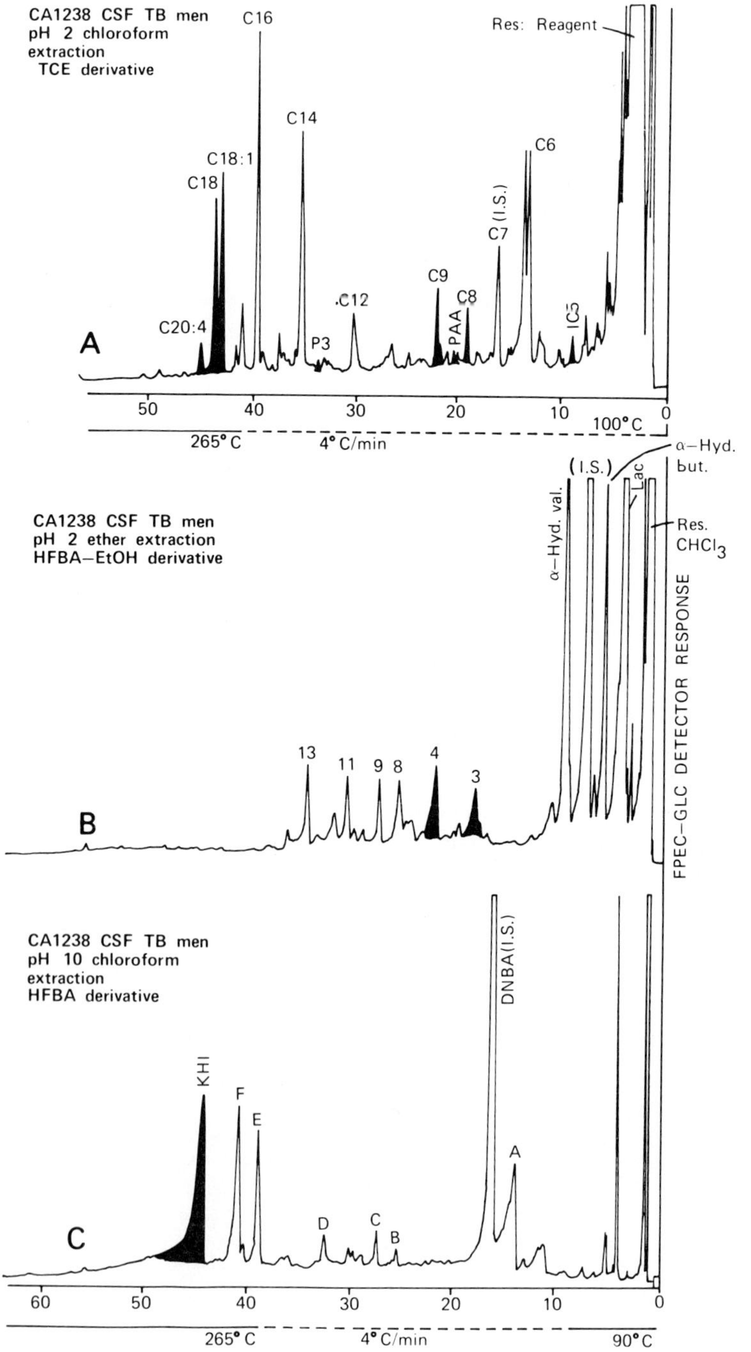

CA1238 CSF TB men
pH 2 chloroform
extraction
TCE derivative
C16
Res: Reagent
C14
C18:1
C18
C6
C7 (I.S.)
C9
C12
C8
PAA
IC5
C20:4
A
P3
50
40
30
20
10
0
100°C
265°C
4°C/min
α–Hyd.
but.
(I.S.)
Lac
α–Hyd. val.
CA1238 CSF TB men
pH 2 ether extraction
HFBA–EtOH derivative
Res.
$CHCl_3$
FPEC–GLC DETECTOR RESPONSE
13
11
9
8
4
3
B
CA1238 CSF TB men
pH 10 chloroform
extraction
HFBA derivative
DNBA(I.S.)
KHI
F
E
A
D
C
B
C
60
50
40
30
20
10
0
265°C
4°C/min
90°C

and these diagnoses were later confirmed. In cooperative studies with Dr. A. J. Nahmias of Emory University (unpublished data) we showed that herpes infection dramatically affected the metabolism of certain tissue culture cells. Studies such as those completed with tuberculous meningitis are still needed to confirm our results with herpes. During our studies of tuberculous meningitis, we found that about 80% of the suspected cases were negative. Among the diagnoses that were later confirmed for these patients were early cancer, Morralet's disease, tertiary syphilis, leptospirosis, aseptic meningitis, and several fungal and parasitic diseases.

B. Arthritis

Among the first body fluids we studied for possible diagnosis by gas chromatography were synovial fluids (Brooks *et al.*, 1974b). The fluids were taken from patients with cases of synovitis caused by joint infections with staphylococci, streptococci, haemophili, and gonococci. Figure 5 shows the type of chromatograms obtained in that study. There is a need for a rapid FPEC–GLC test for identifying some of the noninfectious arthritides. We have conducted preliminary studies of rheumatoid arthritis, Reiters disease, and gout. Our results are still inconclusive. Results of FPEC–GLC studies of patients in both England and the United States indicate a difference in FPEC–GLC patterns among similar types of patients with rheumatoid arthritis.

C. Pleural Effusions

A study involving FPEC–GLC analysis of pleural effusions caused by *Mycobacterium tuberculosis,* other types of sepsis, congestive heart failure, certain malignancies, and uremia showed that specific patterns were present in each of these diseases (Brooks *et al.*, 1978). Several pleural fluids from patients infected with *M. tuberculosis* were studied. After specific treatment for *M. tuberculosis,* the FPEC–GLC patterns changed, as shown in Fig. 6 and 7. All of the FPEC–GLC patterns obtained from the analysis of pleural effusions caused by breast cancer, endometrial carcinoma, congestive heart failure, systemic lupus, certain anaerobic empyema, and staphylococcal empyema were different from each other. The use of FPEC–GLC techniques to aid in the identification of causes of

---

Fig. 3. FPEC–GLC chromatograms of TCE–HFBA, HFBA, and HFBA–EtOH derivatives prepared from chloroform and diethyl ether extracts of cerebrospinal fluid (CSF) taken from a patient with tuberculous (TB) meningitis. LAC denotes lactic acid; 2-Hyd. but., 2- hydroxybutyric acid; 2-Hyd. val., 2-hydroxyvaleric acid; HFBA–EtOH, HFBA–ethanol; I.S., internal standard. For an explanation of the other notation see Fig. 2 and the text. From Brooks *et al.* (1980c) with permission (American Society for Microbiology).

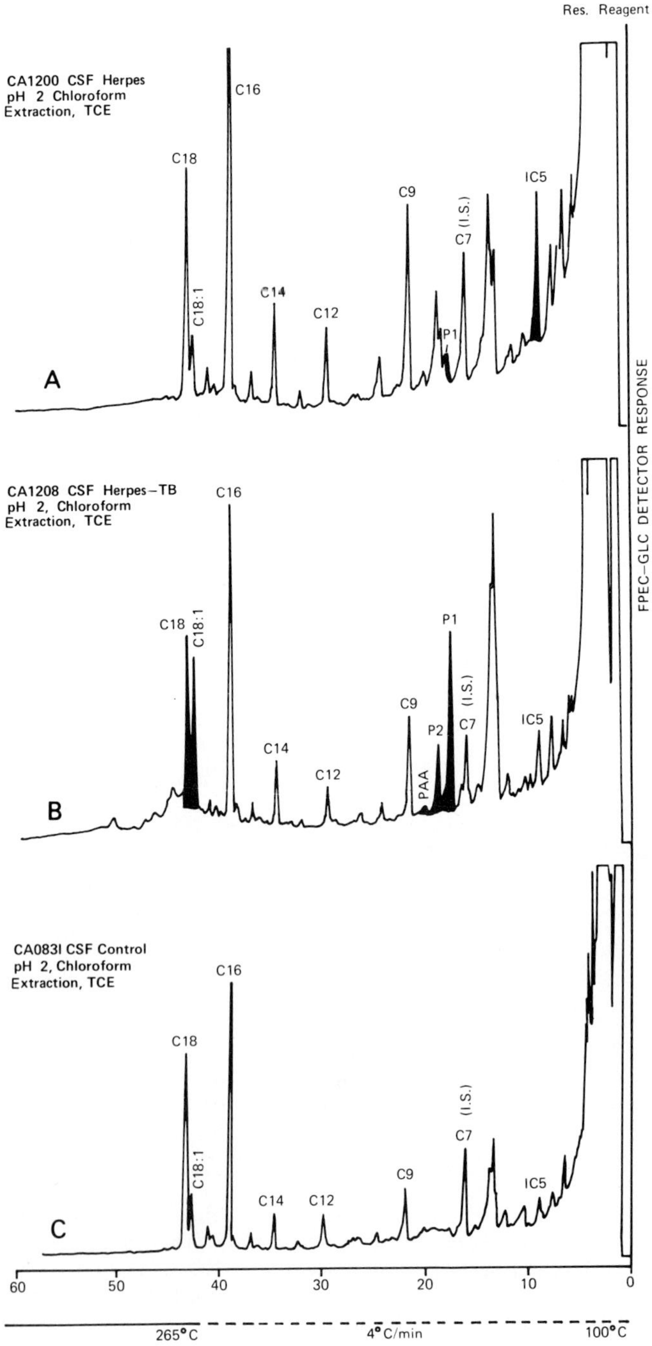

Res. Reagent
CA1200 CSF Herpes
pH 2 Chloroform
Extraction, TCE
C16
C18
C18:1
C14
C12
C9
(I.S.)
C7
P1
IC5
A
CA1208 CSF Herpes–TB
pH 2, Chloroform
Extraction, TCE
C16
C18
C18:1
C14
C12
C9
PAA
P2
P1
(I.S.)
C7
IC5
B
FPEC–GLC DETECTOR RESPONSE
CA0831 CSF Control
pH 2, Chloroform
Extraction, TCE
C16
C18
C18:1
C14
C12
C9
(I.S.)
C7
IC5
C
60
50
40
30
20
10
0
265°C
4°C/min
100°C

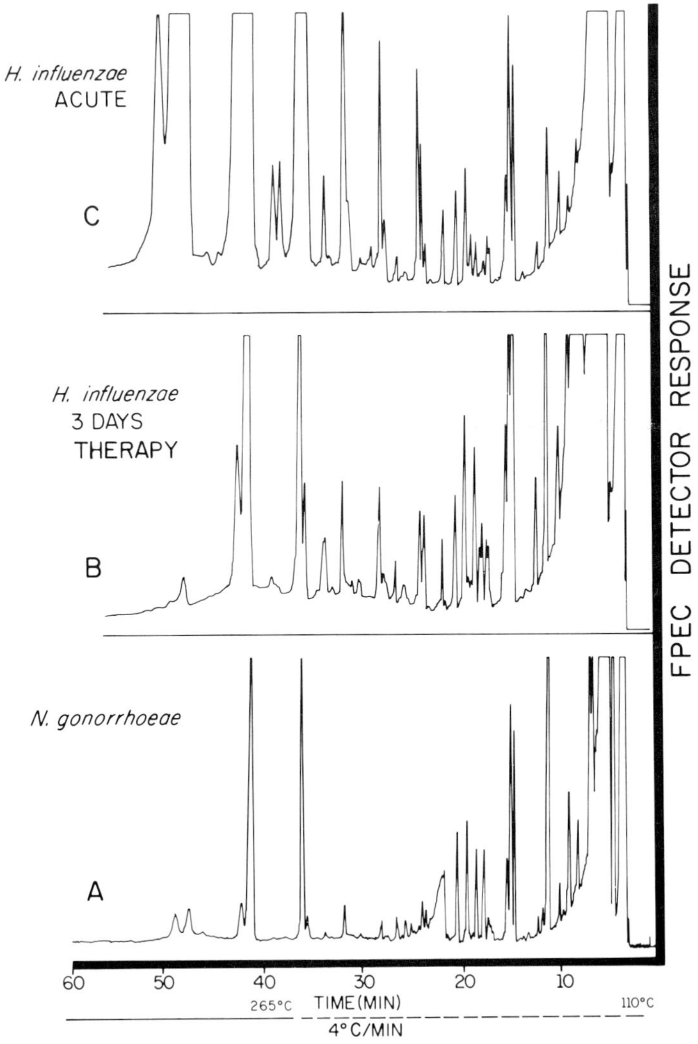

Fig. 5. FPEC–GLC chromatograms of TCE–HFBA derivatives prepared from acidic chloroform extracts of synovial fluids. The type of synovitis is indicated in the figure.

Fig. 4. FPEC–GLC chromatograms of TCE–HFBA derivatives prepared from acidic chloroform extracts of CSF taken from patients with disease, as indicated. For an explanation of the notation see Fig. 2. From Brooks *et al.* (1980c) with permission (American Society for Microbiology).

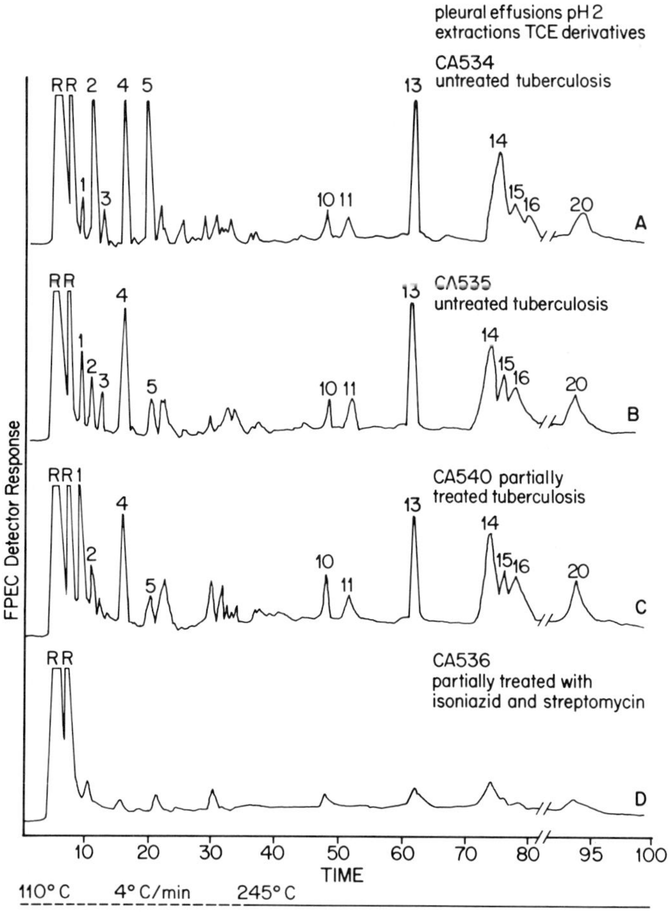

Fig. 6. FPEC–GLC chromatograms of TCE–HFBA derivatives prepared from acidic chloroform extracts of pleural fluids. The pleural effusions were caused by *Mycobacterium tuberculosis*. From Brooks *et al.* (1978) with permission (American Society for Microbiology). All rights reserved.

pleural effusions is now a distinct possibility for hospital and clinical laboratories.

D. Rocky Mountain Spotted Fever and Viral Diseases

So far in this chapter there has been very little discussion about FPEC–GLC analysis of sera. Brooks *et al.* (1981b) used FPEC–GLC to study well-documented cases of Rocky Mountain spotted fever (RMSF). The study included controls of other organisms that produce rashes, namely, chickenpox, measles, enteroviruses, and *Neisseria meningitidis*. In the use of FPEC–GLC fingerprints, comparisons should be kept as restrictive as possible. For example, in the diagnosis of tuberculous meningitis, we restricted our interpretations to cases that had no growth of the major causative agents of bacterial meningitis after overnight incubation and that presented a clinical picture of lymphocytic meningitis. In the case of RMSF we restricted our consideration to rash-producting disease agents that present the physician with a diagnostic dilemma.

We found that samples taken early in the course of the disease (1–3 days) were best for differentiation by FPEC–GLC. The complexity of the chicken pox profile was similar to those obtained for RMSF (Fig. 8). The *N. meningitidis* and

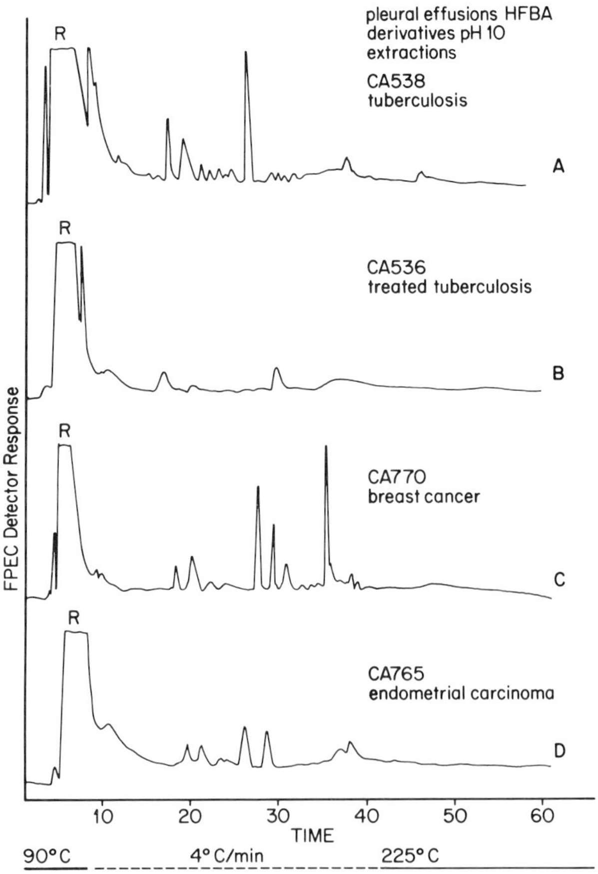

Fig. 7. FPEC–GLC chromatograms of HFBA derivatives prepared from basic chloroform extracts of pleural fluids. The types of pleural effusions are indicated. From Brooks *et al.* (1978) with permission (American Society for Microbiology). All rights reserved.

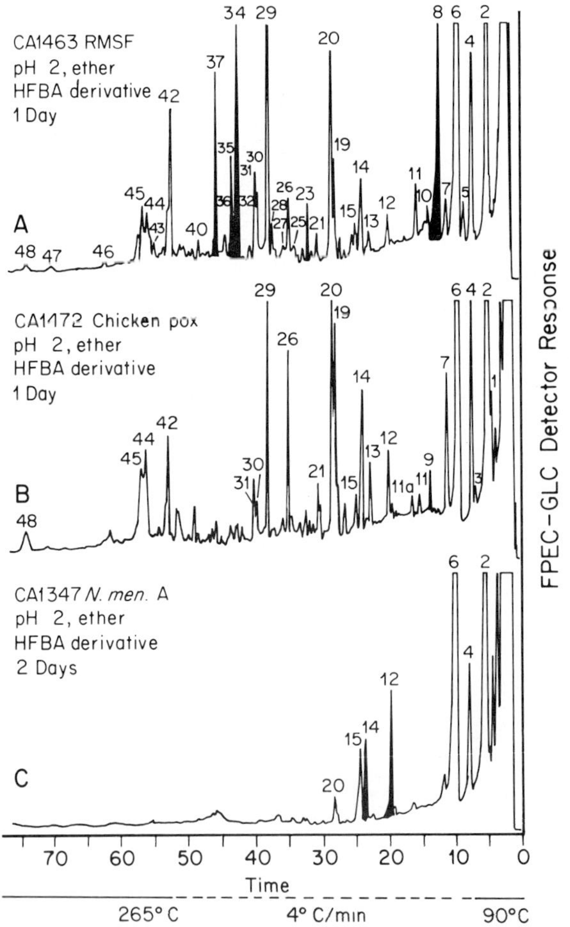

Fig. 8. FPEC–GLC chromatograms of HFBA–EtOH derivatives prepared from acidic diethyl ether extracts (third extraction) of sera. The type of disease is indicated in the figure. Peaks have been blackened to indicate importance in differentiation. From Brooks *et al.* (1981b) with permission (American Society for Microbiology).

measles profiles were relatively simple (Fig. 8 and 9). The enterovirus cases were different among different virus types (Fig. 10), and the FPEC–GLC profiles were intermediate in complexity. The FPEC–GLC profiles in RMSF were different from those observed in chicken pox, *N. meningitidis* infections, measles, and the enterovirus cases studied as illustrated in the figures. The fact that the cases of viral diseases studied gave different FPEC–GLC fingerprints for different types of enteroviruses indicates that FPEC–GLC potentially could be useful for the differentiation of viral diseases.

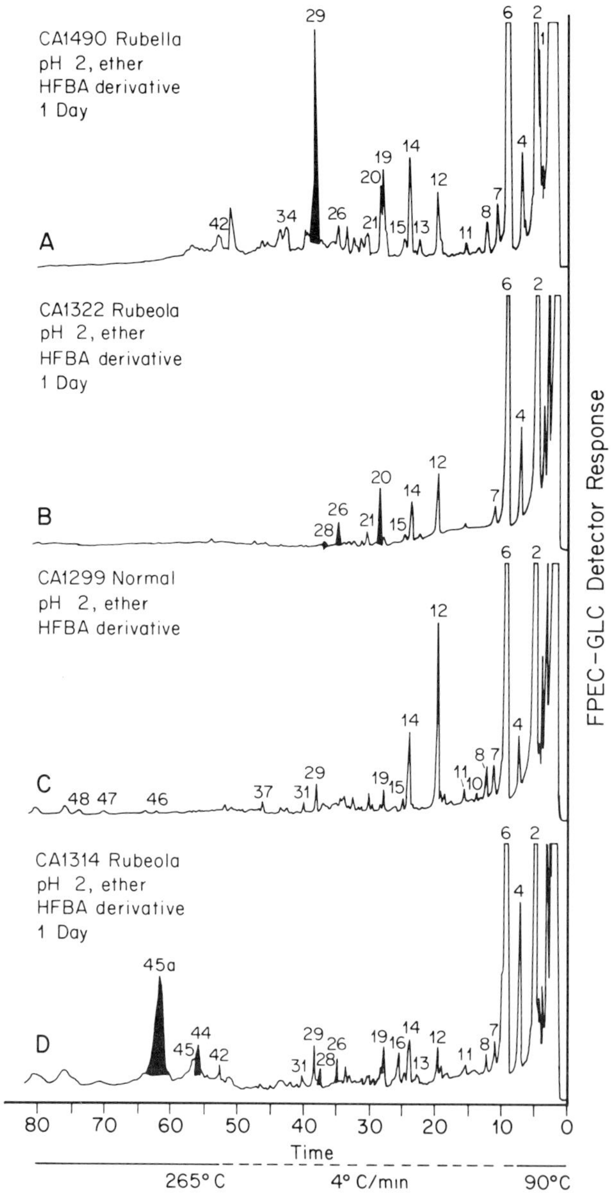

Fig. 9. FPEC–GLC chromatograms of HFBA–EtOH derivatives prepared from acidic diethyl ether extracts (third extraction) of sera. With the exception of the initial serum, the sera were obtained from patients with disease as indicated. From Brooks *et al.* (1981b) with permission (American Society for Microbiology).

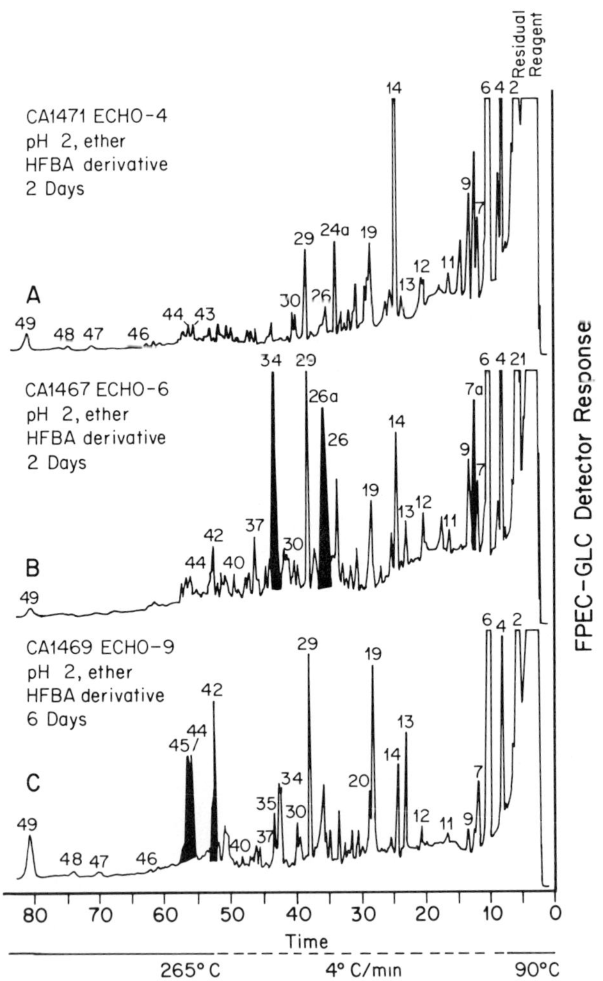

Fig. 10. FPEC–GLC chromatograms of HFBA–EtOH derivatives prepared from acidic diethyl ether extracts (third extraction) of sera. The type of disease is indicated. From Brooks *et al.* (1981b) with permission (American Society for Microbiology). All rights reserved.

VII. IDENTIFICATION OF UNKNOWN PEAKS DETECTED BY FPEC–GLC

Standard mixtures of acids, alcohols, and amines can be prepared, and these standards can be used to identify many of the compounds detected during FPEC–GLC analysis. A typical chromatogram consists of peaks that are identifiable by standard mixtures and those that are unidentified. The unidentified compounds are useful for disease identification purposes because they do form a fingerprint, but an unidentified compound becomes more valuable after identification because it then can be studied for its toxicity and possible physiologic effects on the host. The identification of every unknown compound detected in the analysis is not practical. GLC–Mass spectrometry is the best approach to identification, but a much larger sample is required, as are high-resolution columns for separation.

Much can be learned from the preparatory work on the unidentified compound before FPEC–GLC analysis. The functional group can usually be determined by the extraction and derivatization characteristics of the unknown compound. For example, we knew that 3-(2′-ketohexyl)indoline was an amine by its extraction and derivatization characteristics before we finally determined its structure (Brooks *et al.*, 1977). Unsaturation can be determined by simple techniques of hydrogenation and further analysis by FPEC–GLC (Alley *et al.*, 1979). A good idea of the approximate boiling point of the unknown, and therefore an educated guess of the chain length, can be made by comparing its retention time on OV-101 columns against the retention time of known derivatized compounds. The prime unknown peaks to be selected for identification are those found to be associated with a specific disease or those found to be important in the identification of a particular disease or microorganism.

VIII. INTERPRETATION OF DATA OBTAINED BY FPEC–GLC

As noted earlier one should make the interpretation as restrictive as possible and take advantage of the data already obtained through clinical observations and studies. For example, in the case of RMSF the clinical diagnosis can be confused only with a limited number of rash-causing diseases, and our interpretation was therefore confined to such diseases. In the case of tuberculous meningitis we confined our consideration to lymphocytic-type meningitides that failed to produce overnight growth of a potential etiologic agent. The next step is to look at the chromatogram in its entirety. We took note of the internal standard to ascertain whether it gave a good derivative with the expected FPEC–GLC response. Leaks in the FPEC–GLC system can often be detected by the observation of a change in the elution time of the residual chloroform peak. The size of the residual chloroform peak will also indicate the effectiveness of the evaporation technique for removing chloroform and TCE. After it has been determined that the derivatization process was performed satisfactorily and that the FPEC–GLC analysis was correct, the chromatogram is examined for specifics. Is the chromatogram very complex (as exemplified in Fig. 8A), or are there very few peaks (Fig. 8C)? If one were attempting to determine RMSF or chicken pox and one had a chromatogram similar to Fig. 8C, one could immediately dismiss the possibility of either chicken pox or RMSF on the basis of few peaks. One must remember that multiple diseases and therapy itself can produce conditions that present a confusing chromatogram and that can cause misdiagnosis. If the FPEC–GLC profile of the disease of interest is complex, as is the case in RMSF, and the profile detected by FPEC–GLC is also complex, as in the case of RMSF and chicken pox, then the profiles must be compared in depth. One must look for large quantitative, qualitative, and peak ratio differences, as illustrated in the blackened peaks in Fig. 8A. Qualitative differences are the easiest to detect. It must be kept in mind that one is dealing with living systems and that some differences would be encountered because of differences in the disease state or slight differences in body response to disease among different people. The degree of variation in a disease profile that can be tolerated and still permit the investigator to identify the disease must be determined through research.

IX. POSSIBILITIES FOR AUTOMATION AND COMPUTERIZATION OF DATA OBTAINED FROM FPEC–GLC

A properly equipped gas chromatograph and a person skilled in the interpretation of FPEC–GLC chromatograms, adequate control specimens, and good clinical specimens with histories are essential for FPEC–GLC in research or for diagnostic work. The FPEC–GLC data published to date were obtained without the aid of a computer; however, the complexities of data generated over a period of time suggest that computer analysis could be very beneficial in this area. If automation of the FPEC–GLC technique occurs, computers will be essential to keep abreast of data interpretation. Furthermore, because no two GLC systems produce exactly the same data, standardization of instruments and laboratory procedures is desirable, and computer technology is one means of accomplishing this.

Dependable automatic injectors that permit nonstop utilization of the gas chromatograph without attendants are now available. Current computerized instrumentation holds great possibilities for use in the preparation of derivatives. It is now easy to equip a gas chromatograph with dual columns and dual detectors, which permit two analyses to be made on one programmed run. The recorder can be set up so that it will turn itself off. This operation permits two extra analyses to be made at the end of each day. We have this type of system in operation in our laboratory. It would be possible to design a gas chromatograph that would permit several analyses to be made at once. Presumably, if the demand existed, industry would respond. The use of high-resolution capillary columns can reduce the analysis time by 50%; however, the use of capillary columns for this work has not yet been thoroughly investigated.

We have found that the standardization of data among instruments can be accomplished satisfactorily by the use of standard mixtures and programs. The process involves the analysis of straight-chain carboxylic acids and amines in a range of 1 to 22 carbon atoms. The data are then transmitted from a small computer to a large computer, where spline-fit equations are used to convert the straight-chain carbon retention numbers to straight-chain carbon equivalent numbers (J. B. Brooks and M. Krichevsky, unpublished data). The data can then be manipulated by addition, subtraction, and averaging and used in the final part of our computer development scheme. The current plan is to use standardized data and have the computer create probability profiles of a particular disease and then to compare future analyses with these profiles for best-fit comparison. The software developed for this project could be adapted later for use by computers in the $25,000 range, or a time-sharing mechanism could be set up in a laboratory equipped with a large computer.

X. SUMMARY

A need exists today for the development of new rapid diagnostic procedures in problem areas. The most difficult diagnostic areas are those involving organisms that are slow growing or difficult to cultivate, such as *Mycobacterium tuberculosis,* viruses, fungi, and parasites as well as diseases for which no etiologic agent has been established. The detection of chemical changes in body fluids by the use of highly sensitive selective procedures holds promise for diagnosis. FPEC–GLC in combination with specific electron-capturing derivatives and high-resolution columns has potential as a rapid diagnostic aid.

This chapter describes the development and use of an FPEC–GLC procedure that has been applied to certain diagnostic problems. The areas of study include organisms and conditions that cause lymphocytic meningitis, arthritis, pleural effusions, RMSF, and associated rash-causing diseases. Most of the FPEC–GLC applications have been limited in the number of cases studied, and further studies are needed to determine their true diagnostic potential and defined diagnostic sensitivity and specificity. However, the use of FPEC–GLC in diagnosing of tuberculous meningitis has been extensive, and the data indicate that FPEC–GLC could be a valuable tool for the rapid diagnosis of this disease. FPEC–GLC techniques are amenable to automation and computerization, and it appears that with sufficient consumer demand an FPEC–GLC system would be feasible.

REFERENCES

Alley, C. C., Brooks, J. B., and Choudhary, C. (1976). *Anal. Chem.* **48,** 387–390.
Alley, C. C., Brooks, J. B., and Kellogg, D. S. (1979). *J. Clin. Microbiol.* **9,** 97–102.
Brooks, J. B., Alley, C. C., and Liddle, J. A. (1974a). *Anal. Chem.* **13,** 1930–1934.

Brooks, J. B., Kellogg, D. S., Alley, C. C., Short, H. B., Handsfield, H. H., and Huff, B. (1974b). *J. Infect. Dis.* **129,** 660–668.

Brooks, J. B., Choudhary, G., Craven, R. B., Alley, C. C., Liddle, J. A., Edman, D. C., and Converse, J. D. (1977). *J. Clin. Microbiol.* **5,** 635–628.

Brooks, J. B. Craven, R. B. Schlossberg, D., Alley, C. C., and Pitts, F. M. (1978). *J. Clin. Microbiol.* **8,** 203–208.

Brooks, J. B., Kellogg, D. S., Shepherd, M. E., and Alley, C. C. (1980a). *J. Clin. Microbiol.* **11,** 45–51.

Brooks, J. B., Kellogg, D. S., Shepherd, M. E., and Alley, C. C. (1980b). *J. Clin. Microbiol.* **11,** 52–58.

Brooks, J. B., Edman, D. C., Alley, C. C., Craven, R. B., and Girgis, N. I. (1980c). *J. Clin. Microbiol.* **12,** 208–215.

Brooks, J. B., Kellogg, D. S., Shepherd, M. E., and Craven, D. E. (1981a). *J. Clin. Microbiol.* **13,** 836–842.

Brooks, J. B., McDade, J. E., and Alley, C. C. (1981b). *J. Clin. Microbiol.* **14,** 165–172.

Catlin, B. W. (1974). *In* "Manual of Clinical Microbiology" 2nd ed., pp. 116–123. Am. Soc. Microbiol., Washington, D.C.

Craven, R. B., Brooks, J. B., Edman, D. C., Converse, J. D., Greenlee, J., Schlossberg, D., Furlow, L., Guraltney, J. M., and Miner, W. F. (1977). *J. Clin. Microbiol.* **6,** 27–32.

Edman, D. C., Craven, R. B., and Brooks, J. B. (1981). *CRC Crit. Rev. Clin. Lab. Sci.* **14,** 133–161.

Schlossberg, D., Brooks, J. B., and Schulman, J. A. (1976). *J. Clin. Microbiol.* **3,** 239–245.

Maggs, R. J., Joynes, P. L., Davies, A. J., and Lovelock, J. E. (1971). *Anal. Chem.* **43,** 1966–1971.

25

Challenges in the Development of Automation for the Clinical Microbiology Laboratory

MARY JANE FERRARO

Francis Blake Bacteriology Laboratories
Massachusetts General Hospital
Harvard Medical School
Boston, Massachusetts

During the past several decades the volume of clinical laboratory tests performed in both hospital and private laboratories has skyrocketed; between the years 1970 and 1977 alone, the volume more than doubled. This increase in the number of laboratory tests being performed has not been confined to clinical chemistry and hematology assays but has affected the work load in the clinical microbiology laboratory as well (Finkelstein, 1980). During the same interval, both clinical chemistry and hematology have undergone radical changes with regard to instrumentation and degree of automation (Table I). Such technology has improved turnaround time and made it possible for laboratories to cope with an enormous increase in volume without a commensurate escalation of personnel costs. The result, in general, has been to keep the unit cost of automated tests relatively low. Although a few economic and health care analysts have suggested that the improvement in laboratory automation may itself have been at least partially responsible for the increased demand for tests (Fineberg, 1979; Moloney and Rogers, 1979), most analysts have felt that the increased volume of certain well-established tests cannot be attributed to automation per se; rather, behavioral factors have probably changed independently of technological change (Finkelstein, 1980; Martin *et al.*, 1980).

In contrast to a high level of technological development in clinical chemistry and hematology, automation in clinical microbiology has had a rather bleak history. Although many instruments have reached a developmental phase (John-

The Direct Detection of Microorganisms in Clinical Samples

ISBN 0-12-187780-9

TABLE I

Volume Change in Major Laboratory Subdivisions with Changing Penetration of "Automated" Technology[a]

Subdivision	% of hospitals using "automated" technology 1970	1977	% increase in test volume, 1970–1977
Bacteriology	0	0	150
Chemistry	65	85	108
Hematology	78	98	110
All reported tests			105

[a] Reprinted by permission from S. N. Finkelstein, *Medical Care* **18,** 1048 (1980).

ston and Newson, 1976; Tilton, 1982), few have prevailed in the marketplace. Several of the most successful of these are described in Chapters 26 and 27 of this volume. To date, however, a large proportion of the successful instrumentation has been geared toward analyses performed on pure cultures, that is, susceptibility testing and/or automated biochemical identification of bacteria that have been isolated in pure culture. Such instrumentation includes the AutoMicrobic System (AMS) of Vitek Systems, McDonnell-Douglas, Inc., St. Louis, Missouri; the MS-2 of Abbott Diagnostics Division, Irving, Texas (the first two being the most automated of the systems available); and the Autobac of General Diagnostics, Morris Plains, New Jersey. In addition, there are at least two automated scanners for reading susceptibility and/or identification test results performed either in microdilution trays (the Autoscan, Microscan, Inc., Hillsdale, New Jersey) or in an unique rotor-cuvette (API 3600S, Analytab Products, Plainview, New York). Although many of these innovations have reduced the turnaround time from the traditional 18–24 hr to 4–8 hr for susceptibility and identification results obtainable from pure culture, the same level of "rapidity" can be achieved with manual kits (such as API 20E, Minitek, and Micro-ID) and by the modification of classical methods for susceptibility testing (Lorian, 1977).

Instrumentation geared toward the detection and identification of microorganisms from primary clinical specimens is highly desirable but is still in its infancy. Among the instruments available in the marketplace for the detection and screening of positive specimens are the following: (*a*) for positive urine screen—the Autobac and MS-2 systems, both of which rely on certain principles of densitometry for the detection of microbial growth; (*b*) for positive blood screen—BACTEC (Johnston Laboratories, Inc. Cockeysville, Maryland), which detects microbial metabolism through radiometric monitoring for $^{14}CO_2$ generated by the metabolism of ^{14}C-labeled substrates. The detection stage alone, which is dependent on growth to a concentration approximating 10^5 colony-forming units (cfu) per milliliter requires at least 3–6 hr for prediluted urine specimens and about 1–2 days for the majority of blood cultures. The methods indicate only bacterial presence; they do not distinguish mixed from pure growth or identify the organisms present. After the detection of a presumptively positive specimen using these instruments, the clinical microbiologist must still proceed with the standard methods for identification and susceptibility testing.

An innovative and highly controversial approach to automated microbiological analyses was attempted by National Aeronautic and Space Administration (NASA) scientists and engineers in their development of an automated system for use in the space program. They needed a method for microbial monitoring of other planets, the spacecraft environment, and the astronauts themselves (Gibson, 1982). The system, the prototype for Vitek System's AMS, was designed to provide a method for direct and simultaneous detection, enumeration, identification, and antimicrobial susceptibility testing, all without traditional isolation procedures. This attempt in the late 1960s and early 1970s to identify microorganism groups within a mixed-culture environment was viewed as heresy by the majority of clinical microbiologists. The approach taken involved the use of a combination of selective inhibitors, unique carbohydrate sources, optical enhancement for improved detection of biomass, and miniaturization of optical visualization techniques. The first clinical application of this technique was for urine specimens. The detection, enumeration, and identification steps for most urinary tract pathogens could be accomplished within 4 to 13 hr (Sonnenwirth, 1977). Although far from a perfect instrument, the AMS does an adequate job of detecting and identifying a restricted number of organisms of limited genera and species in primary urine specimens and has been used in several major clinical microbiology laboratories (Isenberg *et al.*, 1979; Smith *et al.*, 1978). With this single exception, there have been no developments in instrumentation since the early 1970s specifically designed to process primary clinical specimens, other than for the simple detection of microorganisms. The reasons for continued encouragement of the development of automated instruments for clinical microbiology are that they may allow for more cost-effective performance of laboratory tests that are either high-volume or require a large number of work units or expensive reagents that can be reduced by the use of automated equipment. Also, they can potentially produce a desired result more rapidly, thereby reducing turnaround time and perhaps making a clinically relevant result available to the physician earlier. Automated tests could allow for better standardization by eliminating subjectivity and/or human error. Finally, automation potentially could be useful in screening or assessing the necessity of processing specimens further by standard methods (Benfari and Kunz, 1979). Why then, although obviously offering the same advantages to microbiologists as to other clinical laboratory specialties, has automation failed to evolve in clinical microbiology? Obviously, the nature of the specimen and the specificity of the constituents sought by clinical chemists and hematologists differ greatly from those with which clinical microbiologists are presented. More often than not, specimens for culture arrive in a variety of physical states ranging from fluid to solid to swab-embedded. They are obtained from body sites that are not normally sterile and are delivered in various collection devices or transport media. These specimens may then contain a variety of flora components, some of which are commensals and some potential pathogens. Usually, they are a mixture of both. The situation is compounded by the fact that a specific microorganism might be considered pathogenic in one instance and avirulent in another, often being dictated by the anatomic site or the underlying illness of the patient from which the specimen was obtained. Pathogens, when present, may exist in such small numbers that they defy many of the available methods for rapid detection or may be significant only when present at a certain numerical threshold. Classical biochemical methods have always required large populations rather than single organisms, and therefore the ultimate identification is usually dependent on growth, a process that takes time. Thus, success in the technological development of clinical microbiology may have been hampered by the attempt to merge classical methods for detecting growth and metabolism with instrumentation that merely mimics and often advances by only a few hours the human observation of such processes. Departures from the use of traditional identification parameters by companies such as Autobac in their elaborate search for compounds with differentiating antibacterial activity (Matsen *et al.*, 1982) or Analytab's enzyme profiles for the identification of certain organisms (Waitkins *et al.*, 1980; D'Amato *et al.*, 1978) have met with skepticism. Certainly, the preformed enzyme profile approach to bacterial identification holds promise, because most bacteria can be detected rapidly and the technology for their automated detection could most likely be adapted from discrete analyzers being developed for clinical chemistry laboratories. Successful, rapid detection, however, sometimes depends on a significant biomass, which may entail growth or subculture. As with biochemical methods, the identification of bacteria is dependent on profiles from multiple substrates; application to primary clinical specimens would be limited to those in which multiple organisms are rarely found. Success in analyzing the primary specimen for more than mere detection or screening for microorganisms may have to await the perfection of several newly emerging technologies. The possible marriage of specific monoclonal antibodies or DNA probes with instrumentation containing more advanced detectors and dedicated microprocessors may allow for direct diagnosis from clinical specimens without intervening growth requirements and/or reliance on traditional biochemically based identification. An excellent example of the potential of such an approach is the use of a DNA-hybridization assay for genes encoding enterotoxins in enterotoxigenic *Es-*

cherichia coli in stool samples (Moseley *et al.*, 1980). Unlike the indirect, classical approach of correlating certain serotypes with toxin production or the current method of detecting the enterotoxins with immunologic, cytotoxic, or bioassay techniques, this novel approach entails the detection of genes that encode for enterotoxin. Labeling of such gene probes with radioactive, chemical, or enzymatic material would potentially allow for automated detection. Although still in its infancy, the field of specific DNA probes for various virulence factors or molecular bases of pathogenicity such as toxins may become an important area in automated microbiology. Available information suggests that the method is potentially very sensitive, detecting as little as 10 toxin-producing colony-forming units per milliliter in pure culture or 1 toxin-producing colony-forming unit per 10^3 or 10^4 non-toxin-producing organisms (Moseley *et al.*, 1982). The apparent detection of specific genes without the interference of DNA from other bacteria or cells makes this approach particularly suitable for use with primary clinical specimens. Less futuristic and closer to fruition is the use of monoclonal antibodies for the detection of specific antigens in clinical specimens using a discrete or multichannel autoanalyzer.

In the meantime commercial ventures might be best concentrated on simple, sensitive, and truly "rapid" low-cost methods for sorting positive from negative specimens (i.e., screening). A very significant portion of the specimens that reach the clinical microbiology laboratory are indeed "negative" in that they either contain no microorganisms (e.g., blood or other sterile body fluids) or the numbers and/or types of organisms present are not clinically significant (Table II). In blood, cerebrospinal fluid (CSF), or other normally sterile body fluids, rapidly distinguishing between positive and negative specimens may be extremely important clinically. In others, such as urines from patients with sus-

TABLE II

Massachusetts General Hospital Distribution of Specimen Types and Approximate Number of "Negatives"

Specimen type	Approximate % of total specimens	Approximate % of "negative" or no growth
Urinary	34	70
Blood	20	90
Genital[a]	11	75
Throat[b]	6	80
Sterile fluids	4	50
Stool[c]	2	90

[a] Examined for *Neisseria gonorrhoeae* only.

[b] Examined for beta streptococci only.

[c] Examined for *Salmonella, Shigella,* and *Campylobacter*.

pected urinary tract infection, the elimination of a large portion of the work load at an early point in the processing could be cost-effective for the laboratory. In order for the screening technique to be acceptable, the instrument used must be at least as sensitive as standard methods (which are often far from perfect) and results must be available rapidly enough (e.g., less than 1 hr) to allow for the discard of negative specimens and further processing of only those that are presumptively positive. Specificity should be such that the additional costs of processing an inevitable false positive group do not outweigh the value of screening costs.

Examples of two such systems that have been designed specifically for the rapid screening of urine specimens are currently being evaluated. The first, the Lumac System (3-M Medical Products Division, St. Paul, Minnesota), makes use of bioluminescence, an assay method that has been recognized since the late 1960s (Chappelle and Levin, 1968). The principle is one of detecting changes in light produced by microbial ATP present in viable cells. The current system supposedly offers improved sensitivity by introducing new purified reagents, a more precise instrument, and a process that eliminates interference from nonmicrobial cells. A screening result is available within 30 min. Although no published data on its clinical performance are available, marketing information from the company indicates that the sensitivity of the system is 96% at 10^4 cfu/ml and greater than 99% at 10^5 cfu/ml. The second system, the Bacteria Detection Device (BDD; Marion Laboratories, Inc., Kansas City, Missouri), is an instrument designed to screen urine specimens within a few minutes to identify those that require processing (Wallis *et al.*, 1981). The screen involves solubilizing 1 ml of urine in a patented solvent, filtration through a special filter, and subsequent entrapment of bacteria, when present, which can be stained with safranine dye. After decolorization, the stained filter's color intensity is compared visually with known standards; the intensity of staining (from pink to red) approximates the number of colony-forming units per milliliter in the specimen. Although filter reading for this system is manual and subjective, a second-generation instrument is planned with an automated, colorimetric reading component. Preliminary data on 630 urine specimens obtained in our laboratory with this device are presented in Table III. Urine specimens giving a BDD filter result of ±, 1+, 2+, 3+, or 4+ were considered positive. Certain filters either became clogged before processing was complete or showed an uninterpretable, pigmented color pattern after staining. Although still in its developmental phase, the BDD performed relatively reliably. As with any screening method, the primary concern is with false negative results. An analysis of the 43 false negative filters (as compared with a stringent criterion of a positive culture result by the loop method of $\geq$ to 10^4 cfu/ml) indicated that 25 of the 43 were from urine specimens that contained "mixed" bacterial species consisting of two or more organisms normally thought not to be potential pathogens (e.g., lactobacilli,

TABLE III

Use of Bacteria Detection Device for Rapid Urine Screening

Total amount of bacterial growth on MGH culture by loop method[a]	BDD filter results					
		Positive				
	Negative	±	1–4+	Clogged	Pigmented	Total
No growth[b]	243	34	16	5	16	314
Rare[c]	33	8	5	1	1	48
Few[d]	32	17	11	1	4	65
Moderate or abundant[e]	43[f]	32	103	17	8	203
Total	351	91	135	24	29	630

[a] MGH, Massachusetts General Hospital.
[b] Certain "no growth" urine cultures were positive by quantitative pour plate count; see text.
[c] Total growth approximates $<10^3$ cfu/ml.
[d] Total growth approximates 10^3–10^4 cfu/ml.
[e] Total growth approximates $>10^4$ cfu/ml.
[f] 25/43 contained "mixed" bacterial species; see text.

diphtheroids, *Staphylococcus epidermidis,* non-beta streptococci). Approximately 75% of these had total growth correlating with between 10^4 and 10^5 cfu/ml. Of the remaining 18 false negative filters, 9 had growth correlating with 10^4–10^5 cfu/ml and 9 with $\geq 10^5$ cfu/ml. In addition, in a subsection of this study approximately 200 urines that gave any degree of positivity, became clogged, or caused pigmentation of the BDD filters (regardless of culture results) were tested for quantitative colony count by standard pour plate methods. Of 60 filter-positive, loop-culture-negative urines tested in this fashion, 14 (23%) revealed growth by quantitative colony count, 9 at $\geq 10^4$ cfu/ml. Although false negative filters were from urines that tended to show evidence of contamination and to have counts below 10^5 cfu/ml (a generally accepted threshold of positivity), the error rate may still be too high in certain medical settings to allow for the comfortable discard of BDD filter-negative urine specimens without further confirmation. This is also a current drawback of the more expensive automated systems with a longer turnaround time. If a slight refinement in sensitivity is possible before this instrument is marketed, it promises to answer the need for an inexpensive, rapid, specific screening method for a large percentage of the clinical microbiology laboratory's work load.

If progress in this area is to continue, it is imperative that generous collaboration between clinical microbiologists and their colleagues in industry be maintained. It is the responsibility of each group to stimulate, support, evaluate, and challenge the other at appropriate stages in the research and development for the benefit of all.

REFERENCES

Benfari, M. J. F., and Kunz, L. J. (1979). *In* "Seminars in Infectious Disease" (L. Weinstein and B. N. Fields, eds.), Vol. II, pp. 29–47. Stratton Intercontinental Medical Book Corporation, New York.

Chappelle, E. W., and Levin, G. V. (1968). *Biochem. Med.* **2,** 41–51.

D'Amato, R. F., Eriquez, L. A., Tomfohrde, K. M., and Singerman, E. (1978). *J. Clin. Microbiol.* **7,** 77–81.

Fineberg, H. V. (1979). *In* "Medical Technology: The Culprit behind Medical Care Costs" (S. H. Altman and R. Blendon, eds.), pp. 144–165. DHEW Publ. No. (PHS)79-3216, Washington, D.C.

Finkelstein, S. N. (1980). *Med. Care* **18,** 1048–1056.

Gibson, S. (1982). *In* "Rapid Methods and Automation in Microbiology" (R. C. Tilton, ed.), pp. 80–89. Am. Soc. Microbiol., Washington, D.C.

Johnston, H. H., and Newsom, S. W. B. (1976). "Rapid Methods and Automation in Microbiology." Learned Information, New York.

Isenberg, H. D., Gavin, T. L., Sonnenwirth, A., Taylor, W. I., and Washington, J. A. (1979). *J. Clin. Microbiol.* **10,** 226–230.

Lorian, V. (1977). *In* "Significance of Medical Microbiology in the Care of Patients" (V. Lorian, ed.), pp. 203–212. Williams & Wilkins, Baltimore, Maryland.

Martin, A. R., Wolf, M. A., Thibodeau, L. A., Dzau, V., and Braunwald, E. (1980). *N. Engl. J. Med.* **303,** 1330–1336.

Matsen, J. M., Sielaff, B. H., and Buck, G. E. (1982). *In* "Rapid Methods and Automation in Microbiology" (R. C. Tilton, ed.), pp. 286–289. Am. Soc. Microbiol., Washington, D.C.

Moloney, T. W., and Rogers, D. E. (1979). *N. Engl. J. Med.* **301,** 1413–1419.

Moseley, S. L., Huq, I., Alim, A. R. M. A., So, M., Samadpour-Motalebi, M., and Falkow, S. (1980). *J. Infect. Dis.* **142,** 892–898.

Moseley, S. L., Echeverria, P., Seriwatana, J., Tirapat, C., Chaicumpa, W., Sakuldaipeara, T., and Falkow, S. (1982). *J. Infect. Dis.* **145,** 863–869.

Smith, P. B., Gavan, T. L., Isenberg, H. D., Sonnenwirth, A., Taylor, W. I., Washington, J. A., and Balows, A. (1978). *J. Clin. Microbiol.* **8,** 657–666.

Sonnenwirth, A. C. (1977). *J. Clin. Microbiol.* **6,** 400–405.

Tilton, R. C. (1982). "Rapid Methods and Automation in Microbiology." Am. Soc. Microbiol., Washington, D.C.

Waitkins, S. A., Ball, L. C., and Fraser, A. M. (1980). *J. Clin. Pathol.* **33,** 53–57.

Wallis, C., Melnick, J. L., and Longoria, C. J. (1981). *J. Clin. Microbiol.* **14,** 342–346.

26

Instrumentation for the Detection of Bacteremia

PATRICK R. MURRAY

Division of Laboratory Medicine
Department of Pathology
Washington University School of Medicine
St. Louis, Missouri

I. Introduction 343
II. Conventional Methods 344
III. Alternative Modifications of Blood Cultures 345
A. Inactivation of Inhibitory Factors 345
B. Removal of Inhibitory Factors 346
C. Removal of Microbes from Blood 346
D. Standardization of Detection Methods 348
IV. Summary 350
References 350

I. INTRODUCTION

Bacteremia represents a potentially life-threatening infection in which the patient's defense mechanisms are overwhelmed by microbes that invade the bloodstream. Although immediate detection and identification of the etiologic agent are desirable, the achievement of these goals is essentially precluded by the limitations of conventional technology and the nature of bacteremia. Bacterial pathogens are normally recovered in blood in small numbers (Salventi *et al.*, 1979; Tenney *et al.*, 1982) and in the presence of a variety of inhibitory factors such as phagocytic cells, antibodies, and antibiotics. Thus, a large volume of blood must be cultured and the inhibitory factors neutralized. These cultures are performed most commonly by inoculating blood into broths supplemented with a neutralizing factor such as sodium polyanetholsulfonate (SPS), and then the broths are examined periodically for evidence of microbial growth. Attempts to

The Direct Detection of Microorganisms in Clinical Samples

ISBN 0-12-187780-9

improve the conventional methods used for detecting bacteremic pathogens have included modifications of the initial processing of the blood specimen and manual or automated methods for detecting microbial growth. These efforts are discussed in the following sections.

II. CONVENTIONAL METHODS

The conventional method of processing blood cultures is to collect a large volume of blood (preferably 10–20 ml) and inoculate at least two bottles of nutrient broth. These bottles are sent to the laboratory, where one bottle is transiently vented and then both are incubated at 35 to 37°C. The bottles are periodically examined for macroscopic evidence of microbial growth (i.e., turbidity, hemolysis, production of gas, discrete colonies), and the visually negative broths are subcultured and stained. Positive broths are subcultured, and the isolated colonies are identified and tested for susceptibility to antimicrobial agents. Many positive cultures are detected within the first 48 hr of incubation (Table I), and definitive identification and susceptibility tests generally require an additional 24–48 hr. Minor modifications of conventional procedures, however, can reduce this time.

The initial detection of positive broths is improved by performing early blind subcultures and stains. The majority of all significant aerobic and facultative anaerobic bacteria are detected in blood cultures with 24 hr of initial incubation if a blind subculture is performed on the day the culture is collected (Todd and Roe,

TABLE I

Time of Detection of Positive Blood Cultures[a]

	Cumulative % positive			
Organism	Day 1	Day 2	Day 5	Day 7
Staphylococcus aureus	55	78	99	100
Staphylococcus epidermidis	19	26	90	96
Streptococcus	59	88	97	99
Enterobacteriaceae	80	94	96	100
Pseudomonas	54	72	98	100
Bacteroidaceae	7	70	90	100
Clostridium	57	100	100	100
Other anaerobes	8	36	50	96
Yeasts	22	29	89	97

[a] Data collected at Barnes Hospital during 1980.

1975; Sliva and Washington, 1980). This is particularly important for isolates that do not grow to a visible turbidity, such as *Haemophilus* and *Pseudomonas*. In contrast, anaerobic subcultures are unnecessary (Murray and Sondag, 1978). The value of routinely staining macroscopically negative blood cultures is controversial. Although one group reported that one-fourth of their positive cultures were initially detected by blind Gram's stains (Blazevic *et al.*, 1976), this has not been corroborated by other investigators. Approximately 10^5 organisms per milliliter of broth must be present for one organism to be seen in each microscopic field (1000 × magnification) of a Gram-stained smear. Thus, the Gram's stain is relatively insensitive. The sensitivity can be improved by the use of an acridine orange stain examined under low-power magnification for fluorescent bacteria (McCarthy and Senne, 1980). However, acridine orange staining is plagued by a high incidence of false positive smears due to nonviable contaminants in culture media and cross-contamination from other slides (Mirrett *et al.*, 1982).

III. ALTERNATIVE MODIFICATIONS OF BLOOD CULTURES

Four general approaches have been used to improve the processing of blood cultures: (*a*) inactivation of inhibitory factors in the blood specimen, (*b*) removal of inhibitory factors, (*c*) removal of microbes from the blood and inhibitory factors, and (*d*) standardization and automation of detection methods. The following sections summarize these approaches.

A. Inactivation of Inhibitory Factors

A potential problem with conventional blood culture procedures is that antimicrobial factors in the blood can delay or prevent bacterial growth. Originally, the deleterious effect of these factors was minimized by diluting blood in broth at a ratio of 1 : 30 to 1 : 50. However, this became less important with the incorporation of SPS in the culture broths, because SPS is antiphagocytic, is anticomplementary, prevents opsonization, and inactivates aminoglycosides and polypeptide antibiotics (Lowrance and Traub, 1969). The inactivation of aminoglycosides and polypeptide antibiotics depends on the polyanionic properties of SPS, which physically binds to the antibiotics and prevents their intracellular penetration into bacteria (Krogstad *et al.*, 1981).

Unfortunately, SPS has no effect on a variety of other antimicrobials including penicillins, cephalosporins, tetracycline, and sulfonamides. Penicillinase can be added to the blood culture broth to inactivate susceptible penicillin antibiotics. However, caution must be used to prevent pseudobacteremias due to contaminated solutions of penicillinase. Thiol blood culture broth inactivates all penicillin antibiotics (e.g., penicillin G, oxacillin, nafcillin, carbenicillin) by an incompletely defined reduction of the β-lactam ring (Murray and Niles, 1982). However, the rate of this reaction is probably too slow to be clinically useful. Sulfonamides are inactive in blood culture broths because thymidine is present in most commercially prepared broths.

B. Removal of Inhibitory Factors

In 1980 Wallis and co-workers reported the development and clinical evaluation of the Antimicrobial Removal Device (ARD; Marion Labs, Kansas City, Missouri), which purportedly removed antimicrobials from blood, thus decreasing the time required to recover microorganisms and increasing the overall incidence of positive cultures. The ARD system consists of a bottle that contains two resins: a polymeric adsorbent resin (Amberlite XAD-4) and a cation-exchange resin (C-249). The resins are suspended in saline supplemented with SPS, which prevents coagulation when the blood is introduced into the ARD bottle. A fine mesh filter is positioned in the neck of the bottle, which prevents the withdrawal of the resins when the blood is removed. Processing blood specimens in the ARD system is accomplished by inoculating a maximum of 10 ml of blood into the ARD bottle, slowly mixing the ARD contents and blood for 15 mins on a mechanical shaker, withdrawing the blood, inoculating it into a blood culture broth system, and then processing the specimen according to conventional culture methods. The ARD removes most commonly used antimicrobials from blood including penicillins, cephalosporins, aminoglycosides, chloramphenicol, tetracyclines, and vancomycin. Clinical evaluations of the ARD and related systems (i.e., BACTEC 16B broth) document a more rapid and increased frequency of detection of bacteremias, particularly with *Staphylococcus aureus* (Wallis *et al.*, 1980; Lindsey and Riely, 1981; Appleman *et al.*, 1982; Appelbaum *et al.*, 1983). However, these results were not corroborated by Wright and associates (1982), who found no increased or faster recovery of any group of organisms after blood was collected from patients receiving antimicrobials. In addition, they reported significantly lower recovery of gram-negative bacilli from blood cultured after processing in the ARD system. An explanation for these discrepancies is not readily apparent, although the lack of effectiveness of ARD in Wright's study may have been due in part to the large volume of blood that they processed. The value of the ARD for processing blood specimens will have to be defined by additional studies.

C. Removal of Microbes from Blood

The first system used for the removal of microbes from blood was the biphasic Castaneda bottle developed for the recovery of *Brucella.* The bottle contains nutrient broth and an agar slant on the side of the bottle. After blood is inoculated into the broth, the organisms are subcultured onto the agar slant by tipping the bottle at the time of macroscopic examination of the broth. Clinical evaluations have documented that significantly more fungi and bacteria are recovered in biphasic blood culture systems (Hall *et al.*, 1979; Pfaller *et al.*, 1982).

Another method for removing microorganisms from blood would be to concentrate the microbes and eliminate the need for broth cultures. Two methods have been developed for doing this: lysis–filtration and lysis–centrifugation.

In the lysis–filtration method (Sullivan *et al.*, 1975a; Zierdt *et al.*, 1977), blood is lysed with a solution of proteases (Rhozyme) and nonionic detergent (polyoxyethylene sorbitan monolaurate, Tween 20) and filtered through a 0.45-μm membrane filter; then the filter with trapped organisms is cultured on agar. Problems with the technique include selective loss of organisms due to the toxicity of the lysing solution, prolonged filtration times due to clogging of the filter with membrane fragments, and a high incidence of bacterial contamination during processing (Sullivan *et al.*, 1975b). Problems with toxicity of the lysing solution and slow filtration times have been overcome (Zierdt, 1982), but the procedure is still plagued by contamination problems. Because the overall rate of recovery of bacteria in blood is improved with this method (Sullivan *et al.*, 1975b; Zierdt *et al.*, 1982), however, there is an impetus for seeking improvements that will simplify the technical manipulations and decrease contamination.

The lysis–centrifugation method is similar in principle to the lysis–filtration method except that after blood from bacteremic patients is lysed, the bacteria are concentrated by centrifugation onto a stabilized density layer rather than by filtration onto a membrane. The organisms are then collected from the density layer and cultured on conventional agar media. This system, developed by Dorn and co-workers (1976a,b, 1978, 1979), consists of a single tube with a high-density hydrophobic cushion (Fluorinert, 3-M Corp.) and an aqueous solution of Dow Corning Anti-Foam B, SOLRYTH, SPS, and sodium thioglycollate. Blood samples are inoculated into the tube, mixed with the lysing–anticoagulant solution, and then centrifuged for 30 min at 3000 *g* in a fixed-angle centrifuge. The supernatant fluid is withdrawn with a syringe and discarded, and the density layer with sedimented bacteria is mixed, removed with a syringe, and cultured on agar media. Clinical evaluations have documented that significantly more organisms are recovered in this system (Isolator, DuPont) in comparison with conventional blood culture methods. Also, there is earlier recovery of organisms, which permits faster definitive identification and susceptibility tests. Problems encountered with the lysis–centrifugation method include a high incidence of contamination, the need for multiple entries into the tubes, the need for prolonged and high centrifugation speeds, and poor recovery of some organisms such as anaerobes and *Streptococcus pneumoniae*. However, these problems can most likely be avoided by modifications of the current system and additional technical experience.

D. Standardization of Detection Methods

The most popular semiautomated instrument developed for detecting bacterial growth in blood culture broths is the BACTEC (Johnston Laboratories, Inc., Cockeysville, Maryland), the radiometric detection system originally developed by DeLand and Wagner (1969). The principle of the system is that bacterial growth in broth can be monitored by measuring $^{14}CO_2$ released during bacterial metabolism of radioactively labeled nutrients. Agitation of the broth cultures increases bacterial replication and release of $^{14}CO_2$ into the air space above the broth. Thus, the BACTEC system is designed to mix the cultures and then to sample periodically the total volume of air in the sealed cultures. A significant increase in $^{14}CO_2$ above baseline levels is considered to represent bacterial growth. This must be confirmed by staining the broths, and the positive broths are then processed in accordance with conventional methods.

Most users of the BACTEC system report detecting 30–50% of their positive cultures within 1 day of incubation and 75–85% within 2 days (Brooks and Sodeman, 1974; Thiemke and Wicker, 1975). However, this is fairly similar to the experience reported by laboratories using a good conventional broth culture method (Table I). Therefore, the major advantage of the BACTEC system is standardization and automation of detecting microbial growth rather than an improvement in the speed of detection. This beneficial effect should not be minimized because the growth of bacteria in broth cultures can be subtle and overlooked by inexperienced technologists.

Technical problems with the radiometric system have included false positive and false negative cultures and cross-contamination of cultures (Griffin *et al.*, 1982). Caslow and co-workers (1974) reported that, of their positive readings in aerobic BACTEC bottles, 36% of the readings during the first 24 hr and 91% between 2 and 7 days were false positive. The false positive readings are most common with leukemic patients and neonates because of the hypermetabolic state of their white blood cells (Bannatyne and Harnett, 1974). These are particularly troublesome with the BACTEC system because the broths must be Gram-stained and microscopically examined, which is difficult because the bottles are continuously mixed during incubation to maximize release of $^{14}CO_2$ into the gas space above the broth (Deland and Wagner, 1969). In addition, the mixed blood cells and cellular fragments make examination of Gram's stains for bacteria, particularly gram-negative bacilli, difficult to interpret. The incidence of false positive results can be reduced by adjusting the reading index that defines a positive culture; however, this would adversely affect the sensitivity of the system.

Many of the false negative results with BACTEC have been corrected by improving the nutritional properties of the broths and supplementing the media with additional labeled substrates. However, *Haemophilus* and many obligate anaerobes still grow poorly in the available broths. It is anticipated that modifications of the broths will correct these problems. A more serious deficiency in the system is the limited volume of blood that can be cultured in each bottle, which in turn adversely affects the overall incidence of positive cultures. Although 3 ml of blood can be cultured in the BACTEC bottles, investigators from Johnston Laboratories (Salventi *et al.*, 1979) estimated that the percentage of positive cultures increases by 13% if the volume of blood is increased twofold. This is consistent with the findings of other investigators who recommended that 20 ml of blood should be cultured routinely (Tenney *et al.*, 1982; Hall *et al.*, 1976). Because cross-contamination of cultures occurred with overfilled BACTEC bottles, an unacceptably large number of bottles per culture would have to be inoculated for optimum recovery of bacteria.

In summary, BACTEC is an innovative system for detecting positive blood cultures by measuring the release of $^{14}CO_2$ during bacterial metabolism of radioactively labeled substrates. The system provides the user with automated examination of blood culture broths for bacterial growth and detects growth at least as rapidly as a good conventional system. The system can save the technologist time, although this will be influenced by the BACTEC system that is used and the number of false positive cultures that must be Gram-stained. After the positive culture is detected, however, the broths must be subcultured and the isolates identified and tested for antimicrobial susceptibility in the conventional manner. False negative results associated with the system are caused by some defective media and the small volume of blood that can be cultured. The system could be significantly improved if a larger volume of blood could be processed in a bottle, although this would most likely involve a costly modification of the instrument by the manufacturer.

Electrical detection systems have also been developed for blood cultures. Metabolically active bacteria can convert complex nutrients in broth cultures to electrically charged end products. Thus, an increase in conductivity of the medium or a corresponding decrease in impedance correlates with bacterial growth. One apparatus (Bactometer, Bactomatic, Inc., Princeton, New Jersey) was developed for the continual monitoring of impedance changes in blood culture broths. Experiments with artificially seeded blood cultures resulted in the detection of most significant blood culture pathogens in 6 to 18 hr (Specter *et al.*, 1977). However, additional studies have demonstrated that the sensitivity of impedance monitoring is between 10^6 and 10^7 organisms per milliliter, a sensitivity that is less than that reported for radiometry (e.g., 10^5–10^6 organisms per milliliter) and just below the level of visual turbidity (Cady *et al.*, 1978). Furthermore, variations in the composition of media affect changes in impedance and thus the sensitivity of the system. Finally, metabolically active nonbacterial cells (e.g., erythrocytes, leukocytes) affect impedance changes, a problem Kagan and associates (Kagan *et al.*, 1977) avoided by including an initial lysis step in the processing of blood cultures. Currently, no efforts are being made to develop the Bactometer for processing blood cultures. Instead, the apparatus is being used for industrial applications.

Another system (Electronic Detection System, Abbott Laboratories, Dallas, Texas), which determines microbial growth in broth cultures by measuring changes in electrical potential across electrodes, was evaluated by Holland and co-workers (1980). They reported that the system detected 96% of the seeded blood specimens and did this an average of 18 hr earlier than with the broth cultures. Despite these initially favorable results, further developments of the system have been slow, and currently the system is not available for clinical use.

IV. SUMMARY

In conclusion, improvements in blood culture techniques are hampered by the small number of bacteria normally present in the bloodstream of bacteremic patients and the antibacterial factors in the specimens. Traditionally, a large volume of blood was cultured in broth to detect a small number of bacteria and to dilute or inactivate the inhibitory factors. Unfortunately, this procedure also delays the detection of positive cultures. Modifications of the processing of blood specimens such as the radiometric or electrical methods have aided in standardizing the detection of positive blood cultures and in some cases improved the speed with which they were detected. However, the inherent delays of systems dependent on microbial growth in broth have not been eliminated. Other innovative methods entail centrifugation and filtration for concentrating bacteria and separating them from the inhibitory factors in blood. The limitations of these procedures are related primarily to contamination from the numerous manipulations involved in processing specimens. Although these are very significant problems, technologic innovation might eventually solve them.

REFERENCES

Appelbaum, P. C., Beckwith, D. G., Dipersio, J. R., Dyke, J. W. Salventi, J. F., and Stone, L. L. (1983). *J. Clin. Microbiol.* **17,** 48–51.

Appleman, M. D., Swinney, R. S. and Heseltine, P. N. R. (1982). *J. Clin. Microbiol.* **15,** 278–281.

Bannatyne, R. M., and Harnett, N. (1974). *Appl. Microbiol.* **27,** 1067–1069.

Blazevic, D. J., Trombley, C. M., and Lund, M. E. (1976). *J. Clin. Microbiol.* **4,** 522–526.

Brooks, K., and Sodeman, T. (1974). *Am. J. Clin. Pathol.* **61,** 859–866.

Cady, P., Dufour, S. W., Shaw, J., and Kraeger, S. J. (1978). *J. Clin. Microbiol.* **7,** 265–272.

Caslow, M., Ellner, P. D., and Keihn, T. E. (1974). *Appl. Microbiol.* **28,** 435–438.

DeLand, F. H., and Wagner, H. N. (1969). *Radiology* **92,** 154–155.

Dorn, G. L., Burson, G. G., and Haynes, J. R. (1976a). *J. Clin. Microbiol.* **3,** 258–263.

Dorn, G. L., Haynes, J. R., and Burson, G. G. (1976b). *J. Clin. Microbiol.* **3,** 251–257.

Dorn, G. L., and Smith, K. (1978). *J. Clin. Microbiol.* **7,** 52–54.
Dorn, G. L., Land, G. A., and Wilson, G. E. (1979). *J. Clin. Microbiol.* **9,** 391–396.
Griffin, M. R., Miller, A. D., and Davis, A. C. (1982). *J. Clin. Microbiol.* **15,** 567–570.
Hall, M. M., Ilstrup, D. M., and Washington, J. A., II (1976). *J. Clin. Microbiol.* **3,** 643–645.
Hall, M. M., Mueske, C. A., Ilstrup, D. M., and Washington, J. A., II (1979). *J. Clin. Microbiol.* **10,** 673–676.
Holland, R. L., Cooper, B. H., Helgeson, N. G. P., and McCracken, A. W. (1980). *J. Clin. Microbiol.* **12,** 180–184.
Kagan, R. L., Schuette, W. H., Zierdt, C. H., and MacLowry, J. D. (1977). *J. Clin. Microbiol.* **5,** 51–57.
Krogstad, D. L., Murray, P. R., Granich, G. G., Niles, A. C., Ladenson, J. H., and Davis, J. E. (1981). *Antimicrob. Agents Chemother.* **20,** 272–274.
Lindsey, N. J., and Riely, P. E. (1981). *J. Clin. Microbiol.* **13,** 503–507.
Lowrance, B. L., and Traub, W. H. (1969). *Appl. Microbiol.* **17,** 839–841.
McCarthy, L. R., and Senne, J. E. (1980). *J. Clin. Microbiol.* **11,** 281–285.
Mirrett, S., Lauer, B. A., Miller, G. A., and Reller, L. B. (1982). *J. Clin. Microbiol.* **15,** 562–566.
Murray, P. R., and Niles, A. C. (1982). *J. Clin. Microbiol.* **16,** 982–984.
Murray, P. R., and Sondag, J. E. (1978). *J. Clin. Microbiol.* **8,** 427–430.
Pfaller, M. A., Sibley, T. K., Westfall, L. M., Hoppe-Bauer, J. E., Keating, M. A., and Murray, P. R. (1982). *J. Clin. Microbiol.* **16,** 525–530.
Salventi, J. F., Davies, T. A., Randall, E. L., Whitaker, S., and Waters, J. R. (1979). *J. Clin. Microbiol.* **9,** 248–252.
Sliva, H. S., and Washington, J. A., II (1980). *J. Clin. Microbiol.* **12,** 445–446.
Specter, S., Throm, R., Strauss, R., and Friedman, H. (1977). *J. Clin. Microbiol.* **6,** 489–493.
Sullivan, N. M., Sutter, V. L., and Finegold, S. M. (1975a). *J. Clin. Microbiol.* **1,** 30–36.
Sullivan, N. M., Sutter, V. L., and Finegold, S. M. (1975b). *J. Clin. Microbiol.* **1,** 37–43.
Tenney, J. H., Reller, L. B., Mirrett, S., Wang, W. L. L., and Weinstein, M. P. (1982). *J. Clin. Microbiol.* **15,** 558–561.
Thiemke, W. A., and Wicher, K. (1975). *J. Clin. Microbiol.* **1,** 302–308.
Todd, J. K., Roe, M. H. (1975). *Am. J. Clin. Pathol.* **64,** 694–699.
Wallis, C., Melnick, J. L., Wende, R. D., and Riely, P. E. (1980). *J. Clin. Microbiol.* **11,** 462–464.
Wright, A. J., Thompson, R. L., McLimans, C. A., Wilson, W. R., and Washington, J. A., II (1982). *Am. J. Clin. Pathol.* **78,** 173–177.
Zierdt, C. H. (1982). *J. Clin. Microbiol.* **15,** 172–174.
Zierdt, C. H., Kagan, R. L., and MacLowry, J. D. (1977). *J. Clin. Microbiol.* **5,** 46–50.
Zierdt, C. H., Peterson, D. L., Swan, J. C., and MacLowry, J. D. (1982). *J. Clin. Microbiol.* **15,** 74–77.

27

Rapid Methods and Instrumentation in the Diagnosis of Urinary Tract Infections

WILLIAM J. MARTIN

Clinical Microbiology Laboratories
New England Medical Center
Tufts University School of Medicine
Boston, Massachusetts

I.	Introduction	353
II.	Rapid Methods	354
	Limulus Amoebocyte Lysate Test	356
III.	Instrumentation	356
	References	360

I. INTRODUCTION

Since Pasteur's time it has been known that freshly voided urine is a good culture medium for the growth of bacteria. Because of this, results obtained from culturing urine were considered to be of little value in determining the presence of urinary tract infections. Not until Marple (1941) applied quantitative bacteriologic principles to urine cultures from catheterized patients was some of the confusion about urinary tract infection eliminated. Fifteen years later Kass (1956) reported rather convincingly that, if proper specimen collection and microbiological principles were incorporated, one could separate insignificant and clinically important bacterial growth in urine. This work ensured that there was a means of evaluating the effect of treatment of urinary tract infection.

Several rapid techniques have been developed for determining the presence of bacteriuria; at present, however, significant bacteriuria is still confirmed primarily by traditional quantitative bacteriologic methods.

The calibrated loop, direct-streak method uses a calibrated bacteriologic loop to inoculate and streak plates of standard or differential culture media (Hoeprich,

The Direct Detection of Microorganisms in Clinical Samples

ISBN 0-12-187780-9

1960). After proper incubation the number of colonies present is estimated and reported as a measure of the degree of bacteriuria. Used properly, this technique is reliable for clinical purposes and is simple and rapid. The delivery volume of calibrated loops may change with use [this can be checked by a procedure outlined by Barry *et al.* (1975)] or may vary with the diameter of the container, angle of the loop, or volume of urine (Albers and Fletcher, 1983). If no growth occurs in 24 hr, the plates should be held an additional 24 hr and, if still negative, reported as "No growth after 48 hr."

Although the pour plate method has been criticized as not entirely reflecting the true bacterial count of a urine specimen (clumps of organisms can give rise to single colonies), it remains the most accurate procedure for measuring the degree of bacteriuria. A Quebec colony counter is used to enumerate the number of colonies in the plate yielding 30 to 300 colonies; this is multiplied by the dilution to obtain the total number of bacteria per milliliter of urine.

Some laboratories consider that a urine culture yielding three or more isolates should be considered a contaminated specimen and therefore do not identify the isolates. In certain instances this is poor practice. For example, such polymicrobic bacteriuria is not uncommon in individuals with indwelling catheters; indeed, such bacteriuria may be of special importance in that it may show a significant association with bacteremia (Gross *et al.*, 1976).

For more detail concerning the use of these conventional procedures, the reader should consult one or more of the following references: Cumitech 2 (Barry *et al.*, 1975), Finegold and Martin (1982), and the *Manual of Clinical Microbiology* (Isenberg *et al.*, 1980).

II. RAPID METHODS

A great deal of attention has been given to chemical tests for the rapid detection of bacteriuria. These include the reduction of triphenyltetrazolium chloride by metabolizing bacteria (Simmons and Williams, 1962); the Griess nitrite test, based on the rapid reduction of nitrate by members of the Enterobacteriaceae and staphylococci; the glucose oxidase test, dependent on metabolism of the small amount of glucose present in normal urine by bacteria (Kunin, 1975); and the catalase test, in which rapid gas production from urine reacting with hydrogen peroxide indicates bacteriuria. However, these tests have been considered generally unsatisfactory due to a substantial number of false-negative results and sometimes lack of specificity as well. Commerical test kits based on these principles likewise have shown equivocal results.

A number of screening culture tests for bacteriuria have been described (Cohen and Kass, 1969; Kunin, 1975). In one such method an agar-coated vehicle (dip-slide, tube, spoon, paddle, pipette, or cylinder) is dipped into a freshly voided, midstream urine specimen; after overnight incubation, colonies are counted or compared with density photographs of known colony counts, and the significance of the culture is determined. A similar procedure utilizes a filter paper strip that is dipped into the urine specimen and then transferred to the surface of either a miniature or a conventional agar plate. After incubation, the number of colonies growing in the inoculum area are counted, and their significance is evaluated. Multiple specimens can be cultured on a single conventional agar plate.

It is the general consensus of most investigators that these methods—chemical or dip-culture—neither are appropriate for nor offer any advantage in the clinical microbiology laboratory; their principal value lies in their use as screening procedures for bacteriuria in a physician's office or for field surveys. A duplicate urine specimen held in the refrigerator can be submitted for conventional culture and sensitivity testing if the screening test proves to be positive.

An inexpensive, rapid, and simple colorimetric test that does not require bacterial growth and has the marked advantage of quantifying bacteria, even when the organisms are present in the urine of bacteriuric patients who are being treated with antibiotics, has been described by Wallis *et al.* (1981). The test is carried out with 1 ml of urine, which is processed through a filter 10 mm in diameter that entraps the bacteria on its surface. Safranine dye is passed through the filter to stain the bacteria and the filter fibers. A decolorizer, which removes the dye from the filter fibers but not from the bacteria, is then passed through the filter. If there are $\geq 10^5$ colony-forming units (cfu) of bacteria per milliliter in the sample, the 10-mm filter disk manifests a pink to red color. If there are $<10^5$ cfu/ml, the filter disk remains white or turns slightly yellow. Of interest is that the entire procedure has been adapted to a semiautomated instrument [Bacteria Detection Device (BDD), Marion Scientific Corp., Kansas City, Missouri], and the time required per test is less than 1 min. Moreover, the results obtained on the test card are considered a permanent record to be filed with the patient's chart. Furthermore, the bacteria can be quickly classified as gram-positive or gram-negative by selective staining of a second milliliter of urine on the filter. Of 441 urine specimens tested by these investigators, 98% were correctly classified as containing more or less than 10^5 cfu/ml; 62 urine specimens were positive by bacterial plating ($\geq 10^5$ cfu/ml), and 59 were positive by colorimetric test. Eight were reported as false positive specimens (colorimetric positive, plate counts $<10^5$) and were from patients receiving antibiotics. Removal of the antibiotics from these urine specimens, with subsequent replating of the samples, indicated the presence of $\geq 10^5$ cfu of bacteria per milliliter in three representative cases tested, suggesting that the results of the colorimetric tests were not false positive but that the plate counts were low due to the inhibition of bacterial growth by residual antibiotics in the urine. Further information on the BDD is presented in Chapter 25 of this volume.

Limulus Amoebocyte Lysate Test

The *Limulus* amoebocyte lysate (LAL) test has been shown to be a reliable, simple, and rapid assay for the detection of significant gram-negative bacteriuria. With culture methods the LAL test has showed 100% correlation, provided that the urine pH was adjusted to 7. Moreover, this test correctly identified 96.2% of urine specimens as containing $<10^5$ or $>10^5$ gram-negative bacteria per milliliter (Nachum and Shanbrom, 1981). Because the majority of bacteriurias are caused by gram-negative organisms, the LAL test may be useful as a rapid urine detection procedure, although bacteriuria caused by gram-positive bacteria and yeast are not detected by this test. A more extensive discussion of the applications of LAL test for urine specimens can be found in Chapter 21 of this volume.

III. INSTRUMENTATION

Other techniques or instrumentation that have been proposed or utilized as urine detection systems include the measurement of bacterial ATP by luciferase, production of radioactive carbon dioxide by the metabolism of labeled substrates, liberation of unique compounds into headspace gas utilizing specialized substrates and headspace gas chromatography, measurement of the potential generated by growing bacteria using a platinum and a calomel electrode, measurement of heat generated by metabolizing microorganisms using a microcolorimeter, measurement of particle size and distribution by electronic pulse height analysis, and detection of bacteriuria by automated electrical impedance monitoring.

Because space does not permit an adequate description of all of these systems, only two of the most promising are discussed here. Particle size distribution analysis proved to be remarkably effective for rapid screening (Dow *et al.*, 1979). With 600 urines, the Coulter counter procedure results agreed with routine culture results 98.8% of the time. Results of negative specimens could be obtained in 5–10 mins. There were no false negative results. It may even prove possible to make a preliminary identification based on the particle distribution profiles (data) being fed directly from the Coulter C1000 Channelyzer to a computer programmed to make the diagnosis. The principle of the electrical impedance has also been applied to the detection of bacteria in urine. As bacteria grow and metabolize, the chemical composition of the supporting medium is altered and metabolic end products are produced; along with this is a corresponding change in the resistance to the flow of an alternating current when a pair of electrodes are placed in the medium. One can determine how many bacteria must be present to produce a given change in impedance in a given period of time. In a study of over 1000 urine specimens by Cady *et al.* (1978) it was found that, by defining an impedance-positive culture as one that gives detectable change within 2.6 hr, 96% of urine cultures tested were correctly classified in terms of having greater than or less than 10^5 organisms per milliliter. The predictive value of a positive test was 88%, and the predictive value of a negative test was 97%. By using a 3.5-hr cutoff, the predictive value positive dropped to 75%, whereas the predictive value negative increased to 98.5%.

Instrumentation that is available commercially for rapid screening of urine specimens and for other purposes include the Abbott MS-2, General Diagnostics Autobac, and Vitek's Automicrobic System. The Autobac has undergone a large trial in the clinical microbiology laboratory setting as part of a system of rapid semiautomated screening and processing of urine specimens (Jenkins *et al.*, 1980). This study confirmed and extended the impressive results obtained in an earlier study by Heinze *et al.* (1979). The Autobac detects positive urines by using light-scatter photometry. At 3 hr, 75% of urine specimens that were eventually positive by culture were positive by the Autobac reading. By 6 hr, the Autobac detected ~96% of the positive specimens; the 4.6% false-negative specimens detected at 6 hours consisted of specimens from individuals receiving antibiotics that would not ordinarily be checked by the Autobac in the proposed system or were positive for slow-growing organisms (probably urethral contaminants). The system proposed by these investigators would exclude from the Autobac screen (these specimens would be cultured directly) people receiving antimicrobial agents, people with indwelling urethral catheters, and specimens obtained by suprapubic aspiration. Under the system, specimens that are positive by Autobac at 3 hr are put into a 3-hr direct susceptibility test scheme and a 4-hr Micro-ID rapid identification test. Thus, the results from these specimens can be completed within an 8-hr work day. The direct 3-hr susceptibility and the direct 4-hr identification were found to be 93 to 94% accurate, respectively, as compared with conventional techniques. In addition to the specimens that were positive by Autobac at 3 hr and therefore could be worked up within an 8-hr day, an additional number of specimens (73% of the total) were negative by Autobac screen at 6 hr. It was the author's impression that negative specimens could be called with a high degree of reliability. Accordingly, no further work was required on these specimens. Furthermore, after dropping the cultures that were mixed by Gram's stain and therefore could not be screened rapidly, 82% of all urine specimens screened by this Autobac procedure were completed within an 8-hr work day.

Originally designed for antibiotic susceptibility testing (Thornsberry *et al.*, 1980), the Abbott MS-2 (Abbott Diagnostic Division, Dallas) is currently being applied to the identification of Enterobacteriaceae (McCracken *et al.*, 1980) and to the screening of urine specimens. The major components of the system are a control module and an analysis module (several analysis modules can be stacked on top of one another). Urine screening requires a special adapter that holds 11 glass ampules, or what the manufacturer calls ampvettes. Each of these is inoculated with 0.1 ml of urine and placed in the adapter, which is inserted into the analysis module. The machine scans each ampvette every 5 min, and data are analyzed by a computer program that compares growth curves with those from control ampvettes. Results are printed for all ampvettes at the end of 5 hr. The major advantage of the Abbott MS-2 system is its level of automation, which is considered to be quite sophisticated. The nearly constant monitoring of growth chambers, speed of detection and interpreting results, and its versatility also are advantages.

The Automicrobic System (AMS) of Vitek Incorporated (McDonnell Douglas Corp., St. Louis) was initially designed for the detection, enumeration, identification, and drug-susceptibility testing of microorganisms in clinical specimens, with all of its functions to be performed simultaneously or consecutively by the same instrument (Alridge *et al.*, 1977; Nicholson and Koepka, 1979; Smith *et al.*, 1978; Sonnenwirth, 1977). It uses a novel array of highly selective media in which a mixture of substrates and inhibitors permits the growth of one or a group of closely related microorganisms, concurrently inhibits the growth of any other organisms present, and enumerates the total number of organisms in the sample. The lyophilized media are incorporated in wells of a sealed, disposable, plastic card, which is inoculated and incubated in an automated instrument equipped with an optical system. The optical system monitors light transmission changes in the card wells and transmits optical measurements to a minicomputer, which compares them with present thresholds and then displays results in 4 to 13 hr.

The AMS is considered to be different from other instrument systems on the market in that it was originally designed to screen urine specimens and identify the organisms present without the usual preliminary pure culture technique. (The instrument has many other uses in identification and susceptibility testing, but these all require pure cultures.) The system consists of a diluent dispenser module, a filling module, an incubator reader, a computer, and a terminal printer. As with the MS-2 and Autobac, it uses a disposable unit, which varies with the type of analysis being made. The urine card that receives the inoculum is about the size of a playing card and 3–4 mm thick. There are 5 wells for quantifying and 15 for identification purposes, of which 13 are actually used. The user simply pipettes undiluted urine into the two plastic tubes of the injector, with 0.5 and 200 μl being diluted, respectively, to 1 : 1000 for enumeration and 1 : 10 for identification with sterile 0.5% saline by the diluent dispenser. This whole unit (card and injector) is placed in the filling module, which is a pressurized system that forces the urine into the respective chambers within the card. The entire procedure takes about 4.5 min, and then the operator discards the two plastic tubes and inserts the plastic card into the incubator reader. From this point on, everything is done automatically. The cards are read hourly, with data being

TABLE I

Urines Received for Culture January 1, 1979 through June 30, 1980

Specimen	Number of cultures	Percentage
Total	28,630	—
No growth	16,081	56.1
No significant growth	3,236	11.3
Total positives	8,042	28.0
Escherichia coli as single isolate, $>10^5$	2,299	28.6
E. coli as single isolate, $<10^5$	440	5.5
Other single isolates, $>10^5$	2,054	25.6
Other single isolates, $<10^5$	1,243	15.5
Multiple isolates, $<10^5$ or $>10^5$ (combination of 2 or 3)	2,006	25.0

transmitted to the computer. At the end of 13 hr all results are printed. Also, one can retrieve results from the AMS at any time after 1 hr. The system has the capacity to identify mixed cultures and also report each of the organisms present, but it will provide only a total count. That is, it will not provide a specific count for each of the various organisms that may be present. Nevertheless, one can obtain a population count and an identification within the same day. Table I summarizes the results obtained with the AMS on midstream urine specimens collected at the UCLA Hospital and Clinics over an 18-month period. Table II presents the identification of single species isolates of $>10^5$ cfu/ml. Of significance is that greater than 95% of the isolates were accurately identified by this instrument system.

Pezzlo *et al.* (1982) compared the Autobac, AMS, MS-2, and Gram's stain for their capacity to detect significant bacteriuria. A total of 1000 urine specimens were evaluated and compared with a semiquantitative culture plate method. When pure pathogens of $>10^5$ cfu/ml were considered, all systems had sensitivities of >95% and could predict a negative urine in >99% of the cases. Moreover, the results of the three automated systems were available in 1 to 13 hr compared with 18 to 24 hr by the standard procedure. When these systems were used for screening, the cost of supplies and technical time with the Gram's stain, Autobac, and MS-2 were comparable and considerably less expensive than for the reference method. Although the AMS was the least expensive system when the cost for identifying probable pathogens was included, this system had the advantage of simultaneously identifying and detecting organisms in positive urines in much less time than the reference method. As a result of their studies, these investigators felt that using an automated system to screen urine specimens

TABLE II

Identification of Single Isolates of >10^5 Organisms per Milliliter

Single isolate	Number of cultures	Percentage
Total	2054	—
Klebsiella pneumoniae	322	15.7
Pseudomonas aeruginosa	312	15.2
Group D enterococcus	264	12.8
Yeast	248	12.1
Proteus mirabilis	243	11.8
Staphylococcus epidermidis	219	10.7
Group B *Streptococcus*	59	2.9
Enterobacter cloacae	48	2.3
Staphylococcus aureus	47	2.3
Viridans group *Streptococcus*	39	1.9
Corynebacterium	35	1.7
Citrobacter freundii	32	1.6
Enterobacter aerogenes	32	1.6
Lactobacillus	20	1.0
Morganella morganii	14	0.7
Serratia marcescens	14	0.7
Providencia stuartii	5	0.2
Klebsiella ozaenae	5	0.2
Providencia alcalifaciens	3	0.1
Other single isolates	93	4.5

provides an overall savings of time to clinical microbiology laboratories. However, each institution should evaluate its needs before making a commitment to any one system.

REFERENCES

Albers, A. C., and Fletcher, R. D. (1983). *J. Clin. Microbiol.* **18,** 40–42.

Alridge, C., Jones, P. W., Gibson, S., Lanham, J., Meyer, M., Vannest, R., and Charles, R. (1977). *J. Clin. Microbiol.* **6,** 406–413.

Barry, A. L., Smith, P. B., and Turck, M. (1975). *In* "Laboratory Diagnosis of Urinary Tract Infections" (T. L. Gavin, ed.), Cumitech 2. Am. Soc. Microbiol., Washington, D.C.

Cady, P., Dufour, S. W., Lawless, P., Nunke, P., and Kraeger, S. J. (1978). *J. Clin. Microbiol.* **7,** 273–278.

Cohen, S. N., and Kass, E. H. (1969). *N. Engl. J. Med.* **277,** 176–180.

Dow, C. S., France, A. D., Khan, M. S., and Johnson, T. (1979). *J. Clin. Pathol.* **32,** 386–390.

Finegold, S. M., and Martin, W. J. (1982). "Bailey and Scott's Diagnostic Microbiology," 6th ed., pp. 92–100. Mosby, St. Louis, Missouri.

Gross, P. A., Flower, M., and Barden, G. (1976). *J. Clin. Microbiol.* **3,** 246–250.

Heinze, P. A., Thrupp, L. D., and Anselmo, C. R. (1979). *Am. J. Clin. Pathol.* **71,** 177–183.
Hoeprich, P. D. (1960). *J. Lab. Clin. Med.* **56,** 899–907.
Isenberg, H. D., Washington, J. A., II, Balows, A., and Sonnenwirth, A. C. (1980). *In* "Manual of Clinical Microbiology" (E. H. Lennette, A. Balows, W. J. Hausler, Jr., and J. P. Truant, eds.), 3rd ed., pp. 52–82. Am. Soc. Microbiol., Washington, D.C.
Jenkins, R. D., Hale, D. C., and Matsen, J. M. (1980). *J. Clin. Microbiol.* **11,** 220–225.
Kass, E. H. (1956). *Trans. Assoc. Am. Physicians* **69,** 56–64.
Kunin, C. M. (1975). *Urol. Clin. North Am.* **2,** 423–432.
McCracken, A. W., Martin, W. J., McCarthy, L. R., Schwab, D. A., Cooper, B. H., Helgeson, N. G. P., Prowant, S., and Robson, J. (1980). *J. Clin. Microbiol.* **12,** 684–689.
Marple, C. D. (1941). *Ann. Intern. Med.* **14,** 2220–2239.
Nachum, R., and Shanbrom, E. (1981). *J. Clin. Microbiol.* **13,** 158–162.
Nicholson, D. P., and Koepka, J. A. (1979). *J. Clin. Microbiol.* **10,** 823–833.
Pezzlo, M. T., Tan, G. L., Peterson, E. M., and De La Maza, L. M. (1982). *J. Clin. Microbiol.* **15,** 468–474.
Simmons, N. A., and Williams, J. D. (1962). *Lancet* **1,** 1377–1378.
Smith, P. B., Gavan, T. L., Isenberg, H. D., Sonnenwirth, A. C., Taylor, W. I., Washington, J. A., II, and Balows, A. (1978). *J. Clin. Microbiol.* **8,** 657–666.
Sonnenwirth, A. C. (1977). *J. Clin. Microbiol.* **6,** 400–405.
Thornsberry, C., Anhalt, J. P., Washington, J. A., II, McCarthy, L. R., Schoenknecht, F. D., Sherris, J. C., and Spencer, H. J. (1980). *J. Clin. Microbiol.* **12,** 375–390.
Wallis, C., Melnick, J. L., and Longoria, C. J. (1981). *J. Clin. Microbiol.* **14,** 342–346.

Index

A

Abbott MS-2, screening of urine specimens and, 357–358
Abcesses, diagnostic microscopy and, 9–11
Absidia, in clinical material, 18
N-Acetyl-L-cysteine
  examination of specimens for fungi and, 19
  liquefaction of sputum and, 93–94
Acridine orange
  detection of bacteremia and, 345
  detection of fungi and, 30
  staining
    cerebrospinal fluid and, 6
    diagnosis of gonorrhea and, 9
    wounds, abcesses, and exudates and, 9–10
Actinomyces
  in clinical material, 33
  immunofluorescence and, 51
Adenovirus
  detection by immunofluorescence, 67, 73
  occurrence and symptoms, 72–73
Affinity chromatography, antiviral antibodies and, 235
Agarose, preparation for counterimmunoelectrophoresis, 91
Agglutination tests
  advantages and disadvantages for antigen detection, 139–141
  for diagnosis of meningitis, 144–152
  future prospects, 141
  principle of, 135–138
  quantification of, 137, 138
  specificity of
    latex particle agglutination, 150–151
    staphylococcal coagglutination, 151–152
Alkaline phosphatase, immunoassays and, 271
Aminotransferase, non-A, non-B hepatitis and, 255–256, 258, 260
Antibodies
  to *Aspergillus,* detection of, 216
  avidity, solid-phase immunoassays and, 235
  clearance of *Haemophilus* polysaccharide from blood and, 122
  clearance of pneumococcal polysaccharide from blood and, 119–121
  covalent linkage to solid phase, 234
  determinants, agglutination tests and, 150
  to hepatitus B surface antigen, 249–251
    assays for, 250–251
  to hepatitus B virus core, 248–249
    assays for, 249
  labels, for solid-phase immunoassay, 226–229
  monoclonal
    detection of antigenemia and, 123
    identification of bacteria and 338–339

immunoassays and, 163
solid-phase immunoassays and, 236
viral diagnosis and, 62
preexisting, inhibition of immunoassays and, 177–180
response, in Legionnaires disease, 188–189
Anticomplement immunofluorescence, 59
usefulness of, 61
Antigens
candidal, preparation of, 213–214
detection in meningitis
background, 161–163
counterimmunoelectrophoresis, 163
enzyme-linked immunosorbent assay, 166–168
latex agglutination, 163
staphyloccal coagglutination, 164–166
diversity, immunoassays for antigenemia and, 177
pneumococcal, detection in clinical samples, 107–109
Antigen determinants
agglutination tests and
cryptococcal, 149–150
group B streptococcal, 148–149
Haemophilus influenzae, 144–145
meningococcal, 146–148
pneumococcal, 145–146
Antigen e, of hepatitis B, 251–252
assay for, 252
Antigenemia
detection by precipitin methods, variables in, 118–123
problems of immunoassays for detection of antigen diversity, 177
nonspecific inhibitors, 177
protein A contamination, 180
rheumatoid factor, 180
specific inhibitors, 177–180
Antimicrobials,
diarrhea following use, 295–296
sodium polyanetholsulfonate and, 345
Antimicrobial Removal Device, blood cultures and, 346
Antisera
for counterimmunoelectrophoresis, 89
to *L. pnemophila* antigen, preparation of, 191–192
nonspecificity and variability of, 162
for rotavirus immunoassay, 269–270
use in therapy, identification of etiologic agent and, 141
to viruses, production of, 61–62
Arabinitol, candidal infection and, 213, 308, 309
Arthritis, GLC and, 323, 325
Aspergillus
antigen
preparation for RIA, 219–220
procedure for RIA, 220
antigenemia, dissociation of immune complexes and, 179
in clinical material, 17, 32
detection of antigens, 215–219
Assay format, solid-phase immunoassays and, 229–233
Autobac, diagnosis of bacteriuria, 357
Automation
challenges in development of for clinical microbiology laboratory, 335–341
GLC and, 332–333
Automicrobic system, screening urine specimens with, 358–359, 360
Autopsy specimens, detection of viruses in, 75–77
brain, 76–77
pulmonary tissue, 77

B

BACTEC, detection of bacteremia and, 348–349
Bacteremia
detection, conventional methods and, 344–345
diagnostic microscopy and, 6
pneumococcal, antigenemia and, 118
rapid diagnosis of
ELISA and, 183–184
radioimmunoassay and, 181–183
use of immunoassays in
rationale, 175–176
technical aspects, 176–177
Bacteria
identification
automation of, 336
bacteria for which immunofluorescence is most commonly use, 41–51
miscellaneous bacteria, using immunofluorescence, 51–52

Bacteria Detection Device, use in bacteriuria, 355
Bacteroides, infections, GLC and 307
Bacteroides fragilis
antigen, ELISA and, 182, 184
cross-reaction with *L. pneumophila* antisera and, 189
identification by immunofluorescence, 41, 42
Bacteroides melaninogenicus, identification by immunofluorescence, 41, 42–43
Bacteriuria, Gram-negative, detection by LAL, 287–290, 356
Bactometer, detection of bacteremia and, 349–350
Bioluminescence, screening of urine samples and, 340
Biopsy specimens, detection of viruses in, 75–76
brain, 76–77
pulmonary tissue, 77
Biotin-avidin system, enzyme immunoassays and, 227
in clinical material, 18
Blastomyces dermatitidis
yeast form, significance of, 14
Blood, peripheral, examination, 6
Blood cultures
for detection of bacteremia, time of detection, 344
alternative modifications of
inactivation of inhibitory factors, 345–346
removal of inhibitory factors, 346
removal of microbes from blood, 346–347
standardization of detection methods, 348–350
Bordetella pertussis, identification by immunofluorescence, 45–47
Body fluids, normally sterile, distinguishing between positive and negative specimens, 339
Brain, viruses in, immunofluorescence and, 76–77
Breast cancer, GLC and, 327
Buffers, for counterimmunoelectrophoresis, 89–90, 107
calculation of ionic strength, 89
1,4-Butanedicarboxylic acid, diagnostic importance of, 307

C

Candida
in clinical material, 19
tissue invasion by, 14
Candidiasis
detection of antigens, 208–213
ELISA protocol, 214–215
RIA protocol, 213–214
immune complexes, dissociation of, 179, 180
Carcinoembryonic antigen, screening for hepatitis and, 261
Carrier state, hepatitis B and, 244–246
antibody to virus core and, 248, 249
e antigen and, 251
Castaneda bottle, for blood cultures, 346–347
Cerebrospinal fluid
concentration of antigen in, 162
detection of pneumococcal antigen in, 109
diagnostic microscopy and, 6
lactic acid in, GLC and, 308
leukocyte count, meningitis and, 160
sample CIE setup for 92
Chemical compounds, volatile, practical methods for recovering from body fluids for derivatization, 314–315
Chemiluminescence, solid-phase immunoassays and, 228
Chicken pox, GLC and, 328
Chlamydia trachomatis, identification by immunofluorescence, 50–51
Chloroform, for extraction of body fluids, 315
Chronic bronchitis, pneumococcal antigens in, 157
Cladosporium carrionii, in clinical material, 19
Clindamycin, antimicrobial-associated diarrhea and, 296, 298
Clinical microbiology laboratory, challenges in development of automation for, 335–341
Clostridium difficile
laboratory procedures
culture, 299–300
cytotoxicity assay for toxin in stool, 300–301
toxins of, 296–299
Clostridium sordellii, antitoxin, *C. difficile* toxin and, 298
Coagglutination tests, in pneumonia, 154–155
comparison to counterimmunoelectrophoresis, 155–156

specificity, 157
Coccidoides immitis
antigen, detection of, 219
infection by, 17, 19, 32, 33
Columns, for GLC, 317
Commercial tests, for rotavirus, 274, 275
Competitive assays, solid-phase immunoassays and, 232–233
Complement components, inhibition of immunoassays and, 177
Computers, GLC and, 306, 332–333
Conjuctival scrapings, detection of viruses in by immunofluorescence, 73–74
herpes simplex, 74–75
varicella-zoster, 75
Controls
agglutination reactions and, 137–138
immunofluorescence tests and, 39–40, 77–78
Coomassie brilliant blue stain, preparation of, 94
Corneal ulcer, detection of Gram-negative bacteria in, 291–292
Coulter counter, diagnosis of bacteriuria and, 356
Counterimmunoelectrophoresis (CIE)
clinical applications of, 94–95
comparison to coagglutination, 155–156
concentration of samples for, 93
for diagnosis of intrapleural empyema, 113–114
methods, 114–115
for diagnosis of meningitis, 163, 164
Escherichia coli, 102
group B streptococci, 101–102
Haemophilus influenzae, 99–100
meningococci, 98–99
pneumococci, 100–101
evaluation in diagnosis of infectious disease
culture-negative diagnosis, 130
discussion, 131–133
identification of pathogens in presence of mixed flora, 130–131
predictive value, 126–129
test procedure, 126
use in early diagnosis, 129–130
methods in pneumococcal infection, 106–107
perspective, 110
principle, 88, 106
variables, 88
antisera, 89
buffers, 89–90
electrophoresis chamber and power supply, 90–91
support systems, 90
Counterstaining, immunofluroescence tests and, 40, 63
C-Reactive protein, meningitis and, 171
p-Cresol, cultures of *C. difficile* and, 297
Criteria, for positive immunofluorescence tests, 40
Cross-reactions
with capsular antigens, 162–163
LPA and, 151
SC and, 151–152
Cryptococcus neoformans
antigen, agglutination tests for, 149–150
in clinical material, 18, 28–29, 30
meningitis and
assessment, 170
background, 168–169
sensitivity of latex agglutination, 169
specificity, 169–170
Culture medium
for *Clostridium difficile,* 296, 299–300
for *L. pneumophila,* 190
spent, application of GLC to detection of chemical changes in, 319, 320
Culture-negative diagnosis, CIE and, 130
Cunninghamella, in clinical material, 18
Cytomegalovirus, immunofluorescence and, 61, 62, 77
Cytotoxicity assay, for *C. difficile* toxin in stool, 300–301

D

Delta antigen and antibody, hepatitis B and, 253–254
assay for, 254
Dematiaceae, infections and, 17
Deoxyribonucleic acid
hybridization
detection of bacteria and, 170
identification of bacteria and, 338–339
Deoxyribonucleic acid polymerase, hepatitis B and, 252–253
Derivatization, practical methods, for GLC, 318
Dermatophytes, in clinical material, 17–18

Detectors, GLC and, 305, 317
Diabetic ketoacidosis, diagnosis of, 307
Diagnosis, of candidiasis, 208
Diagnostic microscopy
  usefulness, 5
    bacteremia: examination of peripheral blood, 6
    eye, ear and nose infections and specimens for examination, 8
    feces, 9
    gonorrhea: exudates and body sites, 8–9
    meningitis: examination of cerebrospinal fluid, 6
    pharyngitis: throat specimens, 8
    respiratory tract infections and specimens for examination, 7–8
    urinary tract infection, 7
  of wounds, abcesses and exudates, 9–11
Diagnostic tests, rapid, selection of, 52
Diarrhea, in patients treated with antimicrobials, 295–296
Dichlorotriazinylaminofluorescein, conjugation to protein, 59
Diethyl ether, for extraction of body fluids, 315
Direct immunofluorescence, advantages and disadvantages of, 59
Dithiothreitol, examination of specimens for fungi and, 19
Double determinant assays, format of, 230–232

E

Early diagnosis, of infectious disease, CIE and, 129–130
Electrical impedance, measurements, diagnosis of bacteriuria and, 356–357
Electron capture detector, GLC and, 317
Electrophoresis, procedure, 92
Electrophoresis chamber, for counterimmunophoresis, 90–91
Encephalitis, GLC and, 321
Endometrial carcinoma, GLC and, 327
Endotoxemia, detection by LAL, 285–286
Endotoxins, *Limulus* amoebocyte lysis and, 284
Enteroviruses, infections, 328, 330
Enzymes, as antibody labels, 226–227
Enzyme-linked immunoassays, detection of antigenemia and, 123, 182–184
Enzyme-linked immunosorbent assay (ELISA)
  detection of candidal mannan and, 210, 211, 212
    protocol, 214–215
  for detecting serogroup 1 *L. pneumophila* antigens
    comparison with RIA, 202–203
    development of, 201
  format of assay, 229–230
  procedure, 166–167
  role in diagnosis of meningitis, 168
  sensitivity, 167
  specificity, 167
Epidemiology
  of hepatitis B, 242–243
  of non-A, non-B hepatitis, 255–258
  of rotavirus, 268–269
Epstein-Barr virus, diagnosis of, 61
Escherichia coli, 144
  counterimmunoelectrophoresis and, 92, 102
  enterophathogenic
    identification by immunofluorescence, 51
    identification of, 338–339
Exudates, diagnostic microscopy and, 9–11
Eye, ear and nose infections, specimens for examination, 8

F

False negatives,
  BACTEC system and, 348–349
False positives
  BACTEC system and, 348
  latex particle agglutination and, 150–151
  Legionella infections and, 45
  staphylococcal coagglutination and, 151
Feces, diagnostic microscopy and, 9
Filters, fluorescent microscopy and, 66
Fingerprinting, GLC and, 308
Fixation, of specimen, for immunofluorescence, 38, 70
Flow chart, for preparation of specimens for GLC, 316
Fluorescein, light absorption and emission by, 58
Fluorescence, solid-phase immunoassays and, 227–228
Fluorescence microscope
  filters and, 66
  light sources for, 65–66

mounting medium and, 66–67
objectives and, 66
Fluorescent antibody, for detection of fungi, 16, 30
Fonsecaea compactum, in clinical material, 19
Formalin, fixation for immunofluoresce and, 38, 76
Fungi
detection in clinical specimens, microscopic techniques, 13–19
examples in clinical material, 31–33
methods for visualizing in clinical material
direct fluorescence, 30
fluorescent antibody technique, 30
processing specimens for examination, 19–26
routine stains, 29
special stains for fungi, 29–30
wet mounts, 28–29
Fusarium, in clinical material, 32–33
Fusobacterium, infections, GLC and, 307

G

β-Galactosidase, immunoassays and, 271
Gas(es), for GLC, 317
Gas–liquid chromatography (GLC)
additional equipment
computers, 306
detectors, 305
mass spectrometer, 306
tubular capillary columns, 305
application to detection of chemical changes in culture media, 319
application to detection of disease specific profiles in body fluids
arthritis, 323
lymphocytic meningitis and encephalitis, 319–323
pleural effusions, 323, 326
Rocky Mountain spotted fever and viral diseases, 326–330
clinical correlations, 307–308
future prospects, 309–310
goals of analytical techniques, 307
identification of unknown peaks detected by, 330–331
interpretation of data obtained by, 331–332
methodology, 304–305
preparative techniques, 305
pitfalls in analysis, 308–309
possibilities for automation and computerization of data obtained by, 332–333
selective, sensitive and high-resolution systems, for body fluid analysis, 315–318
use in fungal diagnosis, 213
Geotrichium, in clinical material, 18
Glycerol, as mounting medium for fluorescence microscopy, 38, 66–67
Gomorri methenamine silver stain, for fungi, 14, 16
Gonorrhea
exudates and body sites, diagnostic microscopy and, 8–9
presumptive diagnosis by LAL, 290–291
Gram's stain
agglutination tests and, 139–140
of CSF, meningitis and, 160, 162
for dection of bacteremia, 345
detection of pneumococci and, 155
disuse of, 4
general uses for, 5
predictive value in meningitis, 129

H

Haemophilus, bacteremia, antigenemia and, 118–119
Haemophilus influenzae
antigens
agglutination tests for, 144–145
ELISA and, 166, 167, 168, 182, 183–184
RIA and, 181, 182
counterimmunoelectrophoresis and, 89, 92, 99–100, 115
identification by immunofluorescence in cases of meningitis, 47–49
immune complexes, dissociation of, 178–179
spent culture medium, GLC chromatogram of, 320
Hemagglutination inhibition, candidal mannan and, 209, 210, 211, 212
Hepatitis, non-A, non-B, 254–255
epidemiology, 255–258
experimental serologic tests, 258–261
Hepatitis B, 241–242
antibody to surface antigen, 249–251
antibody to virus core, 248–249

delta antigen and antibody, 253–254
DNA polymerase, 252–253
e antigen, 251–252
epidemiology, 242–243
nonspecific screening for, 254
surface antigens, 243–248
Herpes virus
encephalitis and, 321, 323, 324
immunofluorescence and, 61, 67, 74–75, 76
1,6-Hexanedicarboxylic acid, diagnostic importance of, 307
Histoplasma capsulatum, in clinical material, 18
Horseradish peroxidase, immunoassays and, 271
Human sera, use in viral diagnosis, 62
2-Hydroxybutyric acid, diagnosis and, 307

I

Immune complexes
dissociation and removal of antibody, detection of antigenemia and, 178–180
formation, staphylococcal antigens, 122
Immunoassays, in meningitis
conclusions and future prospects, 170–172
detection of bacterial antigens, 161–168
fungal meningitis, 168–170
problems in diagnosis, 160–161
Immunofluorescence
diagnosis of legionnaires' disease and, 188–190
viral diagnosis and, 57–58
collection and preparation of specimens, 67–77
fluorescence microscope, 65–67
interpretation of findings, 78–79
reagents, 61–65
staining method, 65
theory of method, 58–61
use of controls, 77–78
Immunofluorescence microscopy
historical background, 35–36
important factors in performance of tests, 36–37
controls and interpretation, 39–40
pitfalls and remedies, 40
reagents and protocol for preparation of smears, 37–39
Immunoglobulin G, iodination of, 194
Immunoglobulin M, agglutination reactions and, 136
India ink mount, detection of fungi and, 28–29
Indicator system, for rotavirus immunoassay, 271
Indirect immunofluorescence, advantages and disadvantages of 59–61
Infections
with *L. pneumophila* serogroup 1, diagnosis with RIA, 194–198
with non-serogroup 1 *L. pneumophila,* RIA in, 198–200
Influenza viruses A and B
detection by immunofluorescence, 67, 72
occurrence and symptoms, 71–72
Inhibitors
nonspecific, in immunoassays, 177
specific, of immunoassays, 177–180
Inhibitory factors, in blood cultures
inactivation of, 345–346
removal of, 346
Instrumentation
for diagnosis of bacteriuria, 356–360
solid-phase immunoassays and, 228–229
Internal standards, GLC and, 304, 318
Intrapleural empyema, diagnosis by CIE, 113–114
methods, 114–115
Iodine
radioactive
as antibody label, 226
disposal of, 200–201

K

3-(2-Ketohexyl)indoline, 331
tuberculous meningitis and, 308
Kits, commercial, for detection of bacterial antigens, 141
Klebsiella pneumoniae, 144
spent culture medium, GLC chromatograms of, 320

L

β-Lactamase, detection of bacteria and, 171
Lactic acidosis syndrome, diagnosis of, 307
Latex particles
agglutination, diagnosis of meningitis and, 163, 165

solid-phase immunoassays and, 234
specificity of agglutination tests and, 150–151
use in agglutination reactions, 136, 139, 140
cryptococcal antigens, 149–150, 169–170
group B streptococcal antigens, 148–149
H. influenzae antigens, 144–145
meningococcal antigens, 146–148
pneumococcal antigens, 145–146

Legionella
antigenicity, fixation and, 38
immunofluorescence tests and, 38–39, 40, 188–189

Legionella bozemanii, infection, RIA and, 200

Legionella longbeachae, infection, RIA and, 200

Legionella micdadei, identification of, 45

Legionella pneumophila
ELISA for serogroup 1 antigen
comparison with RIA, 202–203
development of, 201
identification by immunofluorescence, 43–45
radioimmunoassay for serogroup 1 antigen
development of, 191–194
diagnosis of infections, 194–198
non-serogroup 1 infections, 198–200
problems with RIA, 200–201

Legionnaires' disease
conventional diagnostic tests for
antibody response, 188–189
culture, 190
demonstration of bacilli in clinical specimens, 189–190
individuals at risk, 187–188

Limulus amoebocyte lysate test (LAL)
background and mechanism of, 283–285
clinical applications
detection of endotoxemia, 285–286
detection of Gram-negative bacteria in ocular infections, 291–292
detection of Gram-negative bacteriuria, 287–290
diagnosis of Gram-negative bacterial meningitis, 286–287
other clinical applications, 292
presumptive diagnosis of gonorrhea, 290–291
detection of bacteriuria and, 287–290, 356
meningitis and, 171
in the pharmaceutical industry, 285

Lipopolysaccharide
as antigen in immunoassays, 177
Limulus amoebocytes and, 284

Listeria monocytogenes, 144
immunofluorescence and, 51

Lysis-centrifugation, blood cultures and, 347

Lysis-filtration, blood cultures and, 347

M

Mannan, candidal, detection of, 209–212

Mannose, detection in candidiasis, 213, 308

Mass spectrometer, GLC and, 306

Menningitis
agglutination tests for diagnosis of
cryptococcal antigens, 149–150
group B streptococcal antigens, 148–149
Haemophilus influenzae antigens, 144–145
meningococcal antigens, 146–148
pneumococcal antigens, 145–146

Meningitis
bacterial agents of, identification by immunofluorescence, 47–49
common etiologic agents, 160
cryptococcal, GLC and, 321
diagnosis
antibody determinants and, 150
role of ELISA in, 168
diagnosis by counterimmunoelectrophoresis
Escherichia coli, 102
group B streptococci, 101–102
Haemophilus influenzae, 99–100
meningococci, 98–99
pneumococci, 100–101
diagnostic microscopy and, 6
GLC of CSF and, 308
Gram-negative bacterial, diagnosis by LAL, 286–287
immunoassays in
conclusions and future prospects, 170–172
detection of bacterial antigens, 161–168
fungal meningitis, 168–170
problems in diagnosis, 160–161
lymphocytic, application of GLC to, 321, 322

Metabolic disorders, GLC and, 307

Methylmalonic acidemia, diagnosis of, 307, 309

Microscopic techniques, role in detecting fungi in clinical specimens, 13–19
Measles virus, detection by immunofluorescence, 67, 73, 76, 77
 epidemiology of, 73
Mercury arc burners, fluorescence microscopy and, 65–66, 67
Microbes, removal of, from blood for culture, 346–347
Mucor, in clinical material, 18, 31–32
Mucus, removal from specimens, 70–71, 73
Mycobacterium tuberculosis
 immunofluorescence and, 51–52
 pleural effusions caused by, GLC and, 323, 326, 327
Mycoses, individuals at risk, 207

N

National Aeronautic and Space Administration, automated microbiological analyses and, 337
Neisseria gonorrhoeae, identification by immunofluorescence, 51
Neisseria meningitidis
 antigens
  agglutination tests for, 146–148
  ELISA and, 166, 167
 counterimmunoelectrophoresis and, 89, 92, 98–99
 identification by immunofluorescence, 48, 49
 infection, GLC and, 328
 spent culture media, GLC chromatograms and, 319, 320
Nitrocellulose, solid-phase immunoassays and, 234–235
Nocardia, in clinical material, 33
Nonspecific reactions
 in immunofluorescence, 63, 64, 70–71
 rotavirus immunoassays and, 272–273
Norwalk virus, epidemiology and immunoassay, 274, 276

O

Objectives, for fluorescence microscopes, 66
Ocular infections, detection of Gram-negative bacteria in by LAL, 291–292
Otitis media, detection of Gram-negative bacteria and, 292

P

Papovaviruses, brain and, 76
Paracoccidioides brasiliensis, in clinical material, 18, 31
Parainfluenza virus types 1 and 2, seasonality and symptoms, 72
Parainfluenza virus type 3
 occurrence and symptoms, 72
Parainfluenza viruses types 1–4, detection by immunofluorescence, 67, 72
Particles, solid-phase immunoassays and, 228
Pathogens, identification in presence of mixed flora, 130–131
Penicillinase, blood cultures and, 345
Peptidoglycan, as antigen in immunoassays, 177
Periodic-Schiff stain, for fungi, 14, 16, 29
Petriellidium boydii, in clinical material, 32–33
Pharyngitis, throat specimens and, diagnostic microscopy, 8
Phenylboronic acid, use in CIE, 107, 115
Phenylketonuria, diagnosis of, 307
Phialophora verrucosa, in clinical material, 19
Plasma derivatives
 heptatis B and, 243
 non-A, non-B hepatitis and, 257
Plastics, as support systems for solid-phase immuno-assays, 233–234
Pleural fluid, detection of pneumococcal antigen in, 109
Pneumococcal antigens, detection by CIE
 in other body fluids, 109
 in serum, 108
 in sputum, 108–109
Pneumococcal infections, CIE methods in, 106–107
Pneumococci, *see also Streptococcus pneumoniae*
 identification in sputum, 131
Pneumonia
 coagglutination tests in, 154–155
  comparison to counterimmunoelectrophoresis, 155–156
  specificity, 157
 diagnosis of, 7
Polyanetholsulfonate, detection of bacteremia and, 343, 345

Polyribosylribitol phosphate, immunoassays, sensitivity of, 162
Polysaccharide, pneumococcal
clearance from circulation, 119
secretion rate, 118
Potassium hydroxide, examination of specimens for fungi and, 22, 26, 28
Precipitin methods, for detection of antigenemia
future trends, 123
variables in, 118–123
Predictive value, of CIE tests, 126–129
Presumptive etiologic diagnosis, value of, 141
Procedure, for solid-phase immunoassay, 238–239
Pseudoallescheria boydii, in clinical material, 18
Protein A, contamination, immunoassays and, 180
Pseudomembranous colitis, *Clostridium difficile* and, 295
Pseudomonas aeruginosa, bacteremia, RIA and, 181, 182
Pulmonary secretions, *Aspergillus* antigens in, 218–219
Pulmonary tissue, viruses in, 77
Pus, GLC and, 307–308

Q

Quartz-iodine-tungsten lamps, fluorescence microscopy and, 65, 66

R

Rabies virus, detection by immunofluorescence, 67, 76, 77
Radioimmunoassay (RIA)
for candidal mannan, 210, 211, 212
antigens, 213–214
protocol, 214
detection of antigenemia and, 123
for detection of *Aspergillus* antigens, 216–217, 218
preparation of antigens, 219–220
procedure, 220
for detecting serogroup 1 *Legionella pneumophila* antigens
development of, 191–194
diagnosis of infections, 194–198
non-serogroup 1 infections, 198–200
problems with RIA, 200–201
for rapid diagnosis of bacteremia, 181–183
Reagents
for immunofluorescent diagnosis, 61–65
serum, for rotavirus immunoassay, 269–270
for solid-phase immunoassay, 235–236
Resolving power, of GLC, 306
Respiratory secretions
detection of viruses by immunofluorescence in, 68–71
adenovirus, 72–73
influenza A or B, 71–72
measles, 73
parainfluenza types 1 and 2, 72
parainfluenza type 3, 72
respiratory syncytial, 71
obtaining samples of, 68–69
Respiratory syncytial virus
detection by immunofluorescence, 67, 71
occurrence and symptoms, 71
Respiratory tract infections, specimens for examination,diagnostic microscopy and, 7–8
Retention times, GLC and, 306
Reticuloendothelial system, uptake of bacterial polysaccharides by, 122
Rheumatoid factor
diagnosis of cryptococcal menigitis and, 169–170
in immunoassays for antigenemia, 180, 183
Rhinosporidium seeberi, infection by, 17, 19
Rhizopus, in clinical material, 18
Rhodamine, light absorption and emission by, 58–59
Rickettsia rickettsii
demonstration in clinical material, 37
identification by immunofluorescence, 50
Rocky Mountain spotted fever
GLC and, 326–328
Rotavirus
commercial tests, 274
considerations in establishing solid-phase immunoassays, 269–274
epidemiology and virology, 268–269
Rubella virus
in brain, 76
GLC and, 329
Rubeola, GLC and, 329

S

Safranine
screening of urine samples and, 340–341
detection of bactiuria, 355
Saksenaea, in clinical material, 18
Salmonella, identification by immunofluorescence, 51
Sample(s)
concentration for CIE, 93
derivatization for GLC, 305
Sample preparation, for rotavirus immunoassay, 270–271
Scanners, automated, 336
Screening procedures
for bactiuria, 354–355
nonspecific, for hepatitis B, 254
Sensitivity
of agglutination test
for *H. influenzae* antigens, 144–145
for meningococcal antigens, 146–148
of assays for hepatitis B surface antigen, 246–247
of CIE, 126–127
of ELISA, 166, 167
of solid-phase immunoassays, 237–238
Sephacryl S-200 Superfine gel, preparation of *Aspergillus* antigen and, 219, 220
Septic arthritis, detection of Gram-negative bacteria in, 292
Seroconversion, in Legionnaires' disease, 189
Serologic tests, experimental, non-A, non-B hepatitis and, 258–261
Serum, detection of pneumococcal antigen in, 108, 154–155
Shigella, identification by immunofluorescence, 51
Skin scrapings, detection of viruses in by immunofluorescence, 73–74
herpes-simplex, 74–75
varicella-zoster, 75
Slides, for immunofluorescence microscopy, 70
Smears, preparation for immunofluorescence
choice of test, 37
fixation of specimen, 38
performance of test, 38–39
quality of specimen, 37
Solid-phase immunoassay
in diagnosis of respiratory infections, 225–226
antibody labels, 226–229
assay formats, 229–233
collection of specimens, 236–237
protocol, 238–239
reagents, 235–236
sensitivity of assay systems, 237–238
support systems, 233–235
for rotavirus
calculation of cutoff values for positive results, 273–274
direct or indirect assay, 272
indicator system, 271
nonspecific reactions and selection of controls, 272–273
sample preparation, 270–271
serum reagents, 269–270
solid phase, 270
test conditions, 271
Specificity
of CIE tests, 128
of coagglutination tests in pneumonia, 157
of ELISA, 167
of reagents, solid phase immunoassays and, 235–236
Specimens
for clinical microbiological tests, 338
collection for solid-phase immunoassay, 236–237
of respiratory secretions, treatment in laboratory, 69–70
Sporothrix schenckii, in clinical material, 18, 30
Sputum
detection of pneumococcal antigen in, 108–109, 154–155, 156
liquefaction for CIE, 93–94
Stains
routine, detection of fungi and, 29
special for fungi, 29–30
Staining
of electrophoresis plates, 91, 94
viral diagnosis and, 65
Staphylococcal protein A, indirect immunofluorescence and, 59, 61
Staphylococci
antigens, antibodies to, 177–178
use in agglutination reactions, 136–137, 154–155, 164, 165, 166
specificity and, 151–152, 157
Staphylococcus aureus
antigens, RIA and, 181–183

L. pneumophila antisera and, 195
Streptococcal pharyngitis, diagnosis of, 8
Streptococci
beta hemolytic, identification by immunofluorescence, 51
counterimmunoelectrophoresis and, 89, 92
group B antigens, agglutination tets for, 148–149
Streptococcus agalactiae, antigen, ELISA and, 166, 167
Streptococcus pneumoniae, see also Pneumococci
antigens
agglutination tests for, 145–146
ELISA and 146, 167, 182, 184
counterimmunoelectrophoresis and, 89, 92, 115
identification by immunofluorescence, 48, 49
spent culture medium, GLC chromatogram of, 320
Subtypes, of hepatitis B surface antigen, 247–248
Support systems
for counterimmunoelectrophoresis, 90
for solid-phase immunoassay, 233–235
Surface antigens, of heptatitis B, 243–248

T

Theory, of fluorescence microscopy in viral diagnosis, 58–61
Thermodissociation, of immune complexes, 179 –180
Thiolbroth, blood cultures and, 345–346
Torulopsis glabrata, in clinical material, 18
Toxins, of *C. difficile,* 298–299
Tranfusions
hepatitis B and, 243
non-A, non-B hepatitis and, 255, 260
Treponema, fixation for immunofluorescence, 38
Treponema pallidum, identification by immunofluorescence, 50
Trichosporon, in clinical material, 18
Tuberculostearic acid, GLC and, 308

U

Urinary tract infections
diagnosis
comparison of methods, 359–360
conventional methods, 353–354
instrumentation, 356–360
rapid methods, 354–356
diagnostic microscopy and, 7
Urine
antigen in, concentration of, 162, 170
detection of pneumococcal antigen in, 109, 123
RIA for detection of *L. pneumophila* antigen in, 191
sample CIE setup, 92
screening of specimens, 339–340

V

Varicella-zoster virus, diagnosis of, 62, 67, 75
Viral diagnosis, immunofluorescence in, 57–58
collection and preparation of specimens, 67–77
fluorescence microscope, 65–67
interpretation of findings, 78–79
reagents, 61–65
staining method, 65
theory of method, 58–61
use of controls, 77–78
Viruses
choice of which to stain for, 70
not detectable by immunofluorescence, 67–68
production of antisera and, 61–62, 235

W

Wangiella dermatitidis, in clinical material, 19
Wet mounts, miscroscopic detection of fungi and
India ink mount, 28–29
potassium hydroxide preparations, 28
unstained, 28
Wounds, diagnostic microscopy and, 9–11

Y

Yersinia pestis, identification by immunofluorescence, 51

Z

Zone phenomena, agglutination tests and, 138
Zygomycetes, in clinical material, 18